AF703299

BIOPESTICIDES

BIOPESTICIDES

APPLIED PRODUCTION AND BENEFITS

J.K. Teicher

Kruger Brentt
Publishers
2023

Kruger Brentt Publishers UK. LTD.
Company Number 9728962

Regd. Office: 68 St Margarets Road, Edgware, Middlesex HA8 9UU

© 2023 AUTHOR

ISBN: 9781787150560

Disclaimer:

Every possible effort has been made to ensure that the information contained in this book is accurate at the time of going to press, and the publisher and author cannot accept responsibility for any errors or omissions, however caused. No responsibility for loss or damage occasioned to any person acting, or refraining from action, as a result of the material in this publication can be accepted by the editor, the publisher or the author. The Publisher is not associated with any product or vendor mentioned in the book. The contents of this work are intended to further general scientific research, understanding and discussion only. Readers should consult with a specialist where appropriate.

Every effort has been made to trace the owners of copyright material used in this book, if any. The author and the publisher will be grateful for any omission brought to their notice for acknowledgement in the future editions of the book.

All Rights reserved under International Copyright Conventions. No part of this publication may be reproduced, stored in a retrieval system, or transmitted in any form or by any means, electronic, mechanical, photocopying, recording or otherwise without the prior written consent of the publisher and the copyright owner.

For information on all our publications visit our website at http://krugerbrentt.com/

PREFACE

Biopesticides pose minimal dangers to individuals and the environment, making them a relatively safer alternative to chemical or chemical-derived pesticides. Environmental safety and host specificity are the main benefits of biopesticides in managing agricultural pests. Most farmers attempting to establish a sustainable farming system know that the chemical shed is not their first line of defence against unwanted pests. A "softer" biopesticide or a conventional, synthetic treatment are the farmer's two options when a pest infestation gets too serious, and a chemical application is required. The Integrated Pest Management (IPM) program combines cultural measures, biological controls (such as predatory insects), and chemical control to keep pest populations under control. Biopesticides are more environmentally friendly and do not harm the soil, water supply, or wildlife, including beneficial insects, which is one of the main advantages of introducing them into a sustainable agriculture system. Biopesticides are typically used in rotation with conventional products rather than as a replacement, which reduces the amount of synthetic chemicals used. Insects and diseases develop resilience to synthetic chemicals over time. The effectiveness of the synthetic chemical is increased by alternating it with biopesticides.

Biopesticides are certain types of pesticides derived from such natural materials as animals, plants, bacteria, and certain minerals. Agricultural pesticides, properly used, are essential in supplying the food requirements of the world ever growing population. The use of synthetic pesticides affects the health of human being. The indiscriminate use of pesticides has adversely affected the health of the soil. The residual pesticides in the soil not only affect the soil quality but also the water quality, as they get leached into the ground water. Due to these reasons, role of biopesticides are very important for sustainable agriculture.

The present book contains fourteen chapters covering all related disciplines. These chapters include *Introduction, types of biopesticides, Biopesticides: benefits and their production, Modes of action of biopesticides, Biopesticide for integrated crop management,*

Biopesticides and biofertilizers in organic farming, Biopesticide use in horticultural crops, Nano biopesticide: an emerging technology for pest management for future perspectives, Biopesticides for management of arthropod pests, Bioherbicides for management of weeds, Biopesticides for management of aquatic weeds, Biopesticides for ipm and irm, Resistance to biopesticide, Novel biopesticide formulations for environmental sustainability.

The book deals with the detailed description about a number of biopesticides which are used for management of fungi, bacteria, virus and nematodes. Related terminology is given at the end for ready reference. The book will be of immense help to both the undergraduate and postgraduate students, biologists and agriculturists, who would like to broaden their knowledge and gain substantial experience about biopesticides in agriculture and health, this will enable them to contribute significantly in making the world a safer and healthier place.

We are grateful to all those persons as well as various books, manuals, periodicals, magazines, journals etc. that helped in the preparation of this book. In spite of the best efforts, it is possible that some errors may have occurred into the compilation and editing of the book. Further queries, constructive suggestions and criticisms for the improvement of the book are always welcome and shall be thankfully acknowledged.

J.K. Teicher

CONTENTS

1

INTRODUCTION

Generally, biopesticides are made of living things, come from living things, or they are found in nature. They tend to pose fewer risks than conventional chemicals. Very small quantities can be effective and they tend to break down more quickly, which means less pollution.

Some biopesticides are targeted in their activity, often working on a small number of species. However, users need more knowledge to use biopesticides effectively. This is because they are often most effectively used as part of an Integrated Pest Management approach.

THE NEED FOR BIOPESTICIDES

In the absence of effective alternative management options to tackle pests, smallholder farmers rely extensively on indiscriminate application of synthetic pesticides. These synthetic pesticides are harmful to human health, detrimental to the environment and biodiversity, and lead to rapid build-up of resistance in the target pests while decimating natural enemies of pests, resulting in secondary pest outbreaks.

In addition, presence of pesticide residues on export crops that are above the permissible maximum residue levels of importing nations results in informal trade barriers.

Especially for smallholder farmers who rely on their crops as a primary dietary staple, it is essential to introduce management techniques of lower toxicity. Biopesticides may offer an essential alternative to the indiscriminate use of synthetic pesticides.

Biopesticides are generally less toxic than conventional pesticides and therefore considered safer for human health. Biopesticides often only affect the target pest and therefore do not pose negative effects on the environment. Also, their use does not lead to resistance build-up in target pests.

The global market for biopesticides is valued at 3.0 billion USD, accounting for 5% of the global pesticide market. With an annual compounded growth rate of more than

15%, it is anticipated that biopesticide market share will equal that of synthetic pesticides between 2040 and 2050.

According to the United States Environmental Protection Agency (EPA), biopesticides are pesticides derived from natural materials such as animals, plants, bacteria and certain minerals, and can be classified in three classes:

1. microbial pesticides, which consist of a microorganism as the active ingredient;
2. biochemical pesticides, which are naturally occurring substances that control pests by non-toxic mechanisms such as insect sex pheromones and plant extracts; and
3. plant-incorporated protectants, which are pesticidal substances that plants produce from genetic material that has been added to the plant.

HISTORY OF BIOPESTICIDE

Biologicals are used to control pests, pathogens, and weeds by a variety of means. Microbial biocontrols may include a pathogen or parasite that infects the target. Alternatively, they might act as competitors or inducers of plant host resistance. Biochemical biocontrols can also act through a variety of mechanisms. Some act by inhibiting the growth, feeding, development or reproduction of a pest or pathogen. Still other biocontrols may be used to form a barrier on the host, so as to act as a feeding or infection inhibitor.

Plant extracts were likely the earliest agricultural biocontrols, as history records that nicotine was used to control plum beetles as early as the 17th century. Experiments involving biological controls for insect pests in agriculture date back as far as 1835, when Agostine Bassi demonstrated that white-muscadine fungus (Beauveria bassiana) could be used to cause an infectious disease in silkworm. Experiments with mineral oils as plant protectants were also reported in the 19th century. During the rapid institutional expansion of agricultural research during the early 20th century, an ever-growing number of studies and proposal for biocontrols were developed.

The first, and still most, widely used biocontrols included spores of the bacteria Bacillus thuringiensis (Bt). In 1901, Bt was isolated from a diseased silkworm by Japanese biologist Shigetane Ishiwata. Ernst Berliner in Thuringen, Germany, then rediscovered it ten years later in a diseased caterpillar of flour moth. The Bt pathogen was classified in 1911 as type species Bacillus thuringiensis and remains the most widely used biocontrols to this day. In the early 1920s, the French began to use Bt as a biological insecticide. The first commercially available Bt product, Sporeine, appeared in France in 1938. In the US in the 1950s, widespread use of biocontrols began to take hold as a host of research on Bt efficacy was published.

In the latter half of the 20th century, research and development continued at a low level because of the widespread adoption of cheaper but more toxic synthetic chemical insecticides. During this time, new products were developed and applied; especially in niche markets where petroleum based chemicals were not registered, not effective, or not economical. For example, in 1956, the Pacific Yeast Product Company developed an industrial process known as submerged fermentation, which allowed production of Bt on a large scale.

In 1973, Heliothis NPV was granted exemption from tolerance and the first viral insecticide, Elcar received a label in 1975. In 1977, Bacillus thuringiensis var. israelensis (toxic to flies) was discovered, and in 1983 the strain tenebrion is (toxic to beetles) was found. In 1979, the U.S. EPA registered the first insect pheromone for use in mass trapping of Japanese beetles. In the 1990s, researchers began testing kaolin clay as an insect repellent in organic fruit orchards. It was made commercially available, particularly for use in organic systems, in 1999.

Biological development for the control of plant diseases has undergone a similar transformation. During the early 20th century, studies of soil microbiology and ecology had led to the identification of many different microorganisms that act as antagonists or hyperparasites of pathogens and insect pests.

A number of these were shown to be useful in field-scale inoculations, but few were developed commercially because of the rapid adoption of chemical pesticides during that time period. Commercial success stories from the 1980s and 1990s include products containing Agrobacterium radiobacter for the prevention of crown gall on woody crops and Pseudomonas fluorescen s for the prevention of fireblight in orchards where the streptomycin had been overused and resistant pathogen populations were abundant. In the greenhouse and potting mix industry, products containing a variety of microbes that suppressed soilborne pathogens were introduced into the market.

As the costs of overusing such synthetic chemicals became apparent, there was resurgence in academic and industrial research related to biocontrols development. And with the rapid expansion of organic agriculture during the past decade, adoption rates have rapidly increased. Because of this, development of new and useful biocontrols has continued to increase rapidly since the mid-1990s. In fact, more than 100 biocontrols active ingredients have been registered with the U.S. EPA Biologicals division since 1995. Many of these have been introduced Biologicals division since 1995. Many of these have been introduced commercially in a variety of products. Many of the active ingredients currently approved for use in the U.S.A. can be found in publicly available databases.

Some examples of biocontrols developed in more recent years

Agrobacterium radiobacter Strain K84

Agrobacterium radiobacter Strain K84 is a naturally occurring bacterium found in many soils and in plant root zones. This biocontrols is used in the greenhouse and nursery environment to control crown gall, an important plant disease.

Bacillus spp.

Bacillus licheniformis, B. pumilus, and B. subtilis are naturally occurring soil bacteria with fungicidal properties that together have become one of the fastest growing biocontrols in today's market. Successes include uses as seed treatments or dressings, foliar application and soil-applied control of diseases in a variety of crops.

Coniothyrium minitans

Coniothyrium minitans is a naturally occurring fungus used commercially to control common Sclerotinia plant diseases through parasitism of the resting structures of the pathogen.

Paecilomyces fumosoroseus and P. lilacinus

Paecilomyces fumosoroseus is a naturally occurring fungus used in a greenhouse environment to control several species of insects including whiteflies, thrips, aphids, and spider mites. Paecilomyces lilacinus is used to control nematodes that attack plant roots in field crops including many vegetables, fruit, turf, and ornamental crops.

Trichoderma spp.

Trichoderma spp. is another biocontrols technology developed in the 1990 s that has been widely commercialized in recent years. Trichodermais a genus of fungi that helps to control plant disease by stimulating plant host defenses and growth, and, under certain conditions, parasitizing harmful fungi within the plant root zone.

Azadirachtin

Azadirachtin is an insect growth regulator derived from neem tree seeds. Known to affect some 200 species of insects, azadirachtin disrupts insect feeding and inhibits its ability to molt as it changes from the pupa to adult stage.

Beauveria bassiana

Beauveria bassiana is a naturally occurring soil fungus that grows as white mold. This insect pathogen can be used to control a wide range of target pests, which become infected and develop white muscadine disease, killing the pest within a matter of days.

Cydia pomonella granulo virus (CpGV)

CpGV is a natural pathogen of the codling moth, a major pest of tree fruits such as apples and pears. Developed through research begun in the 1980's, commercial use of CpGV in both organic and conventional systems has gained in popularity over the last ten years as codling moth has displayed resistance to many traditional insecticides.

Dysphania ambrosioides

An extract of the plant Dysphania ambrosioides (syn. Chenopodium ambrosioides) is used to control a number of sucking insect pests such as aphids, leafhoppers, whiteflies, and mites in citrus, grapes, tree nuts, and vegetables. This product breaks down the pest's exoskeleton, adversely affects its respiratory system, and interrupts its ability to navigate (find food).

Biopesticides are classified into three main categories: biochemical, microbials and plant-incorporated protectants.

An increasing number of greenhouse growers in Michigan are discovering biopesticides can be used in conjunction with commercially purchased natural enemies as a part of their integrated pest management (IPM) programs. Biopesticides offer numerous advantages to growers, including lower risk to employees, minimal (or no) re-entry and pre-harvest intervals, and compatibility with biocontrol programs.

In fact, a recent survey of Michigan floriculture growers revealed all of them had used at least one biopesticide in conjunction with their biological control program. While greenhouse growers are now using biopesticides, they often have questions such as: What is a biopesticide? How are they different from each other? Which one should I use?

Biopesticides are pesticides derived from natural materials, including animals, plants, fungi and microbes. The Biopesticides and Pollution Prevention Division of the Environmental Protection Agency, which registers biopesticides in the U.S., classifies classified biopesticides into three categories: biochemical, microbials and plant-incorporated protectants.

Biochemicals

Biochemicals are derived from naturally occurring substances such as plant extracts. This includes insect repellants, insect attractants and repellants, pheromones, and non-pest management class—plant growth regulators.

Examples

- Azadirachtin (broad-spectrum insecticide).
- Capsaicin (compound from chili peppers) (broad-spectrum insecticide, nematicide and fungicide).
- Clove, rosemary and peppermint oil (broad-spectrum fungicide).

Microbials

Microbials are products containing micro-organisms or their fermentation by-products.

Examples

- The bacterium Bacillus thuringiensis for use against caterpillars.
- The fungus Beauveria bassiana for use against whiteflies, aphids and thrips.

Plant-incorporated protectants

Plant-incorporated protectants are pesticidal substances produced by plants as a results of genetic manipulation. There are no plant-incorporated protectant compounds registered for use in greenhouses.

As of April 2016, there were 299 active ingredients in biopesticides registered with the Environmental Protection Agency (EPA). However, only 40 are labeled for use in greenhouses.

BIOPESTICIDES AND THEIR IMPORTANCE IN PLANT PROTECTION

Biopesticides are naturally occurring compounds or agents that are obtained from animals, plants, and microorganisms such as bacteria, cyanobacteria, and microalgae and are used to control agricultural pests and pathogens. According to the US Environmental Protection Agency, biopesticides are 'derived from natural materials such as animals, plants, bacteria and certain minerals'. Products such as genes or metabolites from these biocontrol agents can be used to prevent crop damage. The use of biopesticides is, by far, more advantageous than the use of their counterparts, traditional chemical pesticides, as they are eco-friendly and host specific. The use and application of agro-based chemicals in the agricultural sector to protect crop plants from invading and infecting pests can be greatly improved by employing biopesticides.

MODE OF ACTION OF BIOPESTICIDES

Microbial Biopesticides

Fungicides and bactericides

These biopesticides generally inhibit or disrupt the process of translation and thus protein synthesis in numerous ways, including through binding of 50S ribosomes in prokaryotes, to prevent the transfer of peptides and inhibit chain elongation (such as blasticidin). Sometimes they interfere with the binding of aminoacyl tRNA to 30S and 70S ribosomal subunit complexes and inhibit translation (such as kasugamycin). In the case of streptomycin and mildiomycin, binding with the 30S ribosomal subunit causes abnormal synthesis of protein (nonfunctional) and blocks the activity of peptidyltransferase, respectively.

Metarhizium anisopliae M. anisopliae Sorokin var. anisopliae is an essential entomopathogenic fungus. It propagates worldwide in the soil, demonstrating a wide range of insect host species. This subspecies was first described in 1879 by Metschnikoff, under the term Entomophthora anisopliae, as a pathogen of the wheat cockchafer, and later it was termed M. anisopliae by Sorokin in 1883.

Several entomopathogenic fungi and their derivatives are also used as microbial pesticides. M. anisopliae are hyphomycete entomopathogenic fungi most widely used for insect pest control and are ubiquitous worldwide. This species comprises a huge number of different strains and isolates of various geographical origins and from different types of hosts.

Under natural conditions, Metarhizium are found in the soil, where the moist conditions permit filamentous growth and production of infectious spores, called conidia, which infect soil-dwelling insects upon contact (Fig. 1). M. anisopliae has the potential to be used as a biocontrol agent, particularly for malaria vector species, and is also a suitable candidate for further research and development.

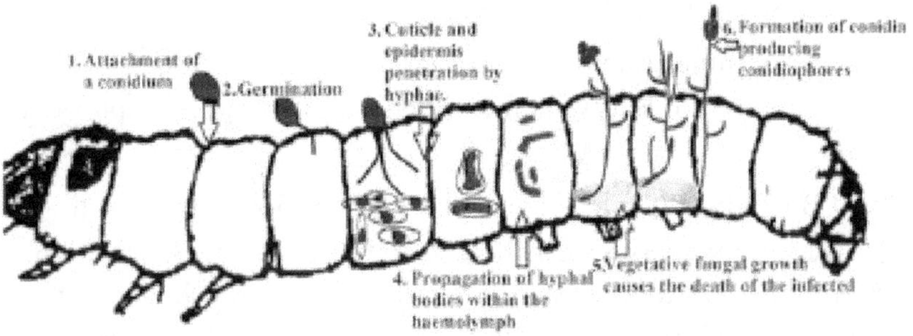

Fig. 1 Mode of action of entomopathogenic fungi against lepidopteran insects

Researchers reevaluated the taxonomy of the genus Metarhizium using sequence data from ITS and 28S rDNA D3 regions and also using RAPD patterns, revealing ten distinguishable clades. M. anisopliae var. anisopliae represents clade 9. These entomopathogenic fungi have been viewed as safe and regarded as an environmentally satisfactory alternative to synthetic chemical pesticides. Recently, these entomopathogenic fungi have been registered as microbial agents and are also under commercial development for the biological control of several pests.

M. anisopliae strains are obtained from different geographical localities, as suggested, among others, by extremely variable toxicities. M. anisopliae have been used on a large scale in countries like Brazil, where 100,000 ha of sugarcanes are treated every year. They are released in the field after thorough assessment of strains. Researchers have reported that the growth and application of fungi as microbial agents for biocontrol of insect pest involve tests with mammalian models to evaluate possible human and animal health risks.

Bacillus thuringiensis

Beginning in the 1980s and continuing to the present, a different molecular approach has been employed to develop market acceptance of biopesticides. Earlier, several efforts were aimed at establishing microbial insecticides, like Bt, which has been used commercially over 40 years. Later, some Bacillus species such as Bacillus thuringiensis israelensis Bti and Bacillus sphaericus 2362 (Bs) were found particularly effective against mosquito and other dipteran larvae. Bti was first discovered to have increased toxicity against mosquito larvae in 1975.

Various bacterial species and subspecies, especially Bacillus, Pseudomonas, etc., have been established as biopesticides and are primarily used to control insect and plant diseases. Most salient among these are insecticides based on several subspecies of Bacillus thuringiensis Berliner. These include B. thuringiensis sp. kurstaki and aizawai, with the highest activity against lepidopteran larval species; B. thuringiensis israelensis, with activity against mosquito larvae, black fly (simuliid), and fungus gnats; B. thuringiensis tenebrionis, with activity against coleopteran adults and larvae, most notably the Colorado potato beetle (Leptinotarsa decemlineata); and B. thuringiensis japonensis strain Buibui, with activity against soil-inhabiting beetles.

Bt produces crystalline proteins and kills few target insect pest species like lepidopteran species. The binding of the Bt crystalline proteins to insect gut receptor determines the target insect pest.

Toxicity of Bti and some other toxic strains is commonly imputed to the parasporal inclusion bodies (d-endotoxins) which are produced during sporulation time. These endotoxins must be assimilated by the larvae to accomplish toxicity. Bt and their subspecies produce different insecticidal crystal proteins (d-endotoxins), and their toxicity was determined. These toxins, when ingested by the larvae, can damage the gut tissues, leading to gut paralysis. After that, the infected larvae stop feeding and finally they die from the combined effects of starvation and midgut epithelium impairment.

Some other microbial pesticides act by outcompeting insect pest organisms. Microbial pesticides need to be continuously supervised to ensure that they do not become capable of injuring nontarget organisms, including humans. In previous studies, the microbial pesticide advance has resulted in a significant decrease of synthetic chemical insecticide usage.

Researchers reported Bt toxins produced by plant growth-promoting rhizobacteria, which also develop bacteriocin compounds of insecticidal attributes. Bt is marketed worldwide for the control of different important plant pests, mainly caterpillars, mosquito larvae, and black flies. Commercial Bt-based products include powders containing a combination of dried spores and crystal toxins. They are applied on leaves or other environments where the insect larvae feed. Toxin genes from Bt have been genetically engineered into several crops.

Biochemical Pesticides

These pesticides are derived from plants. Plants have evolved and developed many compounds, which can help to combat pathogenic microorganisms during the course of infection and attack. These compounds include steroids, alkaloids, phenylpropanoids, phenolics, terpenoids, and nitrogenated compounds. For instance, nicotine was the first insecticide obtained from tobacco leaves in the 17th-century that used to kill plum beetles. Nicotine in tobacco is toxic to most herbivore insects and pesticides derived from them have been regarded as 'green pesticides' with high activity and low toxicity. Researchers have mentioned tobacco to be containing some useful ingredients, such as solanesol and nicotine, which exhibit potent inhibitory activity against Staphylococcus aureus, Bacillus subtilis, and Micrococcus lysodeikticus. Insecticides, such as azadirachtin and nicotine, function by either disrupting respiratory enzymes or inhibiting insect growth regulators, or by binding to sodium channels, while microbicides impair metabolic function and disrupt the integrity of plasma membrane and inhibit conidial formation.

Pheromones

Insects produce chemicals called pheromones to stimulate a certain behavioral reaction from other individuals. These pheromones have numerous effects and are named according to their evoked response, for example, sex pheromones, aggregation pheromones, alarm pheromones, etc. A few pheromones function as sex attractants, permitting individuals to detect and locate mates, whereas others induce trail following, oviposition, and aggregation in other congeners. Pheromones have become essential tools for monitoring and controlling agricultural pest populations, and as such, a huge collection of over 1,600 pheromones and sex attractants has been reported.

In the past decades, there has been a significant volume of literature on insect pheromones and new opportunities have arisen to explore the use of semiochemicals in managing insect pest problems. Insect control by pheromones alone has precincts, but they can be applied in integrated control in combination with other practices. Plant volatiles are recognized as an integral part of the pheromone system of various Coleopteran species studied so far. How the combinations of pheromones and plant volatiles are incorporated into the insects' olfactory systems so that they can discriminate pheromone molecules alone and pheromone plus odor plume strands and react behaviorally to these signals is a question of increasing importance.

Recently, the pest management system has undergone an intentional shift from calendarbased, broad-spectrum insecticide applications to more holistic, integrated, and high-efficacy approaches. Furthermore, environmental conservation, food safety, and resistance management are a few of the key components guiding current pest management policies in commercial agriculture.

Researchers had reported that meeting the necessities of the quickly expanding global population while incorporating sustainability and ecological stewardship is a major challenge facing modern agriculture. In agricultural fields, the application of pheromones and/or allelochemicals as behavioral manipulation tools can replace or complement existing pest management programs, resulting in a decreased rate of broad-spectrum use.

Nowadays, pheromones and other semiochemicals are applied to monitor and control pests in millions of hectares. There are several advantages of utilizing pheromones for monitoring pests, including lower costs, specificity, ease of use, and high sensitivity.

Insect pest monitoring by using pheromone lures can profit management conclusions such as insecticide application timing.

Pheromones produced by insects are highly species specific. Virgin female insects are developing sex pheromones when expecting for a mate and males along the concentration slope for the female producer. Aggregation pheromones are released by insects such as wood-invading beetles to show to others the presence of a good food source.

Some of the alarm pheromones are developed by insects that are beneath approach from a predator and this contributes to a movement of the insect pest aside from the production source and, therefore it becomes dangerous. Plants and its derived attractants are also known that interact with the insects to a valuable food source and, while combine with insect-derived attractants will be developed a potent attraction to some insect pests.

GMO-BASED BIOPESTICIDES

These are produced when genes are transferred into a plant, which allows it to produce compounds, such as Bt toxin, that can be used to combat pests. The delta endotoxins produced by the bacterium B. thuringiensis are broken down into smaller toxins in the insect gut by the action of proteases, which then bind to receptors in the midgut, causing cell expansion, rupture, and ion leakage leading to cell death.

Biopesticides from Algal and Cyanobacterial Sources

Microalgae can be used as an alternative technology to increase productivity in sustainable agricultural systems. A number of microalgae strains produce biologically active compounds that include antimicrobial compounds with the potential to act as biopesticides. The biomass (extracts) can be applied as an alternative to chemical pesticides since it can enhance plant growth and protect agricultural crops. The filamentous cyanobacterium Nostoc piscinale and two single-celled green algae, Chlamydopodium fusiforme and Chlorella vulgaris are reported to have biopesticide activity against certain pathogens. Some important microalgae have been exploited for their beneficial biopesticide activity in the cultivation of spices.

The use of chemical insecticides can result in numerous undesirable effects, including:

i. killing of beneficial and nontargeted organisms and sometimes resurgence;
ii. rapid multiplication of secondary pests;
iii. development of pesticide resistance;
iv. contamination of the environment/ecosystem;
v. accumulation of pesticide residues in food materials;
vi. causing imbalanced ecological processes, such as pollination (pollinators affected by pesticides) and harm to living beings;
vii. carcinogenic and teratogenic effects in nature; and
viii. causing imbalances in hormone systems.

Several microorganisms have been explored for their potential in developing biopesticides. Microalgae have proved to be an excellent source owing to their advantages over traditional chemical pesticides. They produce a plethora of compounds with stimulating activities, including biomass and compounds, which can be used in the preparation of biopesticides, thereby enhancing crop protection. Microalgae can be produced using wastewater, as they require nitrogen, phosphorus, and carbon and ammonium, which are abundant in wastewater, thus representing a nitrogen source. Chlorella vulgaris is generally used in the treatment of wastewater and is able to tolerate ammonium levels effectively.

Researchers assayed the efficacy of C. vulgaris against several phytopathogens, such as Rhizoctonia solani, Fusarium oxysporum, Phytophthora capsica, Pythium ultimum, Clavibacter michiganensis, Xanthomonas campestris, Pseudomonas syringae, and Pectobacterium carotovorum, while observing its antibacterial and antifungal activity, which were higher when cultivated in wastewater.

Researchers argued that rice fields heavily sprayed with synthetic fertilisers to promote better productivity and yield left many detrimental effects on the environment and beneficial soil microflora, including decreased efficiency of fertiliser utilisation by the promotion of rice diseases, inhibition of microbiological nitrogen fixation, and increased nonpoint source pollution; importantly, they were also not cost effective. Furthermore, he added that in developing green rice, Anabaena variabilis could be a potent biofertiliser and biopesticide.

VIRUS

Baculovirus

Baculoviruses are double-stranded DNA viruses present in arthropods, mainly insects. Baculoviruses are usually highly pathogenic and have been used efficaciously in their natural form as biocontrol agents against numerous serious insect pests. . Even so, the application of baculoviruses is still limited in the field of agriculture and horticulture where the thresholds for pest damage tend to be minimized. In Lepidoptera, i.e., the primary group where baculoviruses have been isolated, they only cause mortality in the larval stage.

Baculoviruses need to be ingested by the larvae to initiate infection. After ingestion, they enter the insect's body through the midgut and from there they spread throughout the body, although in some insects, infection can be limited to the insect midgut or the fat body (Fig. 2). Two groups of baculoviruses exist: the nucleopolyhedroviruses (NPVs) and granuloviruses (GVs).

In NPVs, occlusion bodies comprise numerous virus particles, but in GVs, occlusion bodies ordinarily contain just one virus particle. A common feature of baculoviruses is that they are occluded, i.e., the virus particles are embedded in a protein matrix. The presence of occlusion bodies plays an essential role in baculovirus biology as it allows the virus to survive outside the host.

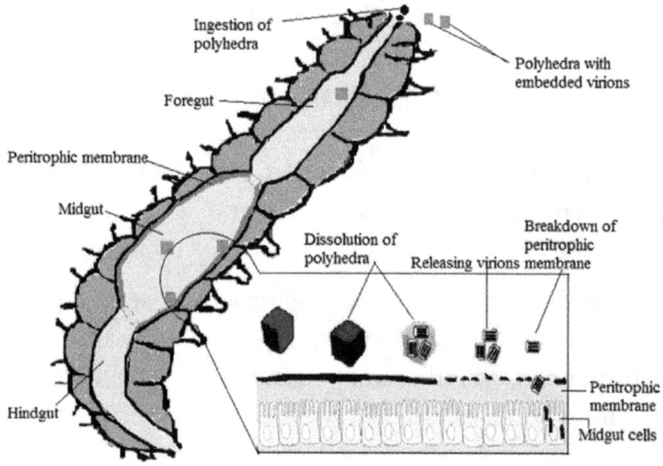

Fig. 2 Mode of action of baculoviruses against lepidopteran insects

Recently, it has been substantially demonstrated that baculoviruses are not infectious to vertebrates and plants. Also within the insects, their host range is restricted to the order from which they were isolated. Genetic engineering of baculoviruses has so far been restricted to isolates from Lepidoptera. Baculoviruses vary in the number of hosts that they can infect; some seem to be particular to one species and hence are very unlikely to be a hazard when genetically modified, whereas others infect a range of hosts.

The majority of new baculovirus recombinants now employ insect-selective toxins. Genetic modification has been mostly carried out on the alfalfa looper, Autographa californica NPV (AcNPV)-the virus for which most molecular information is available, including the complete DNA sequence, which permits precise insertion of foreign genes. Recently, the development of genetically modified baculoviruses has expanded to other strains of commercial or regional interest.

Baculovirus-infected insect larvae have primarily initiated a cascade of molecular and cellular appendages finally the larva enters into the death and the development of huge amounts of polyhedral occlusion bodies comprising rodshaped virions. Lately, a naturally occurring baculovirus (Agrotis ipsilon multiple nucleopolyhedrovirus, family Baculoviridae, AgipMNPV) was indicated to have hope as a microbial insecticide for controlling A. ipsilon in turf.

Numerous viral formulations are available primarily for the control of caterpillar pests. For instance, Certis has recently registered Madex™, an increased-potency codling moth granulosis virus (GV) that also affects oriental fruit moth (OFM). Certis also deals Cyd-X™, which also contains the codling moth GV and which can be an efficient tool for codling moth management. Aside from Madex and Cyd-X, Certis markets Gemstar™, which contains Heliothis zea NPV, and Spod-X™, which contains beet armyworm NPV. Gemstar is also registered for the control of lepidopteran pests, like the cotton bollworm and budworm, caterpillars that are mainly dangerous insect pests of corn, soybean, and other vegetables. Furthermore, Certis has registered a celery looper (Syngrapha falcifera) NPV and an alfalfa looper (Autographa californica) NPV.

NEMATODES

Steinernema (Rhabditida)

One of the new hot products in biopesticide is nematodes. Pest nematodes can be supervised with cover crops, crop rotation, and internalization of organic material into the soil. In the early 1990s, various effective entomopathogenic nematodes from two genera, namely, Steinernema and Heterorhabditis (Nematoda: Rhabditida), were discovered and established as a biocontrol agent against insects.

Insect-parasitic nematodes may encroach upon soil-dwelling stages of insects and kill them within 48 h through the expulsion of pathogenic bacteria. After the host dies, the infectious stages of the nematodes become adults and a modern generation of infective juveniles (IJs) develops (Fig. 3). Entomopathogenic nematodes are commonly available for plant protection from serious insect pests and diseases, and also there have been various efforts to biocontrol insect pest populations in the field by employing IJs via spraying. Nevertheless, little is known about the ability of indigenous nematodes to influence insect pest populations.

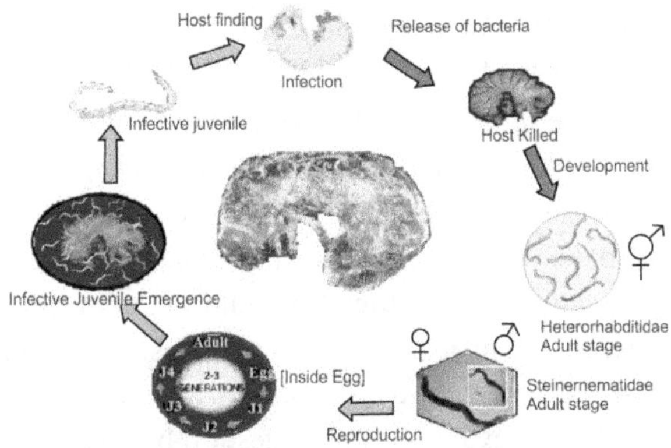

Fig. 4 Mode of action of ntomopathogenic nematodes against lepidopteran insects

In nematodes, the parasitic cycle is initiated by the third-stage IJs. These non-feeding juveniles infest suitable insect host and enter through the insect's natural body openings like the anus, mouth, and spiracles. Once they have entered inside the host, nematodes infest the hemocoel and then release their symbiotic bacteria into the intestine. After that, the bacteria cause septicemia, killing the host within 24–48 h (Fig. 3). The uptake of IJs is rapidly manipulating by the bacteria and decomposed the host tissues. Almost two to three generations of the nematodes are finished within the host cadaver.

In previous research studies, it has been clearly indicated that entomopathogenic nematodes can be effective biocontrol agents against some of the major insect pest families found in stored goods like Pyralidae and Curculionidae. Earlier, researchers had demonstrated the susceptibility of insect pests found in stored products, including Ephestia kuehniella Zeller and Tenebrio molitor L., to an increased concentration of nematodes. Researcher had also advocated a field concentration of >2.5 billion nematodes/ha against some of the major insect pests of row crops, but concentrations few times higher (7–15 billion/ha) are demanded to accomplish the control of pest population.

PROTOZOA

Nosema

Microsporidia, such as Nosema spp., are generally host specific and slow acting, most frequently producing chronic infections. The biological activities of most entomopathogenic protozoa are complex. They grow only in living hosts and some species necessitate an intermediate host. Microsporidia species are among the most commonly observed, and their main benefits are persistence and recycling in host populations and their debilitating effect on reproduction and overall fitness of target insects. As inundatively utilized microbial control agents, some species have been moderately successful.

Some protozoan species are pathogenic like Nosema locustae for grasshoppers, which is the only species that has been registered and established commercially. N. bombycis, the first reported microsporidium, is a pathogen of silkworm pébrine, which persisted in Europe, North America, and Asia during the mid-nineteenth century. Pébrine is still an epidemic and causes heavy economic losses in silk-producing countries such as China.

Almost 1,000 protozoan species, mainly microsporidia, attack invertebrates, including numerous insect species like grasshoppers and heliothine moths. Virtually renowned insect- pathogenic protozoan species are Nosema spp. and Vairimorpha necatrix . Protozoans produce spores, which are the infectious phase in several susceptible insects. Nosema spp. spores are assimilated by the host and develop in the midgut. Germinating spores are released from the sporoplasm and invade host target cells, inducing massive infection and demolishing organs and tissues. Sporulation process begins again from the infected tissues and, upon expulsion and ingestion by a susceptible host, induces an epizootic infection. Naturally, parasitoids and insect predators commonly act as vectors distributing the disease.

BOTANICAL INSECTICIDES

Since ancient times, natural compounds from plants were used, more or less efficaciously to give security from insect pests. In the nineteenth century, these compounds became scientifically established and widely utilized in the earlier period of the twentieth century. Plants and some insects have coexisted on the earth for almost three and a half million years, which has allowed lots of time for both to develop offensive and defensive strategies. Plants have developed many strategies to assist themselves from being assaulted by predators. An example of such plant strategy is developing compounds that are highly toxic to insects.

Neem

In Asia, neem has an extensive history of use mainly against household and storage pests and, to some extent, against insect pests of crops. Nevertheless, a breakthrough in the insecticidal application of neem was attained by the researcher who successfully protected the crops from insects by applying them with low concentration of 0.1% neem seed kernel suspension during a locust invasion. The Indian neem tree (Azadirachta indica) is one of the most important limonoid-producing plants from the Meliaceae family. Several components of its leaves and seeds show marked insect control potential, and due to their relative selectivity, neem products can be recommended for many programs on crop pest management. Neem product activity has been assessed against 450-500 insect pest species in different countries around the world, and from that, 413 insect pest species are reportedly susceptible at various concentrations.

In India alone, neem activity has been assessed against 103 species of insect pests, 12 nematodes, and several pathogenic fungi. Some recent reviews on the potential of neem in pest management. Most works have focused on azadirachtin richly from neem seed extracts which act as both strong antifeedants and insect growth regulators. Azadirachtin affects the physiological activities of insects and does not affect other biocontrol agents. Further, neem products are biodegradable and nontoxic to nontarget organisms.

In several Asian countries, numerous studies have measured neem activity alone or in combination with established insecticides and other biocontrol agents of damaging insect pests in agricultural crop system. In Indian field trials carried out, neem treatments were determined to be effective against some insect species like green leafhopper, yellow stem borer, rice gall midge, rice leaffolder, and grasshopper.

Melia azedarach

The promotion of botanicals as eco-friendly pesticides, microbial sprays, and insect growth regulators has been a major concern amid the presence of other control measures like beneficial insects, all of which demand an integration of supervised insect pest control. Plantbased insecticides are developed naturally from plant chemicals extracted for use against serious insect pests. As a result of concerns about the ecological continuity of synthetic pesticides and their potential toxicity to humans, nontarget beneficial insects, and some domestic animals, there is a regenerated interest in natural products to control insect pests. From this conclusion, the development of biopesticides seems to be a logical choice for further investigation. Meliaceae and Rutaceae species have received much attention due to the fact that they are a rich source of triterpenes known as limonoids.

The Meliaceae plant family is known to hold an assortment of compounds with insecticidal, antifeedant, growth-regulating, insect- deforming, and growth-modifying properties.

M. azedarach, otherwise known as chinaberry or Persian lilac tree, is a deciduous tree that originates from northwestern India, and it has been recognized for its insecticidal properties, which are still to be entirely analyzed. This tree grows in the tropical and subtropical parts of Asia, but nowadays it is also cultivated in other warm places of the world because of its considerable climatic tolerance. The leaves of M. azedarach are used for their insecticidal activity, whereas the fruit extracts of M. azedarach produce a variety of effects in insects, such as growth retardation, reduced fecundity, molting disorders, and behavior changes.

The anti-feedant and insect growth-regulating effects of M. azedarach extracts are known for many insects, the latter effect being the most essential physiological effect of M. azedarach on insects.

As previously mentioned, the Meliaceae plant family has been known as a potential source for insecticide properties. Also, several extracts from neem and other plant seeds and leaves have excellent insecticidal properties against vectors and are at the same time very eco-friendly. The efficaciousness of these neem products on mosquitoes was also demonstrated.

Without a doubt, plant-derived toxicants are a valuable source of potential insecticides. Plants and other natural insecticides may play a vital role in mosquito control programs as well as in other major insect control programs.

BIOCHEMICAL PESTICIDES

Pheromones

Insects produce chemicals called pheromones to stimulate a certain behavioral reaction from other individuals. These pheromones have numerous effects and are named according to their evoked response, for example, sex pheromones, aggregation pheromones, alarm pheromones, etc. A few pheromones function as sex attractants, permitting individuals to detect and locate mates, whereas others induce trail following, oviposition, and aggregation in other congeners. Pheromones have become essential tools for monitoring and controlling agricultural pest populations, and as such, a huge collection of over 1600 pheromones and sex attractants has been reported.

In the past decades, there has been a significant volume of literature on insect pheromones and new opportunities have arisen to explore the use of semiochemicals in managing insect pest problems. Insect control by pheromones alone has precincts, but they can be applied in integrated control in combination with other practices. Plant volatiles are recognized as an integral part of the pheromone system of various Coleopteran species studied so far. How the combinations of pheromones and plant volatiles are incorporated into the insects' olfactory systems so that they can discriminate pheromone molecules alone and pheromone plus odor plume strands and react behaviorally to these signals is a question of increasing importance. Recently, the pest management system has undergone an intentional shift from calendarbased, broad-spectrum insecticide applications to more holistic, integrated, and high-efficacy approaches. Furthermore, environmental conservation, food safety, and resistance management are a few of the key components guiding current pest management policies in commercial agriculture.

Researchers had reported that meeting the necessities of the quickly expanding global population while incorporating sustainability and ecological stewardship is a major challenge facing modern agriculture. In agricultural fields, the application of pheromones and/or allelochemicals as behavioral manipulation tools can replace or complement existing pest management programs, resulting in a decreased rate of broad-spectrum use. Nowadays, pheromones and other semiochemicals are applied to monitor and control pests in millions of hectares. There are several advantages of utilizing pheromones for monitoring pests, including lower costs, specificity, ease of use, and high sensitivity. Insect pest monitoring by using pheromone lures can profit management conclusions such as insecticide application timing.

Pheromones produced by insects are highly species specific. Virgin female insects are developing sex pheromones when expecting for a mate and males along the concentration slope for the female producer. Aggregation pheromones are released by insects such as wood-invading beetles to show to others the presence of a good food source.

Some of the alarm pheromones are developed by insects that are beneath approach from a predator and this contributes to a movement of the insect pest aside from the production source and, therefore it becomes dangerous. Plants and its derived attractants are also known that interact with the insects to a valuable food source and, while combine with insect-derived attractants will be developed a potent attraction to some insect pests.

In previous studies, researchers proposed that all insects produce approximate form of pheromone and companies subsist that synthesize a pheromone for any customer. Recently, 30 mating-disruption pheromonebased products are registered by the US EPA as biocontrol agents of lepidopteran pest species that can cause agricultural damage.

BIOPESTICIDES FROM AQUATIC PLANTS

Duckweed (Lemna minor), muskgrass (Chara spp.), water hyacinth (Eichhornia crassipes), hydrilla (Hydrilla verticillate), water lettuce (Pistia stratiotes), and filamentous algae (Lyngbya wollei) are some common aquatic plants. It is observed that some plants produce allelopathic compounds which have the potential to prevent the growth, germination, survival, and reproduction of surrounding organisms. Neem (Azadirachta indica) extract kills many insects, while Eichhornia crassipes has the ability to inhibit the growth of

Spodoptera litura, a lepidopteran pest. Similarly, Chenopodium album is inhibited by the presence of duckweed and water lettuce. These examples illustrate that similar plants (or weeds) and their allelopathic chemicals have highly potent inhibitory properties against the pathogens and hence can be substituted for conventional chemical pesticides

2

TYPES OF BIOPESTICIDES

Biopesticides are naturally occurring pesticides derived from either plants, animals, microorganisms or minerals. They are non-toxic, environmentally friendly and are a key ingredient in sustainable farming. Canola oil or baking soda are easily obtainable and commonly used as biopesticides for home gardens.

Food security is of critical importance due to the world's growing population. The well-being of humans, as well as global economies, depends upon a stable and reliable food supply. As the population continues to grow, food security will only come under more threat in the future. Sustainable farming and maximizing soil fertility demands non-toxic and environment friendly pesticides. Thus, the need for biopesticides seems poised to grow with time.

TYPES OF BIOPESTICIDES

Biopesticides fall into 3 main categories:

1. Microbial pesticides
2. Plant-incorporated protectants
3. Biochemical pesticides

Microbial pesticides

Microbial pesticides are like security guards hired to protect crops (except they have no labor rights). They are microorganisms, such as viruses, bacteria or fungi that prey on the pests that cause harm to crops. The most common being the bacterium Bacillus thuringiensis (Bt). This bacterium produces a crystal protein that is toxic when eaten by pests, such as caterpillars, moths and worms, proving fatal to them.

They come from naturally occurring or genetically altered bacteria, fungi, algae, viruses or protozoans. Microbial control agents can be effective and used as alternatives to chemical insecticides. A microbial toxin can be defined as a biological toxin material derived from a microorganism, such as a bacterium or fungus. Pathogenic effect of those microorganisms on the target pests are so species specific. The effect by microbial entomopathogens occurs by invasion through the integument or gut of the insect, followed by multiplication of the pathogen resulting in the death of the host, e.g., insects. Studies have demonstrated that the pathogens produce insecticidal toxin important in pathogenesis. Most of the toxins produced by microbial pathogens which have been identified are peptides, but they vary greatly in terms of structure, toxicity and specificity.

These microbial pesticides offer an alternative to chemical insecticides with increased target specificity and ecological safety so that they are used either uniqly or in combination with other pest management programmes. It is a ecologically based pest control strategy that relies heavily on natural mortality factors and seeks out control tactics that disrupt these factors as little as possible. Ideally, an integrated pest management program considers all available pest control actions, including no action, and evaluates the potential interaction among various control tactics, cultural practices, weather, other pests, and the crop to be protected.

These microbials as biocontrol agents present beneficiancy. They have efficiency and safety for humans and other nontarget organisms. They leave less or no residue in food. They are ecologically safe, so that other natural enemies are free of their threatening, leading to preservation of other natural enemies, and increased biodiversity in managed ecosystem. So, microbial agents are highly specific against target pests so they facilitate the survival of beneficial insects in treated crops. This may be the main reason that microbial insecticides are being developed as biological control agents during the last three decades.

Microorganism e.g., a bacterium, fungus, virus or protozoan as the active ingredient can control many different kinds of pests, although each separate active ingredient is relatively specific for its target pest. For example, there are fungi that control certain weeds, and other fungi that kill specific insects. One bacterial species like Bacillus thuringiensis may be more effectiv on Aedes aegypti while one another B. sphaericus strain can be effective on a different types of mosquito like Culex quinquefasciatus.

Advantages of microbial insecticides

Individual products differ in important ways, but the following list of beneficial characteristics applies to microbial insecticides in general.

- The organisms used in microbial insecticides are essentially nontoxic and nonpathogenic to wildlife, humans, and other organisms not closely related to the target pest. The safety offered by microbial insecticides is their greatest strength.
- The toxic action of microbial insecticides is often specific to a single group or species of insects, and this specificity means that most microbial insecticides do not directly affect beneficial insects (including predators or parasites of pests) in treated areas.

- If necessary, most microbial insecticides can be used in conjunction with synthetic chemical insecticides because in most cases the microbial product is not deactivated or damaged by residues of conventional insecticides. (Follow label directions concerning any limitations.)
- Because their residues present no hazards to humans or other animals, microbial insecticides can be applied even when a crop is almost ready for harvest.
- In some cases, the pathogenic microorganisms can become established in a pest population or its habitat and provide control during subsequent pest generations or seasons.
- They also enhance the root and plant growth by way of encouraging the beneficial soil microflora. By this way they take a part in the increase of the crop yield.

Disadvantages of microbial insecticides

Naturally there are also the limitations which are listed below, but do not prevent the successful use of microbial insecticides. These factors just provide users to choose effective microbial products and take necessary steps to achieve successful results.

- Because a single microbial insecticide is toxic to only a specific species or group of insects, each application may control only a portion of the pests present in a field and garden. If other types of pests are present in the treated area, they will survive and may continue to cause damage. Conventional insecticides are subject to similar limitations because they too are not equally effective against all pests. This is because of selectivity indeed and this negative aspect is often more noticeable for both general predators, chemicals and microbials. On the other hand predators and chemicals may be danger for other beneficial insects in threatened area.
- Heat, desiccation (drying out), or exposure to ultraviolet radiation reduces the effectiveness of several types of microbial insecticides. Consequently, proper timing and application procedures are especially important for some products.
- Special formulation and storage procedures are necessary for some microbial pesticides. Although these procedures may complicate the production and distribution of certain products, storage requirements do not seriously limit the handling of microbial insecticides that are widely available.
- Because several microbial insecticides are pest-specific, the potential market for these products may be limited. Their development, registration, and production costs cannot be spread over a wide range of pest control sales. Consequently, some products are not widely available or are relatively expensive.

Plant-incorporated protectants

Plant-incorporated protectants (PIPs) are compounds that are toxic to pests, and are produced by plants that have been genetically modified.

One approach, to reduce destruction of crops by phytophagous arthropod pests, is to genetically modify plants to express genes encoding insecticidal toxins. The adoption of genetically modified (GM) crops has increased dramatically in the last 11 years. Genetically modified (GM) plants possess a gene or genes that have been transferred from a different species.

The production of transgenic plants that express insecticidal ä-endotoxins derived from the soil bacterium Bacillus thuringiensis (Bt plants) were first commercialized in the US in 1996. The expression of these toxins confers protection against insect crop destruction. The lethality of Bt endotoxins is highly dependent upon the alkaline environment of the insect gut, a feature that assures these toxins are not active in vertebrates, especially in humans. These proteins have been commercially produced, targeting the major pests of cotton, tobacco, tomato, potato, corn, maize and rice, notably allowing greater coverage by reaching locations on plants which are inaccessible to foliar sprays.

There are numerous strains of Bt, each with different Cry proteins, and more than 60 Cry proteins have been identified. Most Bt maize hybrids express the Cry1Ab protein, and a few express the Cry1Ac or the Cry9C protein, all of which are targeted against the European corn borer (Ostrinia nubilalis Hubner) (Lepidoptera), a major pest of maize in North America and Europe.

Some recent maize hybrids express the Cry3Bb1 protein, which is targeted against the corn rootworm complex [Diabrotica spp. Coleoptera], also a major pest of maize, especially in North America. Cotton expressing the Cry1Ac protein is targeted against the cotton bollworm [Helicoverpa zea - Lepidoptera].

For example, the gene for the toxic protein produced by Bt is added to such plants, so that they also produce the same protein. This makes the plant toxic for the pests that try to eat it. Once again, the lack of matching receptors prevents any harm to humans from such plants.

Biochemical pesticides

Biochemical pesticides are herbal pesticides—naturally occurring chemicals that possess pest-repelling properties. Examples of biochemical pesticides include hydrogen peroxide (H_2O_2), plant oils like eucalyptus oil, lemongrass oil and rosemary oil.

If a man-made compound is structurally similar to a naturally occurring compound, it is still considered a biochemical pesticide. An example is the biochemical pesticide methoprene. Methoprene is structurally similar to insect juvenile hormone (JH).

Methoprene supplied as a pesticide in unnatural quantities mimics the function of JH and disrupts the biological cycle of the insect, thereby interfering with its development.

While most biochemical pesticides have been used successfully, one problem with this category of biopesticides is that there isn't sufficient scientific research conducted regarding their toxicity or safety. Many of them do possess active ingredients that could prove toxic in high concentrations.

The biopesticides are certain types of pesticides derived from such natural materials as animals, plants, bacteria, and certain minerals. For example, canola oil and baking soda have pesticidal applications and are considered biopesticides. Even at the end of 2001, there were approximately 195 registered biopesticide active ingredients and 780 products. Biopesticides are biochemical pesticides that are naturally occurring substances that control pests by nontoxic mechanisms. They are living organisms (natural enemies) or their products (phytochemicals, microbial products) or byproducts (semiochemicals) which can be used for the management of pests that are injurious to crop plants. Biopesticides have an important role in crop protection, although most commonly in combination with other tools including chemical pesticides as part of Bio-intensive Integrated Pest Management.

They are biological pesticides based on pathogenic microorganisms specific to a target pest offer an ecologically sound and effective solution to pest problems. They pose less threat to the environment and to human health. The most commonly used biopesticides are living organisms, which are pathogenic for the pest of interest. These include biofungicides (Trichoderma), bioherbicides (Phytopthora) and bioinsecticides (Bacillus thuringiensis, B. sphaericus). The potential benefits to agriculture and public health programmes through the use of biopesticides are considerable.

The advantages of using biopesticides (in place of other chemical ones are based on these factors:

- Ecological benefit; inherently less harmful and less environmental load.
- Target specificity; designed to affect only one specific pest or, in somecases, a few target organisms,
- Environmental beneficiency; often effective in very small quantities and often decompose quickly, thereby resulting in lower exposures and largely avoiding the pollution problems.
- Suitability; when used as a component of an integrated pest management (IPM) programs, biopesticides can contribute greatly.

Disadvantages of biopesticides

Biopesticides are not a new invention; in fact, they were used long before the development of chemical pesticides. Now, you may be wondering, if biopesticides are so much better, why did people ever bother coming up with chemical pesticides?

A biopesticide is a natural method of pest regulation, so there are factors that cannot be fully controlled. They may take time to kill the target pest, or the microorganisms may have a complex life cycle, or be difficult to handle. Due to storage and handling constraints, biopesticides are more expensive than synthetic pesticides.

A significant amount of investment is required to research new ways to improve biopesticides; coupled with the probability of low profits and financial rewards, companies are reluctant to invest much in this area. Additionally, there are regulatory constraints that come with the production and distribution of biopesticides, as they are biological agents.

CONCLUSION

Biopesticides are critical to sustainable farming, and major improvements to them will undoubtedly follow. Optimizing production, through lower costs and improved yields, is one way of making biopesticides more financially friendly.

Today, the growing interest in organic, chemical pesticide-free farming has increased the demand for biopesticides. Efforts are being directed towards further research into traditional methods of biopesticide usage so as to reach a better scientific understanding of these time-proven methods.

Currently, the focus of research is on improving their production, commercialization, delivery and awareness amongst farmers.

Overall, biopesticides are environmentally friendly, can target a range of pests, from rodents and insects to microorganisms. They are natural, harmless and bio-degradable (so they don't remain in the environment or the soil). Thus, biopesticides are an excellent solution to tackle both environmental pollution and pest management.

3

BIOPESTICIDES: BENEFITS AND THEIR PRODUCTION

Biopesticides include pesticides produced from living organisms such as plants, animals, and microorganisms (viruses, bacteria, and fungi) to control and provide crops with protection against dangerous plant-damaging pathogens through their non-toxic, eco-friendly mechanism of action. Biopesticides are effective even in low quantities and have no residual effects, which is a major concern for consumers, specifically for edible vegetables and fruits. Table 1 shows the different kinds of plant products used as biopesticides. When using biological pesticides as a source of pest control, the productivity of biological and chemical pesticides on crops (vegetables and fruits) is the same. The importance of biological pesticides increased because the pesticides are effective, biodegradable, non-toxic, different modes of action and available raw materials. The subclasses of biopesticides include botanical, biochemical, and microbial pesticides. Botanical pesticides are derivatives of plants that inhibit growth or kill pests. Plants with pesticidal properties also contain compounds that affect plant pathogens like fungi, viruses, bacteria, and nematodes.

Table 1. Some plant products used as a biopesticide

Plant Product Used as Biopesticide	Target Pests
Limonene and Linalool	Fleas, aphids and mites, also kill fire ants and several types of flies,
Neem	A variety of sucking and chewing insects
Pyrethrum / Pyrethrins	Ants, aphids, roaches, fleas, flies, and ticks
Rotenone	Leaf-feeding insects, such as aphids, some beetles (bean leaf beetle, etc.), caterpillars.
Ryania	Caterpillars (European corn borer, corn earworm, etc.) and thrips.
Sabadilla	Squash bugs, harlequin, leafhoppers, etc.

BIOPESTICIDES - PROS AND CONS

Botanical Pesticides and Mechanism of Their Action

The biologically active substances in botanical pesticides have various modes of action against different pests, including fungi, nematodes, bacteria, and insects. The modes of action include denaturation of proteins, repellent, and inhibition regarding the type of botanical compound and pest. Phenols, alcohols, alkaloids, tannins, and other secondary metabolites found in these pesticides produce toxicity to fungal cell organelles, cell walls, and cell membranes. These metabolites also prevent spore germination, mycelial development, delayed sporulation, and inhibit the production of vital enzymes, DNA, and protein synthesis.

The plant metabolites induce modifications in the structure of hypha and mycelia, which prevents, the production of fumonisin (Fusarium spp.) and aflatoxin (Aspergillus spp.) which are toxins from fungi. The metabolite reduces pathogenicity of mycotoxin producing fungal pathogens. Botanical pesticides contain nematocidal properties that inhibit hatching of eggs and suppress nematode population; its constituent affects the numbers of other microorganisms in the soil, which also affect the survival of nematode eggs and juveniles. Some substances specifically destroy second-stage juveniles, egg masses, cause general larval toxicity, and affect nematode populations in an ecosystem.

For instance, plant extracts derived from Lantana camara and Azadirachta indica prevented the hatching of eggs of root-knot nematode (Meloidogyne incognita) that immobilized and killed the second-stage juveniles. Botanical pesticides have different antibacterial properties, including growth inhibition, and the lack of peptidoglycan cell walls makes Gram-negative bacteria susceptible. For example, methanolic extracts of Aloe vera inhibited the growth of Escherichia coli and Bacillus subtilis, while acetone extracts inhibited Pseudomonas aeruginosa subsequently.

The phytochemicals present in Aloe vera contributes to the antimicrobial activity; destroys microbe proteins, thus disrupting their functionality. The antiviral activity of botanical pesticides occurs through the production of antiviral proteins by manipulation of the host, which inhibits any activities between the virus and the plants.

The mechanism of action includes; inhibits virus penetration and replication in the host cell, enzymatic activity, and hemagglutinations, which are essential for virus attachment. For instance, the cotton-seed oil sludge extracts (acetone) significantly inhibited Tobacco Mosaic Virus obtained from infected tobacco (Nicotiana tabacum) plants under laboratory conditions.

Biochemical Pesticides and Mechanism of their Action

Biochemical pesticides are unconventional pesticides obtained from natural sources such as plants, bacteria, animals, and some minerals. They comprised of naturally occurring compounds that control pests through non-toxic mechanisms. Many biologically functional groups of low-risk act as biochemical pesticides for crops.

Semiochemicals are secretions of plants or animals that change the function of receptor organisms of similar or different kinds. Semiochemicals include; pheromones, which operates, between individuals within a species and allelochemicals act between individuals of different species. Semiochemicals influence insect life situations, including feeding, mating, and egg-laying. Biological control with pheromones or kairomones can identify and monitor insect populations. Another strategy for controlling pests is the use of Semiochemicals as feeding deterrents. The most common mechanism for pest control via Semiochemicals includes attracting, trapping and killing of pest insects.

Enzymes are protein, which serves, as crucial elements for gene expression that catalyses biochemical reactions. Plant protection against insects is mediated by enzymes that impair digestive mechanisms in the insect gut. A potential protective strategy against insects measures the recombinant expression of insecticidal proteins like áamylase and protease inhibitors.

Microbial Pesticides and Mechanism of their Action

Microbial pesticides include a group of biopesticides consisting of naturally occurring bacteria, viruses, fungi, or protozoans that target a specific problem. In certain situations, pesticidal activity may originate from metabolites produced by these organisms. Bacteria target the fly and beetle larvae, caterpillars, fungal, bacterial diseases, and soil-borne pathogens. Most Bacillus thuringiensis strains and subspecies are used in biopesticides.

During the formation of bacterial spores, Bacillus thuringiensis produces a crystal protein called Bt δ-endotoxin. Bt δ-endotoxin can break down gut cells when ingested by susceptible insects. Bt δ-endotoxin is host-specific and can cause the death of the host within 48 hours. Research shows that they are unharmful to vertebrates and safe to beneficial organisms, environment, and humans.

However, over 1000 viruses, which act, as insect pathogens have been isolated in addition to bacteria. Some authors identified different nuclear polyhedrosis viruses (NPVs) that infested 525 insect species. Certain bacteria and fungi can provide defence responses in plants by producing either cell wall components or salicylic acid. Microbial pesticides work differently, and their functionality lies in the type of microbes used.

Though, microbial pesticides interfere with the insect sex pheromones that make them unable to signal other insects for breeding. Microbial pesticides based on fungi control some specific kinds of weed leaving the crops unaffected. Bt microbial pesticide works by targeting the larvae of insects like mosquitoes, flies, and moth larvae. It effectively starves the larvae to death before developing into adults by binding to the larval digestive tract.

Biopesticides have a range of attractive properties for IPM. We list the main benefits here, and we also look at some of the challenges of working with biopesticides.

- In addition to their ability to control pests and diseases, biopesticides produce little or no toxic residue, and partly for this reason they are usually considered to be minimal risk products for human health. Many biopesticides are residue-exempt and are not required to be routinely monitored by regulatory authorities or retailers.
- Re-entry and handling intervals are becoming more important considerations when selecting a plant protection product for use, especially in protected crop and many biopesticides have a zero or low re-entry and handling interval.
- Biopesticides can often be applied with existing spray equipment, and some microbial biopesticides can reproduce on or near to the target pest/ disease, giving a degree of self-perpetuating control.
- As alternatives to conventional chemical pesticides, they can help reduce the selection pressure for the evolution of pesticide resistance in pest populations, and there is good evidence that some microbial biopesticides can stop the expression of resistance once it has evolved.
- The risk of pests and disease developing resistance to biopesticides is often considered to be low, certainly for those agents that have multiple modes of action. However, in principle there is always a potential for a target pest / disease to develop resistance or tolerance, dependent upon the size of the selection pressure, and hence we think it is good practice to adopt an anti-resistance strategy when using biopesticides.
- Biopesticides often have good compatibility both with biological pest control agents (natural enemies) and with conventional chemical pesticides, so they can be readily incorporated into IPM programmes.
- Bioinsecticides can also be useful as a second line of defence or supplementary treatment. In pest management, there are often times in the season when the invertebrate pest population starts to run away from the ability of a predator or parasitoid to control it. In such situations, a bioinsecticide can be used to hold back the population development of the invertebrate pest and allow the predator or parasitoid to "catch up". Having this back-up often makes the difference between success and failure of IPM in protected crops.
- The costs of developing a biopesticide are significantly lower than those of a conventional chemical pesticide, which should encourage companies to develop a wide range of products.
- Product Differentiation: Biopesticides benefit progressive dealers who seek to offer a broad portfolio of crop protection products and serve customers with evolving product preferences. The addition of quality biopesticides to a dealer's inventory differentiates it from outlets that deal solely in traditional chemical pesticides.

- Service Differentiation: Because biopesticides work differently than traditional chemical products, dealers and consultants may provide information services to growers to support product use. These services range from product recommendation to application training to crop assessment.
- Environmental Safety with Biopesticides: The general public's concern for environmental safety is one of the leading drivers of biopesticide usage. As growers focus on this aspect of production and market these practices to their customers, dealers may recommend biopesticides as an environmentally sound and efficacious product. Non-toxic to non-target organisms, including beneficial insects and wildlife, many biopesticides also are biodegradable. They decompose quickly and do not negatively impact surface water and groundwater. Biopesticides typically are effective in small quantities, which eliminates pollution concerns sometimes associated with traditional chemicals. In addition, biopesticides are manufactured from naturally occurring raw materials in an environmentally responsible and sustainable manner.
- Access to Organic Market: Organic farming represents a unique, fastgrowing segment of the food industry. While only 5% of all current biopesticide use is for organic applications,5 many biopesticides are registered for and are effective in certified organic production systems. Dealers who supply biopesticides have access to the organic market and to emerging "no residue" or "pesticide-free" markets.
- Biopesticides Add Value and Yield: Crop quality and yield largely determine a grower's income. Biopesticides provide dealers with products that can markedly improve crop quality and yield by preventing pest damage and promoting physiological benefits in plants, including increased fruit size and enhanced color. Dealers who supply biopesticides and encourage their innovative use are on the forefront of yield-and-profit enhancement practices.
- Biopesticides Partner with Traditional Chemicals in IPM Programs: Sustainable agriculture relies on integrated pest management (IPM) techniques - which combine proven cultural, biological, and chemical methods to control pests - as an environmentally sound approach to crop production. Dealers supply products that support these methods and many biopesticides are specifically designed for use in conjunction with stable traditional chemistries in an IPM program.
- Biopesticides Allow Labor and Harvest Flexibility: Biopesticides provide dealers with a solution to grower requests for products with labor and harvest flexibility. In terms of application, biopesticides generally have short worker restricted entry intervals. Workers can safely return to a field shortly after treatment, allowing them to assess crop conditions and follow up with applications of other crop inputs when needed. In addition, biopesticides generally have short pre-harvest intervals. This allows growers to better manage their labor, harvest and shipping

schedules. In terms of plant growth, Plant Growth Regulators (PGRs) may be used to control and optimize harvest times. PGRs that inhibit the biosynthesis of ethylene in plants, for example, can slow down fruit maturation, ripening, and abscission.

- Biopesticides Make Residue Management Easier: Consumers are increasingly aware of chemical use in food production and the potential for chemical residue on food. Dealers may promote biopesticides, which leave little to no residue on food, to growers as residue management products. Many biopesticides are exempt from residue limits on fresh and processed foods. In addition, biopesticides may be applied in environmentally sensitive areas without residue concerns and may be used to manage residue levels for exported produce.

The downsides to using biopesticides include the following:

- A slower rate of control and often a lower efficacy and shorter persistence compared to conventional pesticides.
- Greater susceptibility to adverse environmental conditions.
- Because biopesticides are not as "robust" as conventional pesticides, they require a greater level of knowledge on behalf of the grower to use them effectively.

CHALLENGES AND OPPORTUNITIES FOR DEALERS

Biopesticide products present several significant opportunities for dealers, including:

- Stocking biopesticides increases overall inventory and meets the needs of a wider audience of growers, including growers from larger operations, who tend to be the most avid users of biopesticide products, and organic growers.
- Dealers who stock biopesticide products today are considered progressive and will be more likely to attract progressive growers. This builds long-term customer base and strengthens reputation.
- The use of biopesticides requires training and related information services. Dealers have the opportunity to sell agronomic services, from technical support to application to assessment, along with products.
- Biopesticides designed for use in conjunction with certain traditional chemical pesticides may be used to spur sales of those chemistries.
- As new biopesticide products are released, dealers may establish and/or strengthen working relations with both large and small-medium manufacturers.

However, there are several significant challenges for dealers, including:

- Biopesticides are relatively new to the market and work in different ways than traditional chemical pesticides. Some growers may be reluctant to alter a program that relies on well-known and affordable traditional chemicals.

- The effectiveness of biopesticides is measured differently than traditional chemicals. Growers are familiar with standard field trial results and may not understand biopesticide success measures.
- There are some common misconceptions about biopesticides, including that they are more expensive to use and less effective than traditional chemicals, that they are only for organic production and that they are difficult to use.

Challenges and Opportunities for Consultants

Biopesticide products present several significant opportunities for consultants, including:

- Biopesticides operate differently than traditional products. Growers need assistance when adopting these new technologies. Consultants may advise growers on such things as proper application timing and frequency; the most effective application methods to ensure thorough crop coverage; target pest identification; and pest/disease pressure and life cycle dynamics.
- Biopesticides provide different benefits depending on the product used and the crop. Consultants may help growers plan for the proper use of biopesticides within an IPM program, monitor pests in-season and evaluate the effectiveness of the product through field performance trials and marketable yield and quality data.

Biopesticides present several significant challenges for consultants, including:

- Biopesticides are relatively new to the market and the need for education of growers is significant. Some growers may be reluctant to be advised on new systems.
- The effectiveness of biopesticides is measured differently than traditional chemicals. Growers may not understand or accept the new measures provided by consultants.

PRODUCTION OF BIOPESTICIDES

Biopesticides are powerful tools for sustainable agriculture products as these are the alternatives to chemical pesticides as they offer a solution to problems such as pest resistance. Biopesticides are certain types of pesticides which are derivative of naturally occurring substances like bacteria, plants, animals and certain minerals. Biopesticides also consist of active creatures that have the ability to repel the agricultural pests. They are quite less harmful as compared to the chemical pesticides. Biopesticides are generally host specific and do not harm the pests that are beneficial for soil health whereas, the chemical pesticides may also mark the organisms which are desired such as mammals, birds, and insects.

Pest problem is one of the major constraints for achieving higher production in agriculture crops. India loses about 30% of its crops due to pests and diseases each year. The damage due to these is estimated to be Rs.60,000 crores annually. The use of pesticides in crop protection has certainly contributed for minimising yield losses. The pesticides, which are needed to be applied carefully, only when the threshold limits of the pest population is exceeded. However, quite often the indiscriminate and unscientific use of pesticides has led to many problems, such as pests developing resistance, resurgence of once minor pest into a major problem besides environmental and food safety hazards.

The problem of insect-pest is acute in case of all the crops and especially so in case of commercial crops. The use of insecticides and pesticides have increased manifolds during the past 3 - 4 decades with the introduction of intensive cropping. The average consumption of pesticides in India is about 570 gms per ha. as compared to developed countries like Japan, Thailand and Germany where the consumption rate is 11 kg, 17 kg and 3 kg per ha, respectively. Though the average quantum of pesticides usage in India is low, the damage caused due to their indiscriminate usage and poor quality maintenance is alarming. Interms of value, much of the pesticide application is accounted for by a few crops. For example, cotton, paddy and vegetable crops account for 80% of the value of pesticides applied in India.

Pesticides or chemicals are meant to control harmful pests such as insects, nematodes, diseases, weeds etc. However, excessive use of pesticides not only leave residues in soil, water and air but also have adverse effects on the non target organisms such as pollinators, parasitoids, predators and wild animals. This has adversely affected the ecological balance resulting in pest resurgence, development of resistance in the pest species and environmental pollution. Development of pest resurgence and resistance has resulted in high cost of production and low income especially to cotton farmers.

In view of the several disadvantages associated with the unscientific use of pesticides in agriculture, there is an urgent need for minimising the use of chemical pesticides in the management of insect pests. Growing public concern over potential health hazards of synthetic pesticides and also steep increase in cost of cultivation/low profit making by farmers has led to the exploration of eco-friendly pest management tactics such as Integrated Pest Management (IPM). IPM aims at suppressing the pest species by combining more than one method of pest control in a harmonious way with least emphasis on the use of insecticides. In simple terms "IPM is the right combination of cultural, biological and chemical measures which provides the most effective, environmentally sound and socially acceptable methods of managing diseases, pests and weeds". The major components of IPM are prevention, observation and intervention. The IPM seems to be the only answer to counter some of the major pests of crops, which have become unmanageable in recent years. The success of IPM largely depends upon conservation of naturally occuring bio control agents.

Production of Biopesticides

Though there are about 140 biopesticide production units existing in the country as on today, they are able to meet the demand of only less than 1% of cropped area. There exists a wide gap, which can only be bridged by setting up of more and more units for production of biopesticides. This requires large scale investment and private participation.

Some of the local small scale industries have already started production and marketing of Trichoderma viride (against few fungal diseases) and Trichogramma (against sugarcane early shoot borer). There is a scope to enhance production and use of biological control agents in the days to come as the demand is on the increase every year.

Location of Biopesticide Units

In order to achieve optimum results, care needs to be taken to set up biopesticide facilities in areas which have appropriate climatic conditions. The production of Biopesticides requires controlled climatic conditions. Temperature control is less costly in locations where there is no extreme conditions. Besides the climatic conditions, the proximity of the location to the market is also important. However, care must be taken that the production facilities are set up at least a quarter of a mile away from farming areas, so as to prevent the contamination of production facilities by insecticides from the farming areas. Also, as air pollution can damage biopesticides, the production should be located away from industrial and urban areas.

Production of Bacillus thuringiensis Biopesticide

Bacillus thuringiensis (Bt) is the most successful commercial biopesticide in the biological control market accounting for 90% of all biopesticides sold all over the world. B. thuringiensis is a Gram positive spore-forming bacterium found naturally in soil, water, dead insects and grain dust. This bacterium is characterized by its ability to produce crystalline inclusions proteins or crystals called endotoxin during sporulation and/or stationary phase. The crystalline inclusions along with the spores have a great potential to control a great number of pest insects belonging to the order Lepidoptera, Diptera and Coleoptera.

The Bt biopesticide preparations are based on endotoxin proteins accumulated as parasporal crystals produced by the bacterial cells. Various agricultural and industrial by-products may be used as raw materials in biopesticide industry, such as maize glucose, soybean flour, peanuts, sugarcane molasses, which also contain N, and liquid swine manure. Researchers also mentioned less expensive complex sources of organic nitrogen such as grains, beans, oilseeds, fish and meat meals. Some carbohydrate sources also contain protein, and vice versa, and could serve as the sole source of nutrients for bacterial culture. Researchers first tested the use of wastewater sludge for the production of Bt based biopesticide, however, the entomotoxicity level reported in the paper was low. Moreover, the low cost of by-products as nutrient source in fermentation media of Bt biopesticide production has received little attention. However, a higher level of entomotoxicity is desired to reduce biopesticide production costs.

Fall armyworm, Spodoptera frugiperda, is one of the most important corn insect pests in Brazil and its damage may reduce yield production up to 34%. The use of chemical insecticides to control S. frugiperda has increased over the years in Brazil, and has generated resistance of this insect to chemicals. Microbial pesticides may become feasible and reliable alternative to control this insect, however Bt based biopesticides is limited due to high production costs. B. thuringiensis strains have a narrow spectrum of insecticidal activity when compared to conventional chemicals. Regarding specificity Bt shows differences even within an order of insects (e.g. Lepidoptera) and dramatic differences in sensitivity are exhibited among species.

Differences in sensitivity were found among species, i.e. Spodoptera spp. are difficult to control with Bt based biopesticide (strain HD1), however Heliothis virescens and Plutella xylostella are not. Researchers also found that Bt was not much efficient against S. frugiperda, but these authors used only one available biopesticide at that time (strain HD1). Nevertheless, the efficiency of B. thuringiensis depends on finding the specific and efficient strain to a specific insect pest. Researchers characterized strain 344, Bt sv tolworthi, very efficient against fall armyworm larvae, and this strain was very stable during the fermentation process.

The use of raw materials in the fermentation media to produce Bt based biopesticide may become a reliable and cheap way to integrate this insect pathogen into the fall armyworm pest management, but these materials may vary due to local availability, cost and in both physical and chemical composition, due to the type of the primary source. The objective of this work was to evaluate the use of industrial raw material and liquid swine manure as the primary carbon and nitrogen sources to produce Bt based biopesticide, using a highly efficient strain against fall armyworm.

Bacterial strain

Bacillus thuringiensis sv tolworthi, strain 344, belonging to Embrapa Maize and Sorghum Microbial Bank was isolated from a soil sample. The sample was mixed with salt solution (0.8%) and the flask was incubated on a rotary shaker for18 hours. One milliliter of this suspension was heated in water bath at 80o C for 30 minutes, cooled in ice for 15 minutes and plated on sporulating medium composed of Luria Bertani (LB) enriched with salts (g.L-1 in distilled water) 0.002 g $FeSO_4$, 0.02 g $ZnSO_4$, 0.02 g $MnSO_4$, 0.3 g $MgSO_4$ and 2 g glucose. The pH was adjusted to 7.2. After 24-72 incubating at 30 ± 1° C, colonies were observed on a phase contrast microscope.

Media preparation

Strain 344 was sub cultured in medium composed of Luria Bertani (LB) enriched with salts ($g.L^{-1}$ in distilled water) 0.002g $FeSO_4$, 0.02g $ZnSO_4$, 0.02g $MnSO_4$, 0.3g $MgSO_4$ and 2g glucose and pH adjusted to 7.5. After sub culturing, strain 344 was streaked on sporulating medium containing the same salts plus 12g of Bacto-Agar (Difco), 8.0g of nutrient broth, and incubated for 24 hour at 30 ± 1°C and maintained at 4°C for future use.

Medium 1 was composed of lab commercial Luria Bertani (LB) plus salts ($FeSO_4$, $ZnSO_4$, $MnSO_4$, $MgSO_4$), and 0.2% glucose; medium 2 was composed of 1.5% glucose, 0.5% soybean flour plus salts; and medium 3 was composed of liquid swine manure (LSM) at 4%. Liquid swine manure (LSM) was collect 32 days after fermentation, under field conditions when the C:N ratio is considered to be stable. LSM was diluted to 4% and enriched with 1.5% soybean flour and 0.5% glucose. All three media were sterilized at 121o C for 30 minutes.

Seed culture preparation

A loopful of strain 344 from Bacto-Agar plates was used to inoculate a 500 mL Erlenmeyer flask containing 150 mL of sterilized LB medium enriched with salts. The flasks were incubated on a rotary shaker at 250 rpm at 30° C for 12-16 hours. The actively growing cells from these flasks were used as a second inoculum for the production in Erlenmeyer flasks.

Growth of B. thuringiensis tolworthi in three different media

The 500 mL Erlenmeyer flasks containing 150 mL of sterilized media 1, 2 and 3 were inoculated with strain 344 grown in the seed culture. A 5% (v/v) inoculum from this flask was used to inoculate 500 mL Erlenmeyer flasks containing 150 mL of sterilized media. The flasks were incubated for 96 hours at 30o C. Samples were withdrawn at different intervals and analyzed for colony forming units (CFU) of VC (viable cell) and heat resistant cells – viable spores (VS) by the serial plate technique on plates. The appropriately diluted samples were plated on agar plates and incubated overnight at 30°C to form fully developed colonies. Two replicates were used for each parameter. Each replicate of spore count, pH and heat resistant spores were checked three times.

Cell mass

Samples of 100 mL were centrifuged at 10000 rpm for 20 minutes. The supernatant was discarded and the pellet was lyophilized. Dry weight was calculated and expressed in gram per liter. The same sample was used for the spore count and toxicity test. Two replicates were used each one measured three times.

Spore count

For spore count culture samples were heat treated at 80° C for 15 minutes, serially diluted and plated on LB plus salts and glucose agar plates. Plates were incubated at 25 ± 1° C for 24 hours to form fully developed colonies that were counted and expressed in colony forming units per mL (c.f.u./mL). Two replicates were used each one measured three times.

pH

The pH of all media was adjusted to 7.2 with NaOH after sterilization, and was measured at regular intervals during 96 hours of fermentation process. Two replicates were used each one measured three times.

Toxicity test

The spore crystal complex produced in different media was assayed against 2 day old larvae (first instar) laboratory-reared S. frugiperda. First instar larvae were raised on an artificial diet for 2 days prior to the bioassay. The bioassays were conducted using corn leaf surface method, where strain 344 produced in three different media were prepared in sterile distilled water, and 18 ìL of each suspension on the corn leaf surface were provided to each larvae. Corn leaves were washed once with 0.5% sodium hypochlorite and two times with distilled water. Each bioassay included 7 doses of 2 replicates each, along with the appropriate control. Larval mortality was scored up to 4 days after larvae infection to check pathogenicity.

FUNGAL BIOPESTICIDES

In the growth of diseases on many agricultural crops, the fungal pathogens show a key part causing serious losses in the productivity of crops1 . Fungal biopesticides can control insects and plant diseases and their method of operation varies and depends on the pest and pesticidal fungus. Two of the most common commercial fungal biopesticides are Trichoderma viride and Beauveria bassiana. In this review, biopesticides are produced using T. viride and B. bassiana.

Trichoderma Viride

T. viride is a fungal antagonist present in almost all farming soils. For multiplying its own spores, Trichoderma rise in the direction of the mycelium of disease producing fungi and then discharge enzymes. Many crop diseases are controlled by Trichoderma which consists of phaseolin, fusarium wilt, root rot and many seed plant diseases.

Banana, brinjal, cardio, tomato, ginger, guava, tobacco, potato, coffee, soybean, chickpea, tea, black gram, cotton, turmeric, pea, groundnut, citrus, sunflower, black pepper, sorghum, sugarcane and sesame are some of field and vegetable crops in which Trichoderma can be used. This can be also used for regulating various fungal diseases which are soil borne. The variety of plant pathogenic fungi for which this biocontrol agent is effective consists of Botrytis cinerea, Pythium spp., Sclerotium rolfsii, Rhizocctonia solani, Slerotinia homoeocarpa and Fusarium spp.

Beauveria bassiana

Beauveria is a type of fungus which is occurring naturally in soils and has been tested for using against soil-borne insects. A number of soil insects have natural tendency to resist these pathogens. Some of the soil insects such as foliar pests do not exhibit this characteristic. Therefore, the development of this fungus is being targeted to resist the foliar feeding pests. B. bassiana is responsible for many diseases such as muscadine disease in insects. The B. bassiana spores start growing in the internal body of the masses as it came in contact with the cuticle of the specified insects. The fungus starts producing toxins as it gets multiplied through the body of the insect and drains out the nutrients from the insect, ultimately slaying them.

The fungal pathogens are not desired to be eaten up by masses in order to be effective, unlike bacterial and viral pathogens. As a white mould, the fungus raises over the cuticle after the death of host and discharge out millions of fresh infectious spores. The rate of killing a host of Beauveria spores is temperature dependent, for example, the plant lice are killed in 3 to 5 days at a constant temperature of 720 F. It can be used against number of insects such as aphids, weevils, whitefly, thrips, mealybugs and psyllids. This can be also useful for plantation crops comprising ornamentals, vegetables, turf, fruits and berries, forestry, and herbs and species.

EXISTING METHODS

Solid State Fermentation (SSF) is a better option as compared to submerged liquid fermentation due to its applications in the production of Biopesticides, enzymes, biopharmaceuticals, aroma compounds etc. and also due to the microorganism's growth similar to their natural habitat. The selection of the substrate and suitable microorganisms, isolation, optimised process parameters are some important factors to be considered.

Substrate particle size should be between optimum as it affects the process at large rate and should between 180 μm to 1.4 mm. The degree of polymerization and crystallinity are the chemical factors that affect solid substrate utilisation. The important parameters to be controlled are temperature, pH aeration and agitation. The temperature varies largely around the trays. pH, control is very difficult in SSF process but pH correcting solutions can help in maintaining it during the process. Water activity is also an important factor which affects the product formation. Better knowledge of the design and process parameters can be helpful in the successful implementation of the SSF process.

In studied a vast variety of compounds comprising fungicides, herbicides, insecticides, rodenticides, nematicides. Plant growth regulators and others are covered by the term pesticides. Insecticides named as Organochlorine (OC) have been successfully used for monitoring many diseases, for example, typhus and malaria. The pesticides are preferred by the farmers due to their fast, simple and cheap resolution for monitoring the pest problems.

Due to these advantages, their extensive use leads to their accumulation in the environment in a significant amount. The deposits of the pesticides have been detected in ground and surface water as well as in air and soil throughout the nations and ushered serious issues. Even the herbicides can contribute to the destruction of the environment. Due to the maximum usage of weed killers, they can also cause severe problems.

So this section concluded that in order to minimize the accumulation of pesticides in the environment and to prevent long-term impacts, the usage of synthetic pesticides should be decreased by using ecofriendly solutions to the problems.

The processes of biopesticides production consist of the genetic integration of DNA which deliberate a shield against damage caused by pests. The acceptance of biopesticides by the farmers can be enhanced by the implementation of organic farming and forcing the use of products which should be chemically free. The paper concluded that in order to register a product in India, the registration committee needs information on the efficiency of the product, its chemistry, poisonousness and labeling and packaging.

A few of biopesticides such as Bacillus thuringiensis var. galleriae, Bacillus sphaericus, Trichoderma viride, T. harzianum and neem-based pesticides are previously registered in India and also being used.

The production of biopesticide from Bacillus thuringiensis using commercial lab media. Accordingly, medium 1 showed the maximum spore concentration during the period of fermentation.

Shelf lives are mostly tested under refrigerated storage conditions but its determination under non-refrigerated conditions is of vital importance particularly under high temperatures. They shelf life of B. bassiana can be increased using optimal conditions.

The improvement in the shelf lives of biopesticides under non-refrigerated conditions is very important. Active packaging should be broadly adopted to enhance the shelf life of B. Bassiana.

The influence of sources of carbon and nitrogen on B. bassiana and T. viride. The study of the mycelium growth and sporulation of both the fungi was done. With the availability of the Czepeck Dox Broth media, the effects of nitrogen and carbon sources were determined. The T. viride is known for its biofungicide whereas the B. bassiana is entomopathogenic fungi. It was concluded that these fungi represent more growth rate on all the nutrient sources considered for the study. In the Czepeck Dox media, the sources of nutrients are potassium nitrate, ammonium sulphate and sodium nitrate. The T. viride showed maximum biomass product in ammonium sulphate whereas B. bassiana showed in potassium nitrate.

The biopesticides are not still able to take over the chemical pesticides in the market due to certain limitations such as low organic carbon in the soil, unstable performance in the fields, poor quality, less mass production and poor shelf life. The metabolites and additives can improve the effectiveness of the product. The biomolecules such as Flavanoids and lipochitooligosaccharides can be helpful in increasing the ability to tolerate the environmental stresses on the crop and will thus increase the yield.

PRODUCTION OF HA NPV AND SI NPV

Introduction

Baculovirus group have a very narrow host range and generally infests the larvae of crop pests. The research aimed at insect pest control is, therefore, confined to nuclear polyhedrosis viruses (NPVs) and granular viruses (GVs).

NPV is a nucleic acid (double standard, circular DNA) enclosed in protein matrix, hence it is called polyhedral occlusion body (POB). NPV infects the nucleus of the cell and multiplies within the nucleus.

In India, extensive research has been conducted on the use of NPVs for tackling two major pests namely Spodoptera litura and Helicoverpa armigera.

Nuclear Polyhedrosis viruses like Ha NPV, SINPV are increasingly being used as alternatives to chemicals. These viruses have distinct advantages over other methods of pest control. NPVs are virulent pathogens of insect characterised by the polyhedral occlusion bodies (POB).

These viruses are highly specific and do not affect beneficial insects like parasitoids and predetors and are safe to fish, birds, animals and man. Considering the usefulness of NPV's there has been a growing demand amongst the farmers for these bioagents.

Major equipment required

The major equipments like centrifuge, laminar flow, magnetic shaker, microscopes, autoclave, coolers, refrigerators, incubator, distillation units etc. are required in addition to glassware, plastic trays, basins, iron racks for mass production of Ha NPV and SI NPV.

Spodoptera litura (Tobacco Caterpillar)

Spodoptera litura commonly known as tobacco caterpillar, is a polyphageous pest. It is a serious pest of tobacco nurseries and also a sporadic pest of cauliflower, cabbage, castor, cotton, groundnut, potato and lucerne. It causes serious crop losses.

SI NPV

The virus is specific and infects only Tobacco Caterpillar. NPV can be successfully multiplied on tobacco caterpillar and the viral extraction can be applied in the field to control the caterpillar. For continuous production of SI NPV, it is necessary to rear Tobacco Caterpillar larvae continuously in a lab condition.

Gram pod borer (Helicoverpa armigera)

It is widely distributed in India and infests/damages a variety of cultivated and wild plants throughout its distribution range. It is a serious pest on commercial crop like cotton; pulses like redgram and bengalgram; vegetables like tomato, bhendi and dolichos bean; oilseeds like sunflower, soybean and safflower and cereals like sorghum and maize.

Ha NPV

Ha NPV is a highly infective microbial biopesticide which can be used to control Gram borer. It is derived from naturally diseased or under laboratory conditions artificially infected larvae of gram borer.

Mass production of Ha NPV and SI NPV

The mass production of Ha NPV and Si NPV involves 3 steps

- Rearing of adult Gram pod borer and Tobacco caterpillar for mass production of eggs.
- Rearing of larvae of the above species either on the host plants like chickpea and castor under semi natural condition or on the synthetic diet in the laboratory conditions. In the model only the later is considered for large scale commercial production of NPV.
- Inoculation of Ha NPV and Si NPV into the larvae of Gram pod borer and Tobacco caterpillar respectively for mass multiplication of viruses and extraction of polyhedral occlusion bodies(POBs) from the diseased larvae, which are used as biopesticide on the crop plants.

Details of mass production

Diet preparation

The larvae of Gram pod borer and Tobacco caterpillar can be multiplied by using chick pea based semi-synthetic diet. The composition of the diet for rearing larvae is as follows:-

	Item	Quantity
'A' fraction:	Chickpea (Kabuli chenna) flour	105.00 gm
	Sorbic acid	1.00 gm
	Streptomycin sulphate	0.25 gm
	10% formaldehyde solution	2.00 ml
'B' fraction:	Agar-agar	12.75 gm
'C' fraction:	Ascorbic acid	3.25 gm
	Yeast tablets	25 tablets
	Multivitaplex	2 capsules
	Vitamin E	2 capsules
	Distilled water	780.00 ml

390 ml of water is mixed with fraction 'A' of the diet in the blender which is run for two minutes. Fraction 'A' and 'C' are mixed and the blender is run again for 1 minute. Fraction 'B' is boiled in the remaining 390 ml water, added to the mixture of A and B and the blender is run for a minute. Formaldehyde solution is added at the end and the blender is again run for a minute.

MASS PRODUCTION OF EGGS

Tobacco caterpillar

The culture of Tobacco caterpillar is initiated by collecting eggs from the fields of castor, cauliflower, lucerne, tobacco etc. These field collected eggs are reared in isolation to eliminate the emerging parasitoids and diseases, if any.

The culture can also be established by collecting the gravid females with the help of light traps. Once the pure culture is established the mass production is commenced under laboratory conditions after the first generation established.

Pairs of newly emerged moths of Tobacco caterpillar are placed in well ventilated plastic containers. The inner wall of the containers is lined with paper to enable the adults to lay eggs. The bottom of the container is lined with sponge covered over by blotting paper. The moths are provided with 50% honey solution and water on two cottons swabs placed in small plastic cups. The eggs which are generally laid in batches on the paper are cut out. Freshly laid egg masses are sterilised by dipping in 10% formalin for 30 minutes, washed in running water for 30 minutes, dried on blotting paper and kept for hatching in sterilised glass vials.

The freshly laid eggs can also be surface sterilised in 0.05 percent solution of sodium hypo chlorite for 5 minutes. These eggs are washed several times in running tap water to remove the traces of sodium hypo chlorite. The traces of sodium hypochlorite could be neutralized by dipping the eggs in 10% sodium thiosulphase solution and again the eggs are washed thoroughly under running tap water. The surface sterilised eggs are kept in plastic tubes (7.5 x 25 cm) on moist tissue paper for continuing the stock culture. After 3 days, the newly hatched larvae are transferred to bouquets of castor leaves and kept in a plastic container with stand for pupation. The pupae are collected 3 days after all the larvae enter the sand. The pupae are sexed and kept on a lid over a wet sponge in adult emergence cage. After 10 days, freshly emerged males and females are collected from their respective emergence cages.

Gram pod borer (Helicoverpa armigera)

The culture of Gram borer is initiated either collecting the adults with the help of light traps. It could be by collection of larvae on a large scale from its host crops in endemic areas. Nucleus culture can also be obtained from the established laboratories. The material thus obtained is reared in the laboratory in aseptic conditions and the healthy progeny is selected and established.

The production starts with the availability of 250 pairs of adults every day, which will yield 10,500 eggs daily. The adults are kept @ 100 pairs in each oviposition cage with a cloth enclosing the frame. A circular plastic mesh (on which cotton swabs soaked in water and honey solution are placed in small containers) rests on a support above the base of the frame. The cloth cover is open at both ends with a 20 cm vertical slit in the centre which can be closed with a zip or cloth clips. The cloth cover enclosing the frame is tied with rubber bands at both ends. It is placed on tray with a sponge at the bottom soaked in water. The temperature inside the cage is maintained at 260 C and humidity at 60 - 90%.

The eggs are laid all over the inner surface of the cloth cover. The egg cloth is removed daily. This cloth is surface sterilised in 10% formalin for 10 minutes, the eggs could also be surface sterilised using 0.2% sodium hypchlorite solution for 5-7 minutes and treated with 10% sodium thiosulphate solution to neutralise the effect of sodium hypo chlorite, rinsed in distilled water. The eggs are later placed on paper towell under laminar flow for drying. The dried cloth pieces containing eggs are kept in 2 litre flasks containing moist cotton. Flasks are plugged with cotton wrapped in muslin cloth and the bottom of the flask is wrapped with aluminium foil.

REARING OF LARVAE ON SEMI-SYNTHETIC DIET

Tobacco caterpillar

Stage - I (rearing of early instar larvae): The rearing unit is prepared by placing a sponge piece on a glass sheet. The sponge is covered with a single layer of soft tissue paper. A small plastic container containing 200 surface sterilised eggs of Tobacco caterpillar is placed in the centre over the tissue paper. A petri dish containing about 200 ml of diet is placed inverted over the tissue paper. The eggs hatch within 25 hr and neonate larvae crawl and spread out on the diet.

Stage - II (rearing of late instar larvae): Late instar larvae are reared in a modified plastic boxes. One window each on the four sides of the box is cut and covered with a fine plastic mesh to provide sufficient ventilation and to prevent moisture accumulation inside the box. A thick layer of sterilised sand is spread at the bottom of the box. A small piece of tissue paper is kept at the centre over the sand.

The diet in the petri dish (containing 200 larvae) is divided into five equal pieces. One piece of diet bearing 40 larvae is kept in plastic box over the tissue paper so that the sand does not soil the diet. In this way, 5 boxes are charged with larvae from 1 petri dish. A plastic grill is fitted into the box in such a manner so that it forms a crest higher than the brim of the box. Thick cake of diet (about 500 gm) in a petri dish is divided into two equal pieces. One such piece is kept on the top of the crest and the lid of the box is then fixed so that the diet and grill crest are opposed to each other just beneath the lid. After consuming the small quantity of diet on tissue paper the larvae crawl and perch on the grill and feed from the ceiling of the box. The boxes are stacked and left intact for 3 days. During this time the diet is almost completely consumed. Now another piece of fresh diet (about 250 gm) is kept on the crest in each box and the boxes are closed and stacked again. During the last 3/4 days of larval stage the food consumption is maximum and so is the fecal matter accumulation on the sand layer. After 20 days from hatching the larvae move into the sand and start pupating. In a period of 25 days, all the larvae, pupate and the chitinisation of pupae is also completed. The boxes are now ready for the pupal harvest. The pupae are collected, cleaned, sterilised and placed in adult emergence cages. The freshly emerged moths are then placed in oviposition cages.

Gram borer

The larvae of gram borer can also be reared on a chickpea based semisynthetic diet as detailed under point 7.2.1.

The diet is poured as per the requirement either on the nylon mesh for rearing 5-7 day old larvae or in tray cells for rearing the older larvae or poured into sterilised petri plates and allowed to solidify. The diet could be stored in the refrigerators for upto 2 weeks. For preparing large quantities of diet, the quantity of diet ingredients to be used should be calculated accordingly and industrial type waring blenders could be used.

The larvae are removed from the top of the aluminium foil wrapped flasks with a brush and then transferred to the diet. 220 larvae are transferred to diet impregnated on nylon mesh and placed in plastic containers or sterilised glass vials. 100 such containers are maintained daily for 5-7 days. Multi-cellular trays with semi-synthetic diet is advantageous for rearing a large number of larvae.

Starting with 10,500 eggs, the total number of larvae available is 10,000 considering an estimated 5% mortality in initial 5 days of emerging and 10% mortality upto first 5 - 7 days. The total number of larvae available for virus production is 8000 (80%). The rest of 20% will be utilized for maintenance of host culture continuously.

The diet requirements at various stages of production of larva are:

- for the young larvae upto 5-7 days will be 2 gms / larva.
- for 5-7 day old larvae for Ha NPV production will be 4gms/larva
- for five to seven day old larvae for continuation of host culture will be 6 gms/ larvae.
- for rearing the field collected larvae for augmenting the nucleus stock will be about 1 kg

In host culture units, larvae start pupating when they are 18-19 days old and the pupation will be over within 2-3 days. The harvested pupae are surface sterilised using 0.2% sodium hypo chlorite solution followed by washing with 10% sodium thiosulphate solution to neutralize sodium hypo chloride and then washed thoroughly with distilled, sterilised water. After washing, the eggs are dried by rolling over blotting paper. The male and female pupae are separated out and placed over moist sponge in adult emergence cages.

The egg, larval, pupal and adult stages of gram borer last 3-4, 18-29, 7-8 and 7-9 days respectively. The oviposition period of the females is about 5 days.

Production of Helicoverpa armigera NPV (Ha NPV) and Spodoptera litura NPV (SI NPV).

For Ha NPV and SINPV production, the synthetic diet prepared is poured at 4gm/cell in the multi-cavity trays and the diet surface is uniformly sprayed with virus prepared in distilled sterilised water at 18 x 106 POBs / ml. Eighty percent of the total 5-7 day old larvae are utilised for Ha NPV and SINPV production.

The trays are incubated at 260 C for 7 days. In case of virus infected larval trays, the diseased larvae dies after attaining its maximum size of 6th instar, where the dead caterpillar will have 2-6 billion poly occlusion bodies (POB) which is in terms of larval equivalent (LE). 1 LE of H.armiegera NPV = 6 x 109 POBs; 1 LE of S. litura = 2 x 109 POBs. The dead larvae have to be harvested, macerated in distilled/sterilised water and filtered through muslin cloth to get the crude suspension of the virus. The extraction is centrifuged to further clarify the solution.

OTHER IMPORTANT ASPECTS

1. General precautions to be followed while maintaining host cultures
    - In production units, keep the host culture in a separate room and the virus production and storage facility should be located in a different facility.
    - In the NPV production units, inspite of best care, 100% larvae are not infected, the larvae which do not turn inactive after 4 - 5 days and keep consuming the normal diet should be culled out regularly from the NPV production unit.
    - Utmost care should be taken to prevent the break in the chain of the production system. This could be achieved only if highly dedicated and disciplined workers are engaged for such production units.
    - Strict hygiene should be maintained in different facilities. The equipments used should be either heat sterilised or sterilised using steam or chemicals. The work place should be thoroughly disinfected with sodium hypo chlorite solution.
    - The host culture should be initiated from a batch of healthy adults.
    - Microbial infection could be avoided if good insect husbandry practices are followed. If infection is detected, the culture or infected part should be destroyed immediately. Besides hygienic conditions, optimum temperature (24o C- 26o C) and humidy (65 - 70%) should also be maintained.
    - The texture and quality of the natural/semisynthetic diet should be good.
    - Entry to host culture unit after visiting virus production unit should be avoided.

2. Mode of action: NPV acts as a stomach poison only to the target host (pest) and hence beneficial insects are not affected. The infected larvae become pale and glossy and tissue get disintegrated and liquified. Most of the body tissues and organs (except gut) get infected by polyhedral occlusion bodies (POBs), which contains the virions. The liquid which oozes out of the infected larvae (which hang upside down) contains millions of POBs. Each POB measuring about one micron in diameter and possessing a characteristic movement can be identified under the microscope.
3. Field application and dosage: Ha NPV is used for controlling H.armigera attacking cotton, redgram, bengalgram, tomato, okra, sunflower, groundnut, chillies, maize, sorgram etc., whereas, SI NPV is used for controlling tobacco caterpillar attacking tobacco, groundnut, soyabean, sunflower, cotton, cabbage, beetroot, cauliflower etc.
4. Directions for use of NPV
    - The recommended dosage is 200 ml of NPV/acre or 500 ml/ha containing 100 and 250 larval equivalent (LE) of NPV respectively as active infective material (one LE = 6 x 109 POBs).
    - 100 ml of NPV could be diluted in 200-400 litres of water when high volume sprayer is used and in 50-70 litres of water in case of power sprayers.
    - Preferable to spray using high volume knap-sack sprayer. Virus should be sprayed during evening hours. Spray should be initiated as soon as some newly hatched larvae are observed or three to five days after a trap catch of 5 moths per pheromone trap. Subsequent sprays should be made at 7-10 days intervals depending upon the pest population.

COMPATIBILITY WITH OTHER INSECTICIDES

The viral pathogens seems to be less sensitive to chemical pesticides. When the combination of pathogen and pesticide is used, sometimes synergistic action is noticed. But is recent years mixing of NPV with insecticides is not advisable due to cross resistance problem.

TECHNOLOGY FOR MASS PRODUCTION OF TRICHODERMA FUNGI

Introduction

Crop losses due to soil borne plant pathogens worldwide are Pythium spp., Furarium oxysporum, Sclerotium rolfsii, Rhizoctonia solani and Phytophthora spp. These fungi pathogens generally cause wilt disease in many crops. Trichoderma, a fungi, which grow saprophytically in soils have proved as an effective biocontrol agent of wilt diseases.

Trichoderma spp. are commonly found in almost any soil and other natural habitats consisting of organic matter such as decaying bark, plant material, etc. They grow trophically towards hyphae of other pathogenic fungi, coil them and degrade their cell walls. This process is called "mycoparasitism", which limits the growth and activity of plant pathogenic fungi. In addition, they produce toxic metabolites which protect the seeds from soil borne pathogenic fungi, by forming a protective coating on them.

Trichoderma spp. are saprophytic fungi that grow best in neutral and acid soils and thrive well in moist conditions.

The important species available for mass production are Trichoderma viride and Trichoderma harzianum

Equipment required

Equipments like fermentor, rotary mixer, auto packer, rotary shaker, laminar flow, water distillation unit, refrigerator, haemo cytometer etc. are required for the production of Trichoderma fungi.

2. Major steps in production process

Inoculation | Fermenter run | Harvesting | Blending | Drying and Packing

Outlines for production of Trichoderma

- The pure mother culture of Trichoderma fungi is being maintained in Agri. Universities, IARI, some ICAR institutions (like PDBC, Bangalore) etc. The mother culture can be purchased from the identified sources. They have to be further sub-cultured and maintained purely for mass production by adopting standard techniques under the supervision of trained microbiologist / pathologist.
- The culture has to be mass multiplied in two levels namely (i) at primary level using shakers in flasks and (ii) secondary stage multiplication in fermenters. The important factor in this is the preparation of growing medium in which the culture is mass multiplied. For Trichoderma Fungi, the growing media used in the model is molasis and protein material.
- After the growing media is formulated and sterilised in fermenter, it is inoculated using the culture multiplied in the flasks.
- The molasis based culture media is continuously aerated by passing sterile air from compressors. After about 3-4 days fermentation period, the culture will be ready for packing in a carrier material.
- While the inoculated culture is gathering ready in the fermenters, the carrier material is sterilised in autoclaves and kept ready for mixing the culture. Talk powder is reported to be the commonly used carrier material for Trichoderma Fungi.
- The cultured (fungi) and sterilised carriers are mixed mechanically in a blender and the material is packed using semi automatic packing and sealing machine.

Dosage

Talc based formulations of the fungal antagonists are applied at the rate of 4gm per kg of seed for controlling soilborne plant diseases. Mix the powder with sufficient quantity of water to make slurry for treating seed before sowing.

Advantages of Trichoderma applicatiom

- Ecofriendly
- Can be used along with organic manure
- Trichoderma spp. are also known to suppress plant parasitic nematodes (roo-knot nematodes).
- Lower cost and longer efficacy than fungicides
- Does not lead to development of resistance in plant pathogens
- No physotoxic effects
- Minimises losses and cost of production and increases yield & profit.
- Promoter plant growth

Application

Soil application

Trichoderma spp. suppress the activity of soil borne fungal pathogens, especially Rhizoctania solani and Pythium spp. and protect transplanted seedlings by colonizing their roots.

Seed treatment

Seed treatment is an alternative approach to introduce Trichoderma spp. into the soil. This method requires smaller amounts of biological material than soil treatment. Unlike chemical fungicides, Trichoderma spp. provide long term protection without any adverse side effects.

SEX PHEROMONE TRAPS OF HELICOVERPA ARMIGERA AND SPODOPTERA LITURA

Introduction

Sex pheromones are single or complex blend of different chemicals released by one insect to attract the opposite sex of the same species. In general, females (especially the moths) emit sex attractants to attract males for mating.

Sex pheromones are artificially synthesized in the laboratories and supplied as sex pheromone lures. Such pheromones are placed in the field to attract trap and kill the males, thus matting is not allowed. Hence, sex pheromone traps can be considered as a key component in Integrated Pest Management (IPM).

Ready-to-use Sex pheromone lures and traps are available for Helicoverpa armigera (attacking crops like cotton, redgram, tomato, okra, sunflower, chillies, maize, sorghum etc.) and Spodoptera litura (attacking crops like tobacco, groundnut, sunflower, cotton, cabbage, beetroot, cauliflower, etc.)

Advantages of pheromone lures

- No harmful effects to beneficial insects, non-target organism or an environment.
- Helps in monitoring & early detection of pests (at moth stage only)
- Helps in scheduling pest control measures
- Reduction in usage of insecticides
- Much simpler

Equipment needed

Only micropippets are required in addition to rubber septas, traps and pouches.

Production of Pheromone Traps

Sex pheromones are insect specific, produced artificially in laboratories and they are generally imported. In India, it is available from National Chemical Laboratory (NCL), Pune. Chemicals obtained from laboratory is diluted to the required dosage and filled into plastic lures with the help of micro pippets and closed with rubber septa. Lures are individually sachet packed and should be stored under refrigerated conditions when not in use.

Field application

Lures containing sex pheromones are placed into insect trap and erected in the field at a recommended spacing. The lure will release the sex pheromane at a constant rate over a period of 2-4 weeks. Male months are attracted and while attempting for matting, fall into a container having pesticide. Thus the female moths in the field are deprived of successful mates and fail to reproduce or lay viable eggs.

Dosage

Timely use of sex pheromone helps in early detection and prompt action against pests. In general, 2-3 traps / acre are recommended for 'monitoring' or more for 'mass-trapping'. These are arranged such that the trap is 1-2 feet above the crop canopy. On the field each lure is effective for atleast 15 days. Change the lures once in two weeks.

CONCLUSION

Organic food indicates huge scope for growth of Bio-pesticides sector in India. At the same time increasing population can be fed by organic farming dependence is a big question and unless organic farming yield can be brought equal to that of conventional farming involving the use of agrochemicals etc, the organic farming may not be feasible at the moment. Rich traditional knowledge base available with the highly diverse indigenous communities in India may provide valuable clues for developing newer and effective bio pesticide. A new more active strain of B.t was produced which has increased the performance and acceptance of commercial products and broadened its use against other insect pests. Commercial biopesticides should be economical to produce, have persistent storage stability, be easy to handle, mix and apply, and provide effective control of target pests.

4

MODES OF ACTION OF BIOPESTICIDES

Most biopesticides are intended for use with traditional pesticides, which has contributed to the rapid increase in their use – the global market for chemical pesticides is experiencing annual growth rates of about 15%, while the global biopesticide market has an annual growth rate of about 5%.

Biopesticides can increase crop yields by working synergistically with chemical pesticides to optimize crop protection strategies, by allowing timely reuptake intervals and through their ability to be used close to harvest. Biopesticide demand is thus also being driven by growers increased knowledge about the mechanisms of action of biopesticides and how to use them.

BIOHERBICIDE MODE OF ACTION

Bioherbicides include microorganisms that infect weeds, and natural compounds with herbicidal activity sourced from microorganisms or plants. Bioherbicides target many of the same plant metabolic processes targeted by conventional herbicides, but they may bind differently to enzymes, or they may have multiple modes of action, making them potentially useful for treating herbicide-resistant weeds. Bioherbicides fall into two major classes:

- Microbial bioherbicides
- Bio-derived (biochemical) bioherbicides

Microbial bioherbicides include bacterial and fungal endophytes (microorganisms which colonize plants without causing disease) as well as viruses.

The modes of action of fungal bioherbicides (mycoherbicides) include the action of plant cell wall degrading enzymes, photobleaching molecules (macrocidins) and phytotoxic molecules such as the diterpene chenopodolin, and growth-regulating plant hormones such as indole acetic acid, IAA. Phoma macrostoma is an example of a mycoherbicide developed for pre-emergence weed control.

Bacterial bioherbicides exhibit phytotoxic activity attributed to specific metabolites as well as the elicitation of plant defense responses such as programmed cell death through the induction of reactive oxygen species..

Bio-derived (biochemical) bioherbicides (phytotoxins) are based on mechanisms used by microbes and plants to exclude invasive plant species from their immediate vicinity.

Allelopathy is a mechanism by which plants produce one or more biochemicals (allelochemicals) that affect the growth and survival of other plants.

The study of allelopathy has, among others, led to the identification of leptospermone, a herbicidal allelochemical produced by the Lemon Bottlebrush, and the subsequent development of its commercial chemical analogue.

Biochemical bioherbicides target many of the same plant metabolic processes targeted by conventional (synthetic) herbicides. However, natural phytotoxins may bind to proteins at different sites than their synthetic counterparts, or may have multiple modes of action.

For example, sorgoleone (from Sorghum bicolor), a D1-protein-binding inhibitor of photosystem II can control weeds resistant to the QB-site PSII-inhibitor, atrazine.

Besides its inhibition of PSII electron transport and the concomitant increase in reactive oxygen species (ROS), sorgoleone inhibits the enzymatic activity of thylakoidal HPPD (hydroxyphenylpyruvate dioxygenase) and thus the formation of vitamin E, deriving the plant of one of its primary antioxidant protective mechanisms. Finally, sorgoleone inhibits mitochondrial respiration by blocking proton-gradient-driven ATP synthesis.

Chloroplast-based metabolic processes inhibited by natural phytotoxins include photosynthetic electron transport, chlorophyll and carotenoid biosynthesis, amino acid biosynthesis as well as lipid membrane function and stability.

Sorgoleone (Sorghum bicolor) and stigmatellin (from Stigmatella aurantica) are inhibitors of photosystem II (and hydroxyphenylpyruvate dioxygenase, HPPD), while pyridazocidin (Streptomyces spp.) is an inhibitor of photosystem I. Aurachins (Stigmatella aurantica) inhibit electron flow between PSII and PSI (Figure 1).

In photosynthesis, the photolysis of water leads to an accumulation of protons in the thylakoids stroma which drive the synthesis of ATP (photophosphorylation) via the enzyme ATP synthase (ATPase). Nigericin (Streptomyces) uncouples the proton gradient, inhibiting photophosphorylation while tentoxin (Alternaria tenuis) blocks photon flow through the chloroplast ATPase.

Gabaculine (Streptomyces toyacaenis) and cyperin (Preussia fleischhakii) are natural phytotoxins inhibiting chlorophyll biosynthesis. Carotenoid biosynthesis may be blocked at different sites by the action of fosmidomycin (Streptomyces lavendulae).

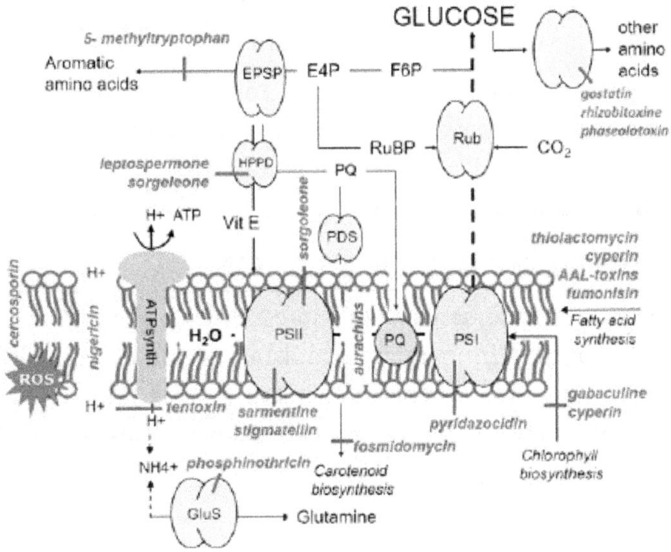

Figure 1: Target sites of natural phytotoxins affecting metabolic processes within the thylakoid membrane.

Leptospermone (Leptospermum scoparium) and sorgoleone inhibit the hydroxyphenylpyruvate dioxygenase (HPPD) enzyme, blocking both carotenoid biosynthesis and the formation of the antioxidant protectant, vitamin E. Like vitamin E, carotenoids have a protective antioxidant function through their quenching or scavenging of oxygen radicals.

Phytotoxins such as sarmentine (Piper longum) and cercosporin (Cercospora kikuchii) lead to loss of membrane integrity through reactive oxygen species (ROS), arising as a result of the over-reduction of the photosynthetic apparatus due to PSII inhibition.

The repair and maintenance of lipid membranes may be inhibited by herbicidal phytotoxins such as thiolactomycin (Streptomyces spp.) and cyperin (Preussia fleischhakii).

Biosynthesis of the aromatic amino acid tryptophan maybe inhibited by the phytotoxin 5- methyltryptophan (Cantharellus cibarius), while a range of phytotoxins such as gostatin (Streptomyces), rhizobitoxine (Bradyrhizobium) and phaseolotoxin (Pseudomonas) inhibit the biosynthesis of other amino acids. Phosphinothricin (Pseudomonas syringae) is a herbicidal phytotoxin inhibiting the synthesis of the amino acid glutamine.

Mitochondria-based metabolic processes inhibited by natural phytotoxins include respiration and lipid biosynthesis (Figure 2). Respiration can be blocked through the inhibition of NADH oxidation by fusicoccin (Fusicoccum amygdali). Syringomycin (Pseudomonas syringae) uncouples the proton gradient through the formation of membrane pores, while sorgoleone (Sorghum bicolor) and juglone (Juglans spp.) block proton-driven ATP synthesis via the H+ ATPase.

Fatty acid biosynthesis in the mitochondria can be inhibited by thiolactomycin (Streptomyces) and cyperin (Preussia fleischhakii).

Thaxtomin (Streptomyces scabies) is a microbial phytotoxin that controls grass, broadleaf and sedge weeds by disrupting cellulose biosynthesis. Additional modes of action for bioherbicides include gene expression and regulation, as well as effects on hormonal processes such as jasmonate mimicking and auxin signalling.

BIOINSECTICIDE MODE OF ACTION

Bioinsecticides include microorganisms that infect insects, and natural compounds with insecticidal activity sourced from microorganisms, plants or animals. Bioinsecticides known not to harm non-target (beneficial) insect species may be considered for use during flowering, a period where many conventional pesticides cannot be used.

Bioinsecticides fall into three major classes:

- Microbial bioinsecticides
- Plant-incorporated protectants
- Bio-derived (biochemical) bioinsecticides

Microbial insecticides include entomopathogenic fungi, capable of acting as pathogenic parasites of insects (and other invertebrates such as nematodes), and typically belong to the divisions Ascomycetes and Zygomycetes.

Besides their parasitic action, entomopathogenic fungi may also produce fungal toxins.

Entomopathogenic bacteria such as Bacillus thuringiensis (BT) are used to control insect pests, typically via their digestive tracts. Entomopathogenic viruses are specific to individual species: Cydia pomonella granulovirus specifically targets the codling moth C. pomonella (a major pest of fruit trees) killing the insect at the larval stage. Beneficial nematodes (e.g. Steinernema), while not micro-organisms, are sometimes classified as such and are used in the biological control of insect pests.

Plant-incorporated protectants (PIPs) are a form of biological insecticide in which foreign DNA is inserted into crop genetic material (GM crops). Bacillus thuringiensis toxin genes may be directly incorporated into plants through the use of genetic engineering, giving these plants tolerance primarily to Lepidopteran insect pests, such as caterpillars and beetles.

RNA interference (RNAi) is an emerging technology, whereby RNA molecules capable of disrupting genes conferring tolerance to insecticides may be used to target resistant insects such as Potato Beetles.

Bio-derived (biochemical) insecticides are naturally occurring compounds, typically produced by plants for defense from predation, but also include microbial extracts that control insects or induce plant defense responses. Insect pheromones are used to attract insect pests to traps, or to disrupt mating cycles.

BT toxin crystals are microbial-derived insecticidal proteins which are activated upon ingestion by lepidopteran insect pests, leading to cell disintegration and insect death. Spinosad is a bioinsecticide based on natural metabolites, and is produced by fermentation of naturally occurring bacteria (Saccharpolyspora spinosa).

Surfactin, a biodegradable lipopeptide biosurfactant isolated from Bacillus species (see below) is a mosquitocide, Chitosan is a polysaccharide derived from fungi or crustacean shells, capable of inducing systemic resistance against insect pests.

Plant extracts such as garlic extracts or canola oil can repel insects, or act as antifeedants, while other plant extracts such as Nicotine (isolated from tobacco), Pyrethrins (isolated from Chrysanthemum plants) or Ryanoids (isolated from e.g. Ryania speciosa) have neural insecticidal activity, leading to hyperstimulation -or paralysis – and death (Fig. 3).

In cholinergic (ACh) neurons, Nicotine competes with ACh neurotransmitters and competitively binds to the ACh receptor on the postsynaptic nerve, resulting in uncontrolled nerve stimulation (hyperstimulation) and death. Pyrethrins disrupt normal nerve function by binding to voltage-gated sodium channels, preventing them from closing normally and giving rise to neural hyperactivity and insect death.

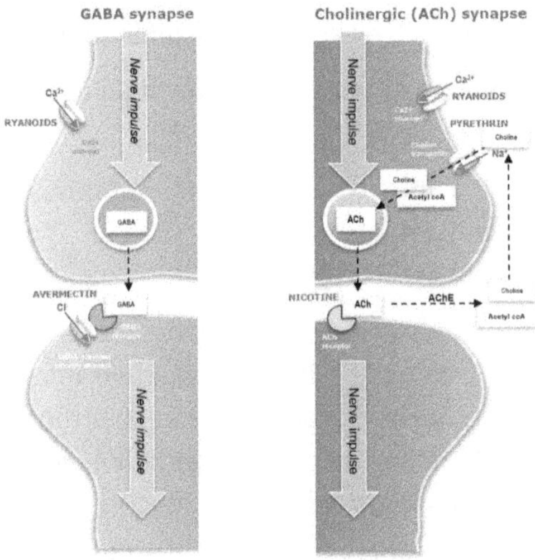

Fig. 3: Mode-of-action of microbial and plant-derived neural bioinsecticides

Avermectins (naturally occurring compounds generated as fermentation products by the soil bacteria, Streptomyces avermitilis) affect insect-specific glutamate-gated chloride channels in GABA neurons, causing an influx of chloride ions (Cl-) into the postsynaptic neuron – leading to hyperpolarisation, paralysis and death.

Ryanoids are insecticidal alkaloids which interact with ryanodine calcium-channel receptors in insects, resulting in paralysis and death.

Other plant extracts such as Rotenone can inhibit mitochondrial electron transport (respiration). Finally, bio-derived insecticides include naturally occurring desiccants, such as borate or silica (diatomaceous earth, the fossilized remains of aquatic diatoms) which absorb the oil and fats from the cuticle of the insect's exoskeleton, leading to desiccation and death.

BIOFUNGICIDE MODE OF ACTION

Biofungicides are formulations of living organisms or natural metabolites used to control the activity of plant-pathogenic fungi.

Many successful microbial biofungicides can colonize and grow in treated soil or on plant leaves, providing a long-term source of microbial and biochemical modes of action.

Biofungicide mode of action is complex and includes:

- antibiosis – microbial metabolites such as antibiotics or fungitoxins which inhibit spore germination or pathogen growth
- plant defense induction – the induction by biofungicides of localized or systemic defense mechanisms within the host plant
- exclusion – competition and displacement of pathogenic fungi from the rhizosphere or on the leaf surface by microbial biofungicides
- mycoparasitism – parasitization of pathogenic fungi by microbial biofungicides
- biostimulation – stimulation of plant growth and resilience

Antibiosis – The genus Bacillus comprises different strains which are commercially available as biofungicides. Bacillus strains such as B. subtilis, B. amylolquefaciens and B. licheniformis produce antimicrobial metabolites such as lipopeptides, which exhibit a broad antimicrobial spectrum and surfactant activity.

Due to their lipid-protein structure, lipopeptides can penetrate lipid membranes such as the cytoplasmic membrane of cells, affecting their integrity and ion permeability. The Iturin and Fengycin families of lipopeptides exhibit strong antifungal activity across a relatively broad range of pathogenic fungi, with fengycins having activity against Oomycetes such as Pythium and Phytophthora.

Another family of lipopeptides, Surfactins exhibit antiviral and antibacterial characteristics. Despite having low inherent antifungal activity, surfactins may act synergistically in combination with iturin.

Surfactins also function as biosurfactants, acting both as a wetter (by decreasing the surface tension of water droplets) and detergent, and are involved in biofilm formation and motility, while inhibiting the biofilm formation of phytopathogenic bacteria, interfering with their attachment to host plants.

Streptomyces is another naturally occurring gram-positive bacteria. It is found in soil, and can produce antimicrobial compounds (the antibiotic Streptomycin).

Induction of localized or systemic plant defense responses – Plants detect invading pathogens and induce defense responses, such as the production of toxic chemicals and cell destruction (hypersensitive responses), to hinder disease development. Plant defense responses are not initiated until pathogens are detected, due to the energy costs of their production and maintenance. Modern biopesticide strategies must thus consider the kinetics of defense response induction prior to pathogen attack.

These strategies include priming and activating plant defense responses prior to pathogenic attack, through application of microbial biofungicides, or through microbial metabolites such as Harpin, a natural protein found in common pathogenic microorganisms.

Systemic Acquired Resistance (SAR) describes plant response to pathogen attack through priming of plant defense responses and compounds. Systemic Acquired Resistance is translocated within plants via the salicylic acid (SA) dependent signaling pathway.

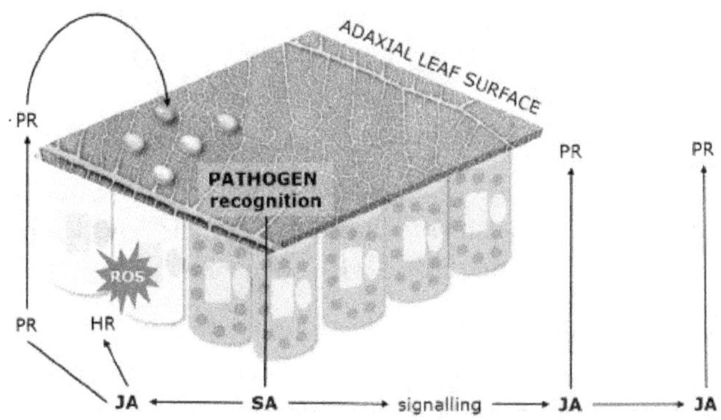

Fig. 4: Mechanisms of Systemic Acquired Resistance (SAR) as plant response to pathogen attack.

Jasmonic acid (JA) mediates the systemic induction of disease defense compounds, including pathogenesis related proteins (PR-Proteins) such as chitinases and peroxidases which degrade fungal cell walls, and antifungal phytochemicals such as phenolics, terpenoids and alkaloids.

Besides direct antifungal strategies, plants may activate a hypersensitive response (HR), also called "programmed cell death," which limits the access of biotrophs that feed on living tissue to water and nutrients.

The hypersensitive response is a mechanism used by plants to restrict pathogen development by inducing the localized, rapid death of cells in the region immediately surrounding an infection (Fig. 5).

Following recognition of the pathogen the activation of resistance genes triggers an oxidative burst of reactive oxygen species (ROS). Reactive oxygen species draw electrons from chloroplast thylakoid membranes (lipid peroxidation), leading to the loss of chlorophyll and associated chlorotic lesions.

Exclusion – Microbial biofungicides may compete with plant pathogens for nutrients and infection sites on leaf surfaces and in the soil. Some microbial biofungicides metabolize root exudates that stimulate pathogen germination, reducing their ability to infest plants.

Microbial biofungicides such as Trichoderma (a genus of soil fungi) may attack and suppress plant pathogens through a combination of displacement and the production of antimicrobial compounds such as cell-wall degrading enzymes.

Fig. 5: Hypersensitive response with associated chlorotic lesions in soya leaves following inoculation with soya rust.

Plant Growth Promotion – Biostimulation and protection against biotic (weeds, pests and diseases) and abiotic stresses have been reported for many agricultural biologicals, including biopesticides, by improving plant growth and root development (improving drought resistance and nutrient uptake). Increased crop vigor ensures optimal natural crop defense responses.

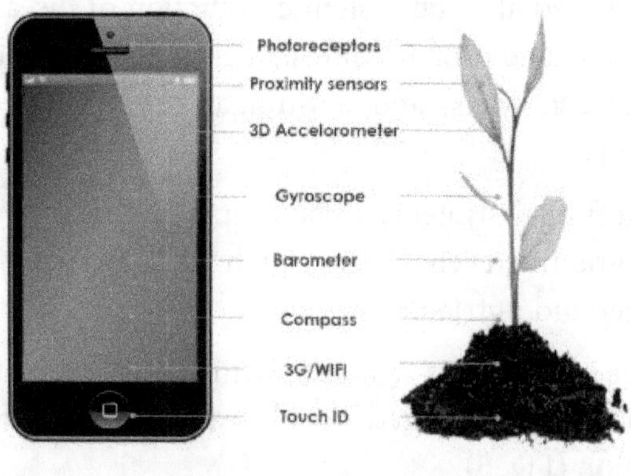

Fig. 6: Comparison of plant environmental, pathogen and pest response mechanisms to smartphone sensors.

Biotic and abiotic plant response mechanisms include photoreceptors, proximity sensors and accelerometers, and may be compared to smartphone sensors. Plants mediate defense strategies through complex communications systems, which may be compared to 3G (volatile plant hormones) and Wi-Fi (root-root and root-mycorrhiza interactions).

Since many biopesticides are slow to control the pest or pathogen, growers may feel that the product has insufficient effect and reapply unnecessarily. Diligent reading of labels and understanding how the product works – their mode of action – is necessary for the optimal application of these products, allowing growers to reap the full benefits and attain consistent results.

Attention is also being given to delivery and application practices, specifically to understanding the impact of plant defense induction kinetics on application timing and placement.

Because the mode of action of a biopesticide typically differs from that of a conventional pesticide, the efficacy of the biopesticide must also be assessed differently – not only through field trials but also by estimating growth, stress tolerance, yield and quality of the final product. It is crucial to understand how a biopesticide works in a cultivation program, and to thoroughly test the product based on all potential benefits.

Efficacy trials demonstrate conclusively that biology-based crop protection strategies can attain an efficiency at least on par with conventional crop protection strategies, when applied with due regard for factors such as plant-response kinetics and disease pressure development.

Reports of inconsistent efficacy lead to either over-optimism or skepticism by corporate decision makers and farmers alike. From a grower / end-user point of view, an important strategic strength of biopesticides lies in their application at specific timings: short preharvest intervals allow the application of biopesticides (especially biofungicides) immediately before harvest, where few (if any) conventional pesticides are available. Crop protection biologists have a unique opportunity to mediate an understanding of biopesticide mode of action, and to ensure the implementation of this knowledge in the development of commercial strategies.

BACILLUS THURINGIENSIS TOXINS IN THE CONTROL OF CATERPILLARS AND STINK BUGS IN SOYBEAN CULTURE

The use of microorganisms has assumed a prominent position among the options that seek to control insect pests without the use of chemicals and with high specific toxicity applied in agroecosystems. The Gram-positive bacterium Bacillus thuringiensis, Bt, stands out representing approximately 95% of micro-organisms used in biological control of agricultural pests in different cultures. Besides the economic aspect and the safety to human health, this bacterium is the most promising for the production of biopesticides and plant resistant to insects, associated with environmental preservation.

Considering the application of Bt in the biological control of insects, Hansen and Salamitou reported that the application of this entomopathogen is estimated at 13,000 tons of commercial formulated in the world, mainly applied against 2 Lepidoptera. Currently the largest single market for European biopesticides is Spain, followed by Italy and France. Despite that the overall market growth of biopesticides does not meet the expectations of the 1990s, the potential remains high and there are opportunities that could increase the total market in 2020.

The efficiency of the use of this entomapathogen is characterized by the production, during the sporulation, of crystalline protein inclusions which are toxic to various insect groups. These proteins are produced in the form of protoxins and transformed into toxic peptides by a number of events which occur in the midgut after ingestion of Bt, killing the susceptible insect.

B. thuringiensis can also serve as a source of toxic genes that can be expressed in plants and thus confer toxic property against different species of insect pests. Genetically modified plants (GMPs) that express the Bt genes, such as rice, corn, potato, cotton, and soybean are associated with the control of pests, especially Lepidoptera. The resistant cultivars result in increased productivity, greater economic value, reduction in the use of chemical pesticides, and benefits in the selectivity of the target pest. Moreover, Bt plants favor the exposure of target insects throughout infestation period, including pests in their most vulnerable stages of development. In another instance, the specificity of toxins of this bacterium is effective for only a limited number of species and the development of resistance to modified plants is more specific than other insecticides.

The main pest, the velvetbean caterpillar, is a defoliator insect that can cause complete destruction of the plant. In some moments in the development of culture, as in phase between flowering and pod production, the attacked plants are able to withstand only 30% of defoliation caused by this insect, without incurring losses in seed production. Chewing insects are susceptible to control with Bt insecticides when applied on the surface of the leaves or when genes are inserted into the plant genome, such as in the case of soybean. Other insect pests, such as the hemipterans, being endophagous, have no larval stage and her nymphs and adults have sucking mouthparts, not favoring the ingestion of the microorganism.

BACILLUS THURINGIENSIS

Bt is an aerobic or anaerobic facultative and sporulating bacterium. It can remain latent in the environment even in adverse conditions for its development. Bt can be found in soil, insects and their habitats, stored products, plants, forest, and aquatic environments.

This bacterium differs from other species belonging to this genus by the presence of a parasporal inclusion body (crystal) of protein origin, formed during sporulation (Fig. 7). This crystal is composed of Cry proteins which are encoded by Cry genes.

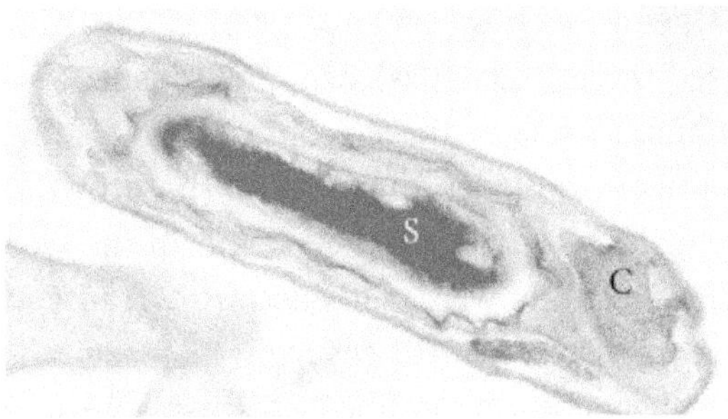

Fig. 7. Transmission electron microscopy of Bacillus thuringiensis, spore (S), and crystal (C).

The crystals of bipyramidal, cuboid, rhomboid, oval, spherical, or even no definite shape of a strain of Bt toxins may contain up to five different protein whit different molecular weight may vary between 30 and 142 kDa as in the case of strain HD-1 (B. thuringiensis subs. Kurstaki) which produces toxins Cry1Aa, Cry2Aa, Cry1Ab, and Cry1Ac Cry2Aa.

Cyt and Cry proteins that form the crystal, with toxic properties to insects, are soluble in water and belong to the δ–endotoxin class of bacterial proteins. Besides these, Bt also produces several other toxins such as a-exotoxin, β–exotoxin, hemolysins, enterotoxins, phospholipases, and chitinase.

The δ–endotoxins form two classes of toxins, the first group: α–helix toxins or α–helical (group of proteins that includes the Cry proteins containing three domains) to which the α–helix region of the protein forms a pore in the membrane and the second group: the β–barrel toxins (includes Cyt proteins), these are inserted into the membrane to form a β–barrel composed of β–sheet hairpins from each monomer.

The amino acid sequences, when aligned, allow the analysis of similarities between the Cry protein classes, revealing the presence of five blocks of conserved sequences located in the internal regions of the protein and in the contact region between domains. In some of these proteins, the blocks vary their position along the amino acid sequence, whereas others may be completely absent.

The relevance of Cry proteins is due to their toxic properties produced after ingestion by insects of different orders and tie order of the toxicity is: Lepidoptera, Diptera, Coleoptera, Hymenoptera, Hemiptera, Isoptera, Orthoptera, Siphonoptera, and Thisanoptera. In the database of Bt, 633 families and subfamilies of Cry proteins with toxic activity including mites, protozoa, and nematodes are described.

Genes that express the delta-endotoxins are called "cry genes" due to the crystalline phenotype. These genes are located on plasmids of large molecular weight. Currently, due to their importance, more than 70 classes of Cry genes are described (cry1 the cry70). These endotoxins have been classified as Cry1–Cry69 and Cyt1–Cyt3 and different subgroups depending on their amino acid sequence.

Of these, some Bt genes such as cry1Ab, cry1Ac, cry2Ab, and cry9C are already being commercially used in GMP such as corn, potato, soybean, and cotton in order to protect against lepidopteran pests of these crops, making it an alternative in reducing the application of chemical pesticides since 1996. The growth scenario of technology adoption could not be different in Brazil, the second in the world in area planted with GMP. In 2011, 160 million hectares were cultivated with Transgenic plants, with a growth of 12 million hectares compared to 2010, ensuring its place in the ranking of countries that use the technology of genetically modified organisms, behind only the United States.

The characterization of Cry genes is important for the differentiation of specific toxicity of Cry proteins active against certain orders of insects. For example, cry1 group has several subclasses identified, each of which is responsible for a specific form of activity against various species of insects. It is widely used in Lepidoptera, whose order focuses most of the studies on the mode of application and action. For some insect groups, such as dipterans or coleopterans, toxicity studies showed that the insecticidal activity of Bt has little relevance.

TOXICITY

Considering the ecology of Bt, studies showed that this microorganism may have a symbiotic relationship with plants,which perhaps explains the production of toxins so specific and efficient against insect pests. However, in the natural environment,several studies indicate that isolates without insecticidal activity are more widely distributed (over 90%) than those with toxic properties.

Currently the demand for new Bt strains will increase the number of toxins available for pest control and management of their resistance. Several isolates have been tested and characterized against insect pests and disease vectors to be used as basis for production of biopesticides or as donors of genes encoding insecticidal proteins.

In Lepidoptera, toxins are consumed through ingestion. Bt toxins classified and studied with insecticidal activity in this group are Cry1, Cry2, Cry9, and Cry15. However only the toxins of the Cry1, Cry2, and Cry9 groups were reported with insecticidal activity to A. gemmatalis.

The association of Cry toxins with specific tissues emonstrated that following ingestion of the Cry1Ac toxin activated with trypsin, the hemipteran L. hesperus has no receptor to this protein, but a wide link to in the microvilli of intestinal cells throughout the intestine for the Cry2Ab toxin.

The low susceptibility of these insects to Bt toxins may be related to similarities between the glycoproteins from insect midgut microvilli and not as a result of direct selection for toxicity. In vivo assays showed that the toxins used showed low activity against the aphids, which may be related to the way the experiments were performed. Once nonsolubilized Bt crystals have been used in feeding assays, which present a disadvantage since the solubilization of the toxin does not occur due to the acidic pH in the stomach of these insects. The proteolytic activation of the Bt toxin ingested in the gut of insects is essential to present toxicity. The differences between the proteolytic enzymes (abundance, type, and its correlation) and the pH of the midgut of hemipterans, as well as other pests, are factors that contribute to toxicity.

The intramolecular proteolytic cleavage is also important for toxicity against insects which have gut with acid or neutral pH, as this favors the creation of Bt plants that express Cry toxins instead of protoxins. The intramolecular proteolytic cleavage increases the solubility of the toxin in the gut, thereby facilitating its binding. These polypeptides can also associate with others and maintain the insecticidal activity. This phase has been extensively studied using Cry1 toxins that involves removal of 27–29 N-terminal amino acids 500–600 and C-terminal amino acids. Hemipterans present acid gut and the proteolytic activities of the membrane are associated with cathepsin of the types G and B.

The toxin ingested by a lepidopteran (third trophic level) induced changes in the distribution of glycogen, lipids, and calcium due to the disorganization of the perimicrovilllar matrix of the gut.

HISTOLOGY AND MODE OF ACTION

Histological studies, using the insects gut, have been the focus of researches to control agricultural pests. This is due the fact that changes in the gut can not only affect their development, but also cause major physiological events, such as changes in nutrient absorption, degenerative transformation, appetite loss and abandonment of food, gut paralysis, physiological disorders, and total paralysis. These are the most common symptoms observed from the moment the susceptible insects ingest the Bt spores and crystals, leading to insect death, when larvae show a blackened color, a characteristic symptom of infections caused by this microorganism.

In lepidopterans, the chewing mouthparts promote the ingestion of Bt toxins both in a product form and in the form of a toxin-containing GMP. Their digestive tract is divided into three regions: preintestine (front), midgut, and hindgut and is one of the most important areas of contact between the insect and the environment. This has been the subject of research aiming to develop alternative control methods.

In caterpillars, the midgut epithelium, pseudostratified columnar, consists of four types of cells, which are involved in the processes of absorption and secretion of enzymes, which are columnar and caliciform cells with ionic homeostasis function, endocrine cells with endocrine function, and regenerative cells involved in the renewal of the epithelium. All of them are coated by a peritrophic membrane, which serves to protect the epithelium from mechanical damage and also as a barrier against harmful chemicals and toxins.

The columnar cells are in larger quantities, tall, and present long and numerous microvilli on the apical portion. The basal portion, invaginations form the basal labyrinth.

The caliciform cells have a large central cavity delimitated by cytoplasmic projections filled with mitochondria. The regenerative cells present an electron-dense cytoplasm and few organelles. The endocrine cells are characterized by the presence of electron-dense secretory granules concentrated in the cytoplasm of the basal cell.

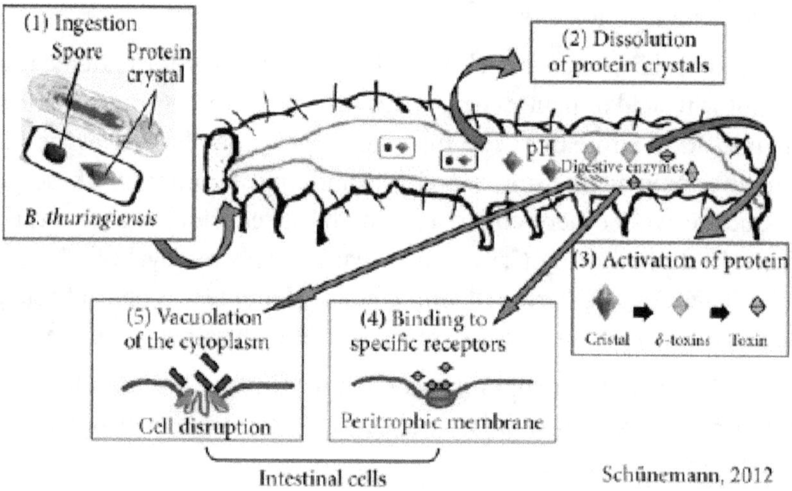

Fig. 8. Mode of action of Bacillus thuringiensis in Lepidoptera: ingestion of bacteria (1); solubilization of the crystals (2); activation protein (3); binding of proteins to the receptors (4); membrane pore formation and cell lise (5).

Several histopathological and ultrastructural studies have investigated the interaction between Bt toxins in the midgut of larvae of these insects. In Fig. 8, the series of events that occur after ingestion of Bt in Lepidoptera are shown.

After ingestion, the crystals are solubilized in the alkaline (pH 9 to 12) midgut environment. The proteins specific for Lepidoptera are soluble at pH above 9.5. The pH of the midgut of insects has great influence on the specific activity of Cry toxins. Some toxins are activated under alkaline conditions (CryII1A) and others are activated under conditions of neutral to acid pH (Cry1b). Cleavage of Cry toxins is a crucial step in the activation of the toxin and also in its specificity in different insects.

When toxins are solubilized, protoxins are released through the action of proteases resulting in active proteins of 60–70 kDa. Their toxic fragments were described as connected to the N-terminal region of the polypeptide chain, which is later removed.

The protoxins are activated by digestive enzymes in the midgut and bind to specific receptors in the microvilli of the apical membranes of the columnar cells of the lepidopterans gut. The binding of Cry toxins to the apical microvillus of the membrane vesicles of the insect determines the specificity of the Cry toxins.

Different proteins have been identified as receptors for Bt in Lepidoptera, and we can highlight with Lepidptera, including aminopeptidases, "cadherin-like protein", and alkaline phosphatase.

The binding method may be described as a biphasic step consisting of reversible and irreversible stage. The interaction between the toxin and its binding site is a basic requirement for the toxicity. The degree of toxicity is determined when the insertion in the cell membrane is considered irreversible binding [24].

The molecules for binding of Cry proteins are a cadherin type protein (CADR), an aminopeptidase anchored to a glicosilfosfatidil-N-inositol (GPI), an alkaline phosphatase anchored to a glicosilfosfatidil-inositol (GPI), and a 270 kDa glycoconjugate. Other experiments have shown that glycolipids may also be involved as receptor molecules in some insects and nematodes. Studies suggested that the toxicity may be related to G-protein-mediated apoptosis upon binding to the receptor.

The Cry toxins cross the peritrophic membrane by binding to specific receptors on the apical membranes of intestinal cells causing opening or pore formation followed by vacuolation of the cytoplasm by osmotic imbalance between the intracellular and extracellular environments and cell disruption.

This destroys the microvilli, causing the insect to stop feeding, leading to its death. A second model to describe the mode of action of Cry toxins has been proposed and refers to signal transduction. Although being only studied in insect cells, it does not differ from the first model to the stage of binding the protein formed by cleavage of the protoxin in intestinal microvilli. Regarding stink bugs, these present needle-like sucking mouthparts that are formed by two mandibular and two maxillary stylets and a narrow channel which injects saliva in the salivary plant tissue.

The digestion of these insects is extra-oral, secreting saliva in the food, which digests sap proteins using proteases present in saliva. After sucking predigested nutrients, they are completely digested by proteases in the gut for absorption of nutrients. The midgut of Hemiptera varies in its morphology and functional activity due to dietary habits of these insects. It is divided into an anterior dilated region, a median tubular region, and posterior dilated one.

The anterior region is responsible for electrolyte balance and involves the transport of ions and water and digestion of carbohydrates and is an important location for the storage of lipids. The process of digestion enzyme secretion is generally observed in the middle and posterior regions. Histologically, the pentatomids midgut has a simple epithelial layer and consists of digestive cells and regenerative distributed of all throughout the midgut. The hemipterans have no peritrophic membrane, therefore they are unprotected by a chitin and protein membrane that prevents the action of pathogens and excretion of digestive enzymes.

APPLICATION OF BT TECHNOLOGIES

Bt Products

Despite the importance and viability of microbial agents for pest control, only 2% of the insecticides used worldwide are based on the application of biopesticides, in which Bt represents approximately 95% of microorganisms used. For over 50 years, Bt has been used in formulations for the biological control of many agricultural pests and vectors of human diseases and based on more than 90% of commercially available microorganism products.

Since the first product has been launched in France in 1938, over 100 formulations were placed in the world market, with the American continent being responsible for 50% of this market. From this fraction, the United States and Canada account for 90% and the countries of Latin America ranking with only 8–10% of consumption concentrated, mainly in Cuba and Mexico, to control pests in cotton, banana, potato, citrus, vegetables, tobacco, corn, and pasture. In Brazil Bt-based products are used in approximately 150,000 acres for the control of about 30 pests with agricultural importance.

There are different commercial Bt products developed for control agricultural insect pests and also against mosquito species. Most of the spore-crystal formulations are obtained from different strains. These include B. thuringiensis var. kurstaki (Btk)-isolate HD1 (contains Cry1Aa, Cry1Ab, Cry1Ac, and Cry2Aa proteins); B. thuringiensis var. kurstaki (Btk)-isolate HD73 (contains Cry1Ac); B. thuringiensis var. aizawai-isolate HD137 (contains Cry1Aa, Cry1B, Cry1Ca, and Cry1Da); B. thuringiensis var. San Diego and B. thuringiensis var. tenebrionis, (which contains Cry3Aa); B. thuringiensis var. israelensis (containing Cry4A, Cry4B, Cry11A, and Cyt1Aa) toxins.

Bt Plants

A selection of different soybean cultivars with resistance against lepidopterans and other insect classes. This occurs through genetic improvement of plants which express chemical characteristics that involve the production of toxins such as isoflavonoids, that act as repellents, feeding and oviposition suppressors, and digestibility reducers of these insects, is also a way to reduce the damage caused by these insects. The use of these plants with other techniques of integrated pest management (IPM) can keep the pest population levels below the levels of economic damages.

The insecticidal effect of GMPs on natural enemies is a controversial topic. Even after rigorous risk assessment, planting these cultivars has aroused concern over environmental impacts such as gene escape and effects on biodiversity. Considering soybeans, the risk of horizontal gene transfer is remote because of the lack of compatible wild species. The gene escape to the conventional crops can occur but can be avoided by isolation of the cultures.

Plants with insecticidal activity, since they are rapidly degradable, reduce environmental impact, in addition to maintaining the beneficial fauna. Their resistance may contribute to the reduction of the insect population below the economic injury level, do not cause imbalances in agroecosystems, and have a cumulative and persistent effect. They do not promote increases in production costs and are compatible with other control tactics.

Transgenic soybean is the main culture that currently occupies the crops of GMPs in the world, occupying 47% of the 160 million hectares of GM crops currently grown in the area in 2011. The soybean transformation was first reported in 1988. To date, the transformation is not yet considered routine due to the complexity of combining techniques for its transformation and regeneration. Transgenic soybean is an important tool for achieving economic growth of this crop in Brazil; however, it is essential to know how to use this technology and associate it with other tools.

Bt soybean is the second largest group of GMPs within this culture. The cultivars are characterized by expressing Cry toxins with insecticidal properties to groups of insect pests of this crop. The toxins are expressed by cloned Cry genes from Bt and hence comes the name of the cultivar: Bt soybean.

FINAL CONSIDERATIONS AND FUTURE PROSPECTS

There are increasingly records of Bt isolates showing toxic activity to some insect pests. But in the market, there are few toxins of this microorganism used in formulated products and GMPs and may quickly cause the emergence of resistant populations. This fact occurred in different populations of Lepidoptera exposed to Cry toxins.

It is clear that cultures that express Bt toxins have had a very important beneficial impact on global agriculture due to the reduction in the number of pests and hence the total application of chemical insecticides used for its control, as well as the final production.

The large-scale cultivation of genetically modified crops over several years may increase the selective pressure on the pest species, which may result in the development of resistance. Resistance to insect control practices is a threat to development, implementation, and maintenance of IPM practices. The emergence of resistance is often related to the reduction of pest populations by the increased mortality or reduced fecundity in insects. Thus, when there are differences in survival or fecundity between the specimens that form the population of a pest species after use of Cry proteins, the selection of resistant insects to protection techniques recently applied may occur.

Therefore, when there is significant variability in insect population with any physiological or behavioral characteristics that contrast the protection to Cry proteins insecticides, it is formed an inevitable selection of adapted insect pests. Thus, it essential to observe proper practices in the integrated management (IPM) of this pest, avoiding a continuous selection of resistant specimens, reducing the frequency of resistance that may compromise the protection to crops by pests.

Due to the large potential that pest populations have to develop resistance, some strategies have already been implemented in other cultures. As an example, using vegetative insecticidal proteins (VIPs) and using genes from plants or animals which encode immunosuppressive proteins.

In the cultivation of Bt soybean, the most recommended technique is the use of refuge areas. This consists of partially planting non-Bt plants in the area of soybean containing Bt proteins. The importance is to keep population of the target pest without exposure to Bt crops and thus sensitive to the proteins. This way, pest individuals that develop in the refuge can mate with any resistant individuals that may have survived in the culture. This would convey susceptibility to Bt for future generations of the pest, generating resistant and susceptible individuals within the population, preserving the Bt technology.

Thus, assessing the mode of action of Cry toxins is a complex process that involves interaction with the receptor molecules that lead to different mechanisms of membrane insertion and lysis of intestinal cells. However the characterization of the mode of action of Cry toxins in other susceptible organisms may be important for understanding the mode of action of this protein family in different insects, which often do not show insecticidal activity with in vivo toxicity tests. The analysis of interaction between insects and bacteria is essential for effective management of the target species, and steps for evaluation may not be the same for both (Fig. 9). Advanced studies provided detailed techniques in each step of interaction between insect and pathogen, which can reveal new ways of using Bt toxins.

However, there is no information about the digestive physiology of hemipterans associated with Bt toxins, and no literature regarding the mechanism of resistance or susceptibility of these insects to Cry toxins is available. Thus, in Brazil where these species have an agricultural importance, it is necessary to evaluate the in vitro action of Cry proteins in the gut of these insects, testing the hypothesis of degradation or activation of Cry toxins in the gut of these pentatomids.

This knowledge may be through the identification of novel molecules and the development of bioinsecticides. In this context, it is of great importance the discovery of bacterial strains with increased activity, adapted to environmental conditions and with high specificity. Thus, the mapping of new Cry proteins to non-susceptible insects and specificities of the receptors may be critical for the development of new products for the control of these insects.

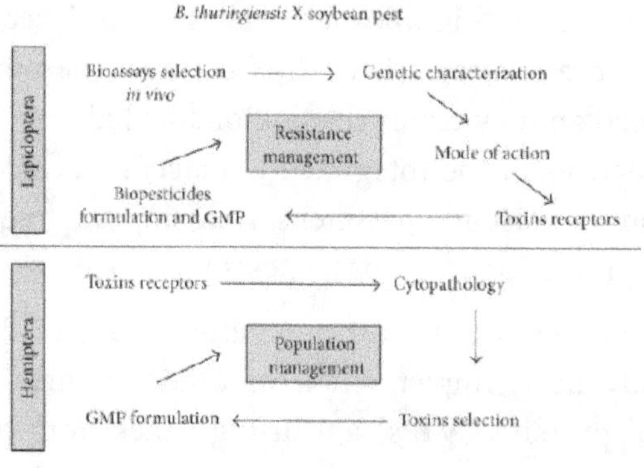

Fig. 9. Comparative analysis of studies of the evolution of application of Bacillus thuringiensis to control caterpillars and stink bugs in agroecosystems.

Studies on the identification of receptor molecules and binding sites, especially in Hemiptera and Lepidoptera, are important for the development of strategies to control insect resistance. In addition, knowledge of the target molecules for insecticides for agricultural pests is also essential to use in GMPs in order to control the different susceptible targets to the toxins used.

5

BIOPESTICIDE FOR INTEGRATED CROP MANAGEMENT

To feed the ever-growing global population, we need to produce more food and livelihood opportunities from less per capita arable land and available water. Providing ample food is only the first part of challenge, the second and more important challenge is to produce this in a safe and sustainable manner.

Most of the cultivated crops/varieties have reached their yield plateau, hence protection of crops to harvest maximum is one of the ways to meet the increasing demands of food and to attain global food security on sustainable basis. There are about 67,000 pest species that damage agricultural crops, and a significant portion of agricultural inputs is required for pest management to protect the crop. Pesticides are extensively used in the intensive agriculture to control pests, diseases, weeds, and other crop enemies to reduce yield losses and maintain the product quality. Insect-pests management in high yielding varieties by extensive use of synthetic pesticides has certainly provided protection to crops; but it has also raised concerns about pesticide residues in the food and the environment.

This pest management strategy adversely affects even beneficial organisms, leave harmful residues in food, feed and fodder, and causes environmental pollutions. Human exposure to pesticides occurs primarily through contaminated food, feed and drinking water.

Their adverse effects depend on toxicity of pesticides, method of application, the dosage applied, their adsorption on soil colloids, the weather conditions prevailing after their application, and how long the pesticides persist in the environment. Therefore, assessment of risks of pesticides either on human health or on the environment is not an easy and accurate process because of differences in the periods and levels of exposure, and the types of pesticides used. Hence, the need of the day is to produce maximum from the decreasing availability of natural resources without adversely affecting the environment.

Biopesticides are products and by-products of naturally occurring substances such as insects, nematodes, microorganisms, plants as well as semiochemicals. Based on the nature and origin of the active ingredients, biopesticides fall into several categories such as botanicals, antagonists, compost teas, growth promoters, predators and pheromones. Plants and microorganisms are the major sources of biopesticides due to the high components of bioactive compounds and antimicrobial agents.

The active compounds in plants include phenols, quinones, alkaloids, steroids, terpenes, alcohols and saponins. Different plant families have varied antimicrobial bioactive compounds which include oil components such as α– and β–phillandrene, limonene, camphor, linalool, β-caryophyellene and linalyl acetate depending on the plant family. Microbial biopesticides include bacteria species such as Pseudomonas, Bacillus, Xanthomonas, Rahnella and Serratia or fungi such as Trichoderma, Verticillium and Beauveria species. Biopesticides exhibit different modes of action against pathogens such as hyperparasitism, competition, lysis and predation.

Plant growth promoting rhizobacteria protect plants from biotic and abiotic stresses and they also enhance plant growth and enhance formation of root hairs. The most common species of plant growth promoting rhizobacteria include Agrobacterium, Ensifer, Microbacterium, Bacillus, Rhizobium, Pseudomonas, Chryseobacterion and Rhodococcus. They colonize the environment around the plant roots, fix nitrogen, increase phosphate solubilisation and result in general increase in plant yield. Species of Pseudomonas and Bacillus have been used as biofertilizers with reports showing increase in plant growth, yield and phosphorous and zinc content in fruits and soils.

Natural enemies including predators, pathogens and some insects are also used as biopesticides in management of insect pests. Parasitoids, wasps, beetles, lace wings, bugs and lady birds are used in management of destructive pests such as boll worms (Helicoverpa armigera) in important crops such as cotton. Compost teas are filtrates of compost extracts and are similarly used as biopesticides.

This review discusses the current status of knowledge on biopesticides including their sources, production, formulation, commercialization, role in sustainable agriculture and their limitations. It also brings together the different types of biopesticides and evidences of their use against important pests in different crops.

BIOPESTICIDES AND INTEGRATED CROP MANAGEMENT

Biopesticides are formulations made from naturally occurring substances that control pests by non-toxic mechanisms and in an ecofriendly manner. They may be derived from animals, plants, microorganisms and include living organisms, their products or byproducts which can be used for the pest management. Biopesticides are used primarily as preventive measures, so they may not perform as quickly as some synthetic chemical pesticides do. Biopesticides are generally less toxic than chemical pesticides, often target specific pests, have little or no residual effects hence pose less risks to human health, the environment and have acceptability for use in the organic farming. Biopesticides may be categorized into three major groups: plant-incorporated protectants (PIPs), biochemical, and microbial biopesticides. While microbial biopesticides use microorganisms (bacteria, fungi, viruses or protozoans) as active-ingredient, biochemical pesticides are naturally occurring substances from plants and animals. PIPs are produced naturally on genetic modification of a crop plant, such as Bt cotton. Such transgenic plant produces biodegradable protein with no harmful effect on animals and human beings, and thus curtails the use of hazardous pesticides. PIPs may be more effective and economical strategies in the developing countries to help produce more food, feed and forages in an environmentally safer manner.

Integrated Crop Management (ICM) is a pragmatic approach to crop production which includes Integrated Pest Management (IPM) focusing on crop protection. ICM is based on understanding the intricate balance between the environment and agriculture, and is a whole-farm approach in achieving a proper balance. Its basic components are crop management, nutrient management, pest management and ultimately financial management. One of the main objectives of ICM is reduction of external farm inputs, such as inorganic fertilizers, pesticides and fuel by means of farm produced substitutes.

Although complete replacement of these inputs is not possible without significant loss of yields, but partial substitution of the inputs can be achieved by the use of natural resources. ICM assumes a broad palette of available disease and pest control methods, such as biological, cultural and physical controls, host plant resistance, and decision support tools. In recent decades, the focus on crop production has moved from yield to quality and safety, then more recently sustainability. ICM and IPM strategies combine a range of complementary methods to reduce pest populations below economic injury level while minimizing impacts on other components of the agro-ecosystem and environmental conditions of the area. While considering pest and disease management, the use of biological control methods is considered to complement physical and cultural methods.

In ICM, synthetic pesticides are treated less as a blanket solution to crop protection and more as a tool to be used selectively in ways that complement other methods and thus minimize the chances of resistance development in pests. Evidences show that biopesticides can be valuable components of ICM for promoting sustainable agriculture. Biopesticides have gained lot of interest in the last decade particularly in view of the growing demands for organic foods.

BIOPESTICIDES AND ENVIRONMENTAL ISSUES

Although pesticides are developed through very strict regulatory processes to function with reasonable certainty and minimal impact on human health and the environment, increasing public concern about the potential adverse effects of synthetic agrochemicals prompts search for the technologies and products safer for the environment. Moreover, due to the problems of pesticide resistance and withdrawal of some pesticides for either regulatory or commercial reasons, a fewer chemical pesticides are available in the global market. Hence, ICM approach needs to be deployed to counteract degradation of the agro-ecosystem due to the ongoing intensive agriculture. This would include the use of biofertilizers, biopesticides, IPM, soil and water conservation practices, biodiversity conservation etc.

A better understanding of genes from micro-organisms and crop plants has allowed isolating genes effective against particular pests, and they are being deployed to control insect pests and diseases of crop plants. Benefits of the use of biopesticides in agriculture and public health programs are considerable. They do not have residual effects which is a matter of significant concern for consumers, particularly in case of fruits and vegetables. When used as a component of IPM, efficacy of biopesticides can be equal to the conventional pesticides.

Interest in biopesticide is gaining because of its advantages associated with environmental safety, target-specificity and efficacy in very small quantity, natural decomposition and suitability in IPM. Thus, biopesticides are one of the promising alternatives to manage environmental pollutions. They can replace some of the hazardous chemical pesticides.

Though potential of biopesticide for promoting sustainable agriculture has been known for years, it has gained interest in view of the growing demands for safe and healthy organic food. Although use of agrochemicals is indispensable to meet the ever growing demands of food, feed and fodder, opportunities do exist in selected crops and niche areas where biopesticides can be used as a component of IPM.

Increasing demands for residue-free crop produce, growing organic food market and easier registration than chemical pesticides are some of the key drivers of the biopesticide market. There are rising concerns for conservation of biodiversity and threats to some of the endangered species, set against the requirement to increase agricultural production without excessive reliance on chemical pesticides. Development of biopesticides has largely followed a chemical pesticide model that does not exploit fully the favorable biological properties of the biological agents.

While there is commercial pressure from the manufacturing side to develop products, based on a single strain that are broad spectrum in order to control a range of pests on different crops and may not be endemic to the areas of application; the environmentalists want narrow spectrum products based on strains from the area of use. To reconcile these divergent demands, biopesticides in the market have been maintained at minimal negative impact, if any, on the environment.

BIOPESTICIDES OF BOTANICAL ORIGIN

Based on the method of extraction, botanical pesticides can either be plant extracts or essential oils. They are obtained from plants parts such as leaves, barks, flowers, roots, rhizomes, bulbs, seeds, cloves or fruits which are either fresh or dried. Dried plant parts are preferred as this reduces water concentration resulting in higher yield of active ingredient. A Gas Chromatography-Mass Spectrometry (GC-MS) analysis was carried out on Citrus sinensis and d-limonene and myrcene were reported as the major constituents of the oil component. The products were tested against a cereal leaf beetle (Oulema melanopus) on wheat and a mortality of up to 85% was reported on larvae observed in 48 hours. Aqueous fruit extracts of Withania somnifera were tested for activity against Fusarium oxysporum f.sp. radicis-lycopersici, the causal agent of fusarium crown and root rot disease in tomatoes. At a concentration of 2% the extracts inhibited growth of the fungal pathogen by up to 56% in vitro.

Table 1 shows examples of plants reported to have potential as sources of botanical pesticides and respective target pests. In vitro experiments involving several ethanolic plant extracts, turmeric (Curcuma longa), lemon (Citrus limon), garlic (Allium sativum), pepper (Capsicum frutescens) and ginger (Zingiber officinale) were reported to be significantly effective against Alternaria solani, Pythium ultimum, Rhizoctonia solani and Fusarium oxysporum f. sp. lycopersici. Turmeric (Curcuma longa) was reported to be the most effective with a growth inhibition of up to 73% against Alternaria solani. Pseudomonas syringae p.v. tomato was effectively managed in vitro by Rhus coriaria, Eucalyptus globulus and Rosmarinus officinalis.

Eucalyptus globulus was reported efficacious in preventing the bacterial speck of tomato (Pseudominas syringae p.v. tomato) by up to 65% under greenhouse conditions. At 5% concentration, a mortality rate of up to 78% was reported on juveniles of root knot nematodes (Meloidogyne sp) by extracts of Nerium oleander. When the concentration was increased to 10% a mortality of between 65% - 100% was observed on second stage juveniles treated with extracts of Eucalyptus sp, Cinnamomum verum, Nerium oleander, Azadirachta indica, Zingiber officinale and Allium sativum. The most common and already commercialized botanical pesticides are derived from neem (Azadirachta indica), pyrethrum (Chrysanthemum cinerariifolium), sabadilla (Schoenocaulon officinale) and tobacco (Nicotiana tabacum).

The quality of extracts and oils is highly dependent on the solvent used and method of extraction. The solvents should be of low toxicity, able to dissolve as many compounds as possible, evaporate easily at low temperatures and should preferably possess preservative properties. The choice of solvent is dictated by the target active compounds. Although water is the universal solvent, it extracts less antimicrobial compounds compared to other solvents. Organic solvents such as ethanol and methanol yield better extracts and their results are consistent. Other extraction solvents include dichlomethane, acetone and hexane. According to a study methanol and hexane extracts from Tarchonanthus camphoratus gave higher growth inhibition capacity against Escherichia coli, Salmonella typhi, Staphylocccus aureus, Proteus mirabilis, Candida albicans and Klebsiella pneumoniae. Methanol was also reported to be efficient in extracting seed components of Morinda citrofolia and the extracts showed antibacterial activity of up to 63% against Escherichia coli and Pseudomonas spp. The differences among the different solvents are attributed to their polarity. The plant parts from which the extracts are obtained also influences the quality of the extracts.

Table 1. Plants with potential as botanical pesticides and respective target pests.

Source plant	Target pest	Host
Azadirachta indica	*Aphis craccivora*	
	Aphis gossypii	
	Amrasca devastans	
	Myzus persicae	
	Sitobion avenae	
	Lipaphis erysimi	*Vigna unguiculate;*
	Bemisia tabaci	*Gossypium hirsutum;*
	Sciothrips cardamomi	*Solanum tuberosum;*
	Rhizopus sp. Aspergillus sp.	*Triticum sp.;*
	Monilinia fructicola	*Brassica sp.;*
	Trichothecium roseum	*Prunus salicina;*
	Pythium aphanidermatum	*Lycopersicon esculentum;*
	Alternaria alternata	*Capsicum chinense;*
	Helminthosporium sp.	*Cardamomum sp.;*
	Vibrio cholerae	
	Bacillus subtilis	
	Meloidogyne javanica	
	Meloidogyne ingognita	
Allium sativum	*Curvularia lunata*	
	Aspergillus niger	
	Candida albicans	
	Trichophyton rubrum	
	Drechslera tritici-repetis Bipolaris sorokiniana	*Triticum sp.;*
	Alternaria raphanin	*Lycopersicon esculentum;*
	Fusarium flocciferum	*Abelmoschus esculentus;*
	Fusarium graminearum	*Panax sp.;*
	Rhizoctonia solani	*Capsicum sp.;*
	Colletotrichum sp.	*Human and animal sp.;*
	Bacillus subtilis	*Oryza sp.;*
Source plant	*Target pest*	*Host*

Source plant	Target pest	Host
	Salmonella senflenberg	*Gossypium hirsutum;*
	Staphylococcus aureus	*Stored grain products;*
	Staphylococcus epidermidis	*Vigna unguiculate;*
	Sitotroga cerealella	*Brassica oleracea;*
	Spodoptera littorals	
	Tenebrio molitor	
	Callosobruchus maculatus	
	Plutella xylostella	
	Brevicoryne brassicae	
Euphorbia sp.	*Pseudomonas aeruginosa*	
	Enterobacter aerogenes	
	Escherichia coli	*Human and animal sp.;*
	Salmonella typhi	*Arachis hypogaea;*
	Aspergillus flavus	
Curcuma longa	*Tribolium castaneum*	
	Bactrocera zonata	
	Trichoplusia ni	*Triticum aestivum;*
	Alternaria solani	*Prunus persica;*
	Streptococcus pyogenes	*Brassica oleracea;*
	Streptococcus mutants	*Solanum lycopersicum;*
	Ralstonia solanacearum	*Human sp.;*
	Escherichia coli	*Animal sp.;*
	Listeria monocytogenes	
	Bacillus subtilis	

BIOPESTICIDES FOR ORGANIC FARMING

Organic produce perhaps does not mean to be pesticide-free or chemical-free. In fact, under the laws of many countries, organic farmers are allowed to use a wide variety of biochemical sprays and powders on the crop. If pesticide is to be used in the organic farming, it must be derived from natural sources and must not be synthetic chemical in nature. It is also important that the biopesticides must be applied using equipment that has not been used to apply any synthetic chemicals and the land used must not had any treatment with synthetic chemicals in the past three years. Primarily different agricultural practices, such as crop rotation, cover crops, disease resistant varieties and good seed bed preparation etc. are used to control pests and diseases in organic farming. However, these practices are not always sufficient to protect crops from losses. Hence, organic farmers turn towards biopesticides to ensure and enhance quality of their organic products. The biopesticides that can be used by organic farmers include microbial and biochemical pesticides only.

Biopesticides represent just 1% of the global market for agrochemicals. Ninety per cent of the microbial biopesticides are derived from just one entomopathogenic bacterium, Bacillus thuringiensis. As of early 2013 there were approximately 400 registered biopesticide active ingredients, and more than 1250 registered biopesticide products. Data on microbial biopesticide agents from Agriculture and Agri-Food Canada and the US Environmental Protection Agency (EPA) indicates that more than 200 products are being sold in the US, compared to only 60 comparable products in the EU. There are several reasons why adoption of biopesticides has been higher in the US than elsewhere. US has a separate Biopesticides and Pollution Prevention Division which promotes the use of biopesticides as components of IPM programs and coordinates the Pesticide Environmental Stewardship Program. This voluntary program forms partnerships with pesticide users in order to reduce the potential environmental hazards and health risks associated with pesticide usage and implements pollution prevention strategies. EPA tests biopesticides for safety but not for efficacy, as efficacy testing may result in higher cost of biopesticides than chemical pesticides

MICROORGANISMS AS SOURCES OF BIOPESTICIDES

Microorganism-based biocontrol agents form the bulk of commercialized biopesticides and they include bacteria, viruses, fungi, nematodes and protozoa. There are up to 175 reported microbial based biopesticide active agents and they have been used in management of pathogens, weeds, insects and nematodes.

Examples are illustrated in Table 2. Majority of the microbial biopesticides are used to manage soil borne pathogens. Bacterial species that have been utilised as biopesticides include Bacillus, Pseudomonas, Burkholderia, Xanthomonas, Enterobacter, Streptomyces, Serratia and these are either obligate facultative or crystalliferous. Fungi used as biopesticides include species of Trichoderma, Beauveria, Metarhizium, Paecilomyces, Fusarium, Pythium, Penicillim and Verticillium. Steinernama and Heterarhabditis are nematode species used to make biopesticides. The mechanisms of action exhibited by microorganisms against plant pathogens include hyperparasitism, competition, secretion of volatile compounds, antibiosis and parasitism.

The major sources of microorganisms with pesticidal activity are agricultural fields where they co-exist with other microorganisms including pathogens and beneficial species. The rhizosphere is usually concentrated with various classes of important microorganisms. Other rich sources of microorganisms include hay, manure, cow shed, as well as straw. Formulation of the microbial pesticides has a great contribution to the effectiveness of the resultant product and it is usually dependent on the substrate used. A study showed that a formulation of Trichoderma harzianum prepared in black gram bran, peat soil and water had a high level of activity against damping off of eggplant seedlings caused by Sclerotium rolfsii. The activity was attributed to the high number of spores produced by the fungus.

Table 2. Examples of microorganisms with potential as biopesticides.

Microorganism	Target pest	Host
Trichoderma spp.	*Penicillium stekii*	*Lycopersicon esculentum;*
	Fusarium spp.	*Phoenix dactylifera;*
	Ceratocystis radicicola	*Zea mays;*
	Pythium spp.	*Gossypium hirsutum;*
	Alternaria spp.	*Solanum lycopersicon;*
	Rhizoctonia spp.	*Solanum tuberosum;*
	Verticillium spp.	*Capsicum annum;*
	Mcrophomina phaseolina	*Solanum melongena;*
	Botrytis cinereal	
	Phytotphtora spp.	
	Meloidogyne spp.	
Beauveria spp.	*Dendrolimus tabulaeformis*	*Pinus tabulaeformis;*
	Spodoptera litura	*Phaseolus vulgaris;*
	Uvarovistia zebra	*Brassica oleracea;*
	Cyclocephala lurida	*Avena sativa;*
	Frankliniella occidentalis	*Triticum aestivum*
	Hylobius abietis	
	Oryzaephilus surinamensis	
	Tuta absoluta	
	Polyphylla fullo	
Paecilomyces spp.	*Meloidogyne javanica*	*Ssolanum lycopersicon;*
	Phytophthora palmivora	*Vigna mungo;*
		Theobroma cacao;
Bacillus spp.	*Macrophomina phaseolina*	*Glycine max;*
	Fusarium graminearum	*Triticum aetivum;*
	Colletotrichum spp.	*Capsicum annuum;*
	Phaeomoniella chlamydospore	*Vitis vinifera;*
	Fusarium solani	*Vicia faba;*
	Meloidogyne incognita	*Vigna mungo;*
	Spodoptera frugiperda	*Solanum lycopersicon;*
	Xanthomonas campestris pv. campestris	
Metarhizium spp.	*Musca domestica*	*Citrus sinensis;*
	Spodoptera litura	*Polyphylla fullo;*
	Ceratitis capitate	*Phaseolus vulgaris;*
	Polyphylla fullo	*Solanum lycopersicon;*

Microorganism	Target pest	Host
	Tetranuchus urticae	
	Maruca vitrata	
Microorganism	*Target pest*	*Host*
	Tuta absoluta	
	Callosobruchus maculatus	
Pseudomonas spp.	*Vibrio harveyi*	*Mangifera indica;*
	Fusarium oxysporum	*Solanum lycopersicon;*
	Alternaria alternata	*Solanum melongena;*
	Sclerotium rolfsii	*Vitis vinifera;*
	Botryodiplodia theobromae	*Oryza sativa;*
	Rhizoctonia solani	*Triticum aetivum;*
	Aspergillus aculatus	
	Dreshlera biseptata	
Verticillium spp.	*Trialeurodes vaporariorum*	*Phaseolus vulgaris;*
	Bemicia tabaci	*Euphorbia ulcherrima;*

The exposure of Helicoverpa armigera to conidial suspension of Beauveria bassiana resulted to antifeeding habits, blackening of the body and the larvae becoming sluggish and morbid. The fungus finally consumes the entire larval tissues resulting in its death. Researchers reported that isolates of Bacillus showed antagonistic activity against rice pathogen, Xanthomonas oryzae p.v. oryzae, and the activity was attributed to production of a bacteriocin by the bacterium. Treatment of wheat and rice plants with concentrations of Chaetomium globosum reduced the severity of wheat rust (Puccinia recondite) and rice blast (Magnaporthe grisea) by up to 80%. Late blight (Phytophthora infestans) on tomato was also controlled by Chaetomium globosum by up to 50% while mycelial growth of Pythium ultimum was inhibited in vitro in well diffusion assays. The activity of the fungus was attributed to production of two types of chaetoviridins, A and B. These studies indicate that microbial biopesticides can be incorporated in integrated pest management for sustainable agriculture.

PREDATORS AND PARASITOIDS AS BIOPESTICIDES IN AGRICULTURE

A predator kills and feeds on prey while parasitoids grow on or inside their hosts and eventually kill them. The predators include beetles (Carabidae), ladybirds (Coccinellidae), spiders, lacewings (Chrysopidae) and true bugs while parasitoids mainly consist of wasps and other hemipterans. Examples of predator organisms used in management of crop pests are presented in Table 3. These natural enemies are mainly found in the environment and are not evenly distributed. In order to have them in large numbers, they are either reared under controlled conditions and released into the fields or are multiplied in open fields containing the prey. The most common way of rearing these predators is by growing them on their preferred hosts. This is either done in screen houses or growth chambers where the host plants are first grown and then exposed to pest infestation.

Table 3. Examples of predator organisms used in management of crop pests.

Predator	Target pest	Host
Phytoseiulus spp.	*Tetranychus urticae*	*Phaseolus vulgaris;*
	Tetranychus evansi	*Solanum lycopersicon;*
		Fragaria ananassa;
Neoseiulus spp.	*Tetranychus urticae*	*Vicia faba;*
	Oligonychus perseae	*Capsicum annuum;*
		Persea americana;
Amblyseius swirskii	*Scirtothrips dorsalis*	*Capsicum sp.;*
Aphidius colemani	*Aphis gossypii*	*Dendranthema grandiflora;*
Diglyphus isaea	*Liriomyza huidobrensis*	*Solanum tuberosum;*
Encarsia formosa	*Bemisia tabaci*	*Solanum lycopersicon;*

The predators are then introduced and they are maintained by growing on the prey. Alternatively, the predators can be grown in cylinders where they are supplied with the prey and all other necessary conditions for growth are provided. An example is the mass rearing of Phytoseiulus persililis on Tetranychus urticae Koch. The optimum growth of predator mite, Neoseiulus californicus, was observed when grown on an artificial diet supplemented with eggs of Ephestia kuehniella, Artemia franciscana cysts and maize bran.

Such artificial diets are important in reproduction, development and survival of the predators during rearing as well as reduction of production costs. Predators can also be grown on egg masses of their prey or other suitable hosts which gives them a longer storage capacity. This has been employed in management of mealy bugs using parasitoids. The predators can also be grown on other feeds such as rice bran as long as it provides the necessary nutrients to the insects.

Due to economic concerns, these organisms are reared on artificial media with carefully evaluated nutritional needs and requirements. Their growth media ranges from beef and liver to crushed lepidopteran pupae. These provide a combination of hormones and nutrients needed by the predators for growth. Artificial media provides as good nutrients as the host plants and reduces the cost of growing the plants.

The artificial media is mostly used in laboratories and has been used for rearing Trichogramma and Anastatus spp. An in vitro study showed that Amblyseius swirskii predated on Frankliniella occidentalis and Tetranychus urticae nymphs with preference to their first instars. The predation rate on T. urticae was 4 - 6 nymphs in 12 hours. The authors reasoned that the outcome of the predation is highly dependent on several factors among them being the host plant traits. A synergistic effect on predation between

Amblyseius swirskii and Phytoseiulus persimilis against two-spotted spider mite (Tetranychus urticae) with a mortality rate of up to 86% was reported.

Introduction of Amblyseius californicus and Amblyseius degenerans into a population of Tetranychus urticae under laboratory conditions recorded a mortality of up to 72% within 15 days.

ICM AND IPM: OPPORTUNITIES AND CHALLENGES FOR BIOPESTICIDES

Integrated crop management (ICM) is a holistic approach to crop management, integrating agricultural and environmental management practices such as optimal site selection, crop rotation, and optimization of crop nutrition, soil, and water management, as well as environment, energy, and waste management and integrated crop protection strategies, in accordance with local conditions, climate, and requirements. Driven by legislation, as well as consumer and industry demands, ICM strategies integrate local knowledge and experience with research and technological innovations to ensure sustainable agricultural production.

Integrated Pest Management (IPM) is a part of Integrated Crop Management (ICM) and the concept was introduced more than 50 years ago as a response to the preponderance of broad-spectrum, mainly neurotoxic conventional insecticides available to the market. Currently, the definition of "Pest" in IPM includes vertebrate and invertebrate pests (e.g., birds, insects, mites, and nematodes), plant pathogens (e.g., fungi, bacteria, and viruses), as well as weeds.

The objective of IPM is to maintain pest populations below levels that would cause economic loss while minimally impacting the environment, and integrates several strategies including chemical, biological and mechanical control, precision technology, and decision support tools (e.g., big data, including crowd-scouting and degree-day models for pest prediction) and the activation of natural plant defense responses, as well as the use of resistant cultivars.

The aim is to reduce the use of chemical pesticides and manage the development of resistance in target pests to both chemical pesticides and biopesticide toxins. By introducing unique modes of action, biopesticides are used in integrated strategies with traditional pesticides to increase crop yields by working synergistically with chemical pesticides, to extend application timings and allow timely re-uptake intervals, as well as to provide resistance strategies.

While the adoption of more "biologically-based" IPM strategies in the field is still low, in part due to a perception of lower and variable efficacy and greater dependence on environmental conditions on the part of end users, biological control plays an important role in the production of greenhouse crops, primarily due to their effectiveness under controlled environmental conditions, short persistence, and low operator risk.

With advances in the development of biopesticides and technological developments in the use and application of conventional and biological pesticides, we now have a real opportunity to harmonize biological and chemical crop protection and realize the concept of IPM as a truly sustainable and holistic solution to current crop protection challenges.

Production, Formulation and Commercialization of Biopesticides

Botanical pesticides are prepared from plants and plant parts obtained from the environment, natural or man-made. The materials are cleaned of dirt or foreign materials and then extracted either using solvents or distillation to obtain extracts or essential oils, respectively. The resultant extracts are then subjected to screening for activity in vitro against different pests using different methods such as disc diffusion, agar well diffusion, agar dilution and poisoned food technique.

The most active botanicals are then evaluated for efficacy in managing pests and diseases under field conditions. The active constituents of the selected extracts are then identified for optimum formulation. Intensive laboratory and field trials conditions are carried out to ensure that the most efficacious combination of the active compounds, carrier materials, emulsifiers, surfactants and other components used in pesticide development are optimized.

The efficacy report from the laboratory and field trials is used to request for registration of the product from the pest control products body. Production of microbial pesticides follows the same procedure as botanicals, except the that the antagonistic microorganisms are collected from sources like the cowshed, hay fields, rhizosphere, compost and manure. They are isolated in to pure cultures in the laboratory and maintained in agar slants. In vitro efficacy trials are carried out following methods such as dual culture, agar discs diffusion and agar well diffusion. The active microorganisms are multiplied on a suitable substrate in the laboratory and mixed with carrier materials, enhancers and stabilizers for field application.

Repeated laboratory and field efficacy trials are conducted until registration process begins. Before the natural products are commercialized, they are usually tested in the laboratory and under field conditions for efficacy against the target pests. The active compounds are also identified using techniques such as thin layer chromatography (TLC), high-performance liquid chromatography (HPLC) and gas chromatography-mass spectrometry (GC-MS). Stabilizers and carrier materials are always added to the active compounds to enhance their applicability and longevity.

Formulation of the active compounds should improve the stability of the compound as well as increase its efficiency and applicability. It should also reduce its degradability due to climatic factors such as heat, water and acids. Carrier materials majorly used include petroleum distillates, corn starch, talc, clays, and water.

Emulsifiers such as soap are also added to the compounds during formulation and they are optimized to ensure effectiveness is not lost. Biopesticides of plant origin that have been formulated and commercialized for agricultural use include neem and pyrethrum while microorganisms include species of Bacillus and Trichoderma.

MODES OF ACTION OF BIOPESTICIDES

Each type of biopesticide exhibits varied modes of action. Microbial pesticides act on pathogens by antagonism, hyperparasitism, antibiosis and predation. Botanical pesticides inhibit growth of pathogens, modify their cellular structures and morphology and exhibit neurotoxicity on insects. Botanicals also repel insects, suppress oviposition and feeding. Predators mainly kill the prey through parasitization or injection of toxic substances which eventually kill the prey. Semiochemicals are used to lure the target pests and they can then be managed through other means such as sterilization or death.

Extracts from plants belonging to Asteraceae family have been reported to inhibit hyphal growth and induce structural modifications on mycelia of plant pathogenic fungi. Asteraceae plants contain compounds such as flavanoids, coumarins alkaloids and terpenoids which could lead to absolute fungal toxicity. Some compounds lead to changes in the cell wall as well as the morphology of celleular organelles.

In some instances, the bioactive compounds cause partitioning of fungal cell membranes making them permeable leading to leakage of cell contents. Plant bioactive compounds also lead to separation of the cytoplasmic membrane which leads to damage of the intracellular components and swelling of cells leading to eventual death. Compounds such as allicin found in garlic (Allium sativum) bulbs lead to suffocation of the pest due to effects on receptors of neurotrans-mitters. Phenolics and terpenoids build hydrophobic and ionic bonds which attack a multiple of proteins in the insects leading to physiological malfunction.

Compounds in plant extracts and essential oils also interfere with receptor cells leading to malfunctioning of the nervous system and failure of coordination leading to death of the insect. Different classes of microorganisms have different modes of actions. Hyperparasitism is one of the most reported mode of action on many biocontrol agents. The antagonist kills the pathogen or its propagules while some attack the sclerotia or the hypha of the fungal pathogen. A single pathogen could be attacked by a number of biocontrol agents. Pasteuria penetrans is an example of a biocontrol agent that parasitizes on root-knot nematodes of Meloidogyne spp.

Species of genus Trichoderma exhibit a predation mode of action by producing enzymes that directly kill cell walls of the pathogens and colonize the environment therein. Some microorganisms produce compounds that kill other microorganisms, a mechanism called antibiosis. This is most common with bacteria belonging to species of Pseudomonas, Agrobacterium, Bacillus, Burkholderia, Pantoea and it has also been reported in the fungus Trichoderma spp.

Sufficient quantities of antibiotics need to be produced for enhanced biocontrol. Some microbial species such as Bacillus cereus produce multiple compounds that could suppress more than two pathogens and this is effective in crop disease management. Other classes of microorganisms such as Lysobacter and Myxobacteria produce lytic enzymes which hydrolyze compounds leading to suppression of pathogens. Beauveria bassiana inhibits chitin development in insects by conidia attaching to the body of insects.

After germination the hypha penetrates through the cuticle and grows throughout the insect body eventually killing it. Semicochemicals such as female sex pheromones are used to lure the male insect pests which are then sterilized thereby decreasing their effectiveness. Upon mating with the sterile male insects, the females lay unfertilized eggs thereby reducing harmful insect populations.

Host location pheromeones lure insects into sites with mass traps from where they may be sterilized or starved to death. Predators may feed on the prey or a particular life stage of the prey such as nymphs or larvae. The predator prey ratio is of importance in balancing the populations of the pests as well as biodiversity.

EFFICACY OF DIFFERENT TYPES OF BIOPESTICIDES

Synthetic pesticides are considered more effective than biopesticides in managing crop pests. Their effectiveness sometimes has nonetheless no much significance in managing a particular population of pest as would the biopesticides. Biopesticides in other instances perform better than synthetic pesticides when applied in the right regimes, concetrations and appropriate frequencies. Research reports across the world have presented different plants, microorganisms and predators with potential as biopesticides. Natural enemies predate on insect pests which balances their population in the ecosystem. Such predators are important in agricultural systems. The mechanisms used by predators to lure insects include scents and other attractants. Some of these scents, called pheromones, have been commercialized and are being used in management of important crop pests such as Tuta absoluta. The commercial pheromones are baited to aid in luring the adult insects and then deactivating them by sterilization or startvation to death.

Certain plants contain compounds which they use to protect themselves against pests and this ability has been explored by researchers in an effort to manage different crop pests. Some plants have been found to contain compounds that are effective against several pests including fungi and nematodes. Some species of microorganisms have antagonistic properties towards other species and are therefore effective as biopesticides.

Researchers reported that extracts of neem (Azadirachta indica) and Mexican sunflower (Tithonia diversifolia) inhibited growth of rotting disease pathogens of tomato, Aspergillus niger, Fusarium oxysporum and Geotrichum candidium by up to 100%. Extracts of castor seeds (Ricinus communis) effectively inhibited growth of post-harvest pathogens Penicillium oxalicum and Aspergillus niger of yams (Dioscorea alata) in a dose dependent poisoned food technique experiment. Similar effects were reported on post-harvest fungi including Fusarium solani, Rhizopus arrhizus and Sclerotium rolfsii after using extracts from Duranta erecta and Lasonia ineruis. Methanolic extracts of Chenopodium ambrosioides exhibited antifungal activity against Fusarium oxysporum f.sp. ciceris a pathogen that causes wilt of chick pea (Cicer arietinum) by up to 50%.

A biopesticide formulation containing onion (Allium cepa) and ginger (Zingiber officinale) was evaluated for efficacy against tomato fruit worm (Helicoverpa armigera) and registered a 70% - 80% control. During the study, yield increment was also observed on plants treated with the formulation compared to the untreated controls. Researchers reported over 50% reduction of rape aphids (Brevicoryne brassicae) and tomato red spider mites (Tetranychus evansi) by powder extracts of Lippia javanica and Solanum delaguense. Populations of Megalurothrips sjostedti were reduced by extracts of Piper nigrum, Cinnamomum zeylanium and Cinnamomum cassia and were reported to be strong repellents. Number of larvae and pupa of Helicoverpa armigera were effectively reduced by extracts of Curcuma longa, Allium sativum and henge (Ferula assa-foetida) in a study. Extracts of Artemisia herbaalba, Eucalyptus camaldulensis and Rosmarinus officinalis soaked on leaves of broad bean (Vicia faba) caused a mortality of 60% - 100% of green peach aphid (Myzus persicae) after 24 hours of exposure in dose dependent in vitro experiments. In another study, topical application of Azadirachta indica, Mangifera indica, Polyalthia longifolia, Annona squamosa and Ficus benghalensis caused a 100% mortality of bed bugs (Cimex lectularius) after 19 seconds of contact. The effectiveness of plant extracts on insects is credited to the solvents used and their ability to extract major compounds with insecticidal properties.

Bacillus subtilis, Pseudomonas putida and Pseudomonas aeruginosa were evaluated against Fusarium oxysporum f.sp. ciceris and reported to have better control in seed treatment and resulted in an increment in growth parameters. Species of Bacillus have been reported to produce compounds effective against important fungal pathogens including Rhizoctonia solani and Xanthomonas oryzae pv. oryzae. Compounds from Chaetomium globosum have also been reported effective against important fungal pathogens of rice such as Magnaporthe grisea and Puccinia recondita.

Researchers reported reduced severity of fusarium wilt of tomato after using Streptomyces griseus. In a seeded media experiment, Stenotrophomnas maltophilia, Bacillus subtilis and Pseudomonas aeruginosa exhibited antagonism against Erwinina carotovora. A formulation consisting of compost tea extracts and poultry litter reduced severity of bacterial wilt (Ralstonia solanacearum) of brinjals (Solanum melongena). Higher efficacy was observed when the compost tea extract was applied as a soil drench and the poultry litter applied on the soil which resulted in healthy plants and improved yield.

A similar formulation reduced the incidence and severity of late blight (Phytophtora infestans) of potato (Solanum tuberosum) when the compost tea was applied as a foliar spray. Research showed that a compost tea formulation containing wood chips improved yield of lettuce (Lactuca sativa var. gentilina) and Kohlrabi or German turnip (Brassica oleracea var. gongylodes) when applied as a foliar spray.

Semiochemicals have been employed in management of insect pests. As reported, mass trapping by use of sex pheromones with water traps has been used in management of Tuta absoluta by delaying initial attacks on tomato plants. Similarly, fruit flies (Rhagoletis cingulata) have been managed through semiochemicals such as sex pheromones, oviposition, host location and mating pheromones.

These semiochemicals could be insect-plant induced or insect induced and the end result is enhanced parasitizing of the insect populations.

Predators of insect and microbial nature have been effectively used in management of insect pests. Species of Amblyseius swirskii have been used in the management of thrips, Frankliniella occidentalis and Scirtothrips dorsalis and spider mites (Tetranychus urticae). Phytoseiulus persimilis is an effective predator mite against spider mites (Tetranyshus spp.)

WHY BIOPESTICIDES IN SUSTAINABLE AGRICULTURAL PRODUCTION

Biopesticides are as effective as synthetic pesticides in management of crop pests. Natural products are also eco-friendly since they are easily biodegradable and therefore do not pollute the environment.

Consumer tastes and preferences fluctuate over time and following the demand for organically produced food, this makes biopesticides suitable alternatives to synthetic pesticides. Biopesticides have very short pre-harvest intervals and are therefore safe to use on fresh fruits and vegetables. They are also target specific and hence do not affect the beneficial organisms such as the natural enemies.

They are effective in small quantities and their use promotes sustainable pest management and hence contribute towards sustainable agriculture. Natural pesticides do not cause resistance build up among pests.

Availability of their source materials makes them inexpensive to attain since they are found within the natural environment and some of them are used for other purposes like food and feed. Biopesticides are safe products both for the applicant and the consumer since they have no toxicity.

Therefore, biopesticides can suitably be incorporated in integrated pest management (IPM) which helps reduce the amounts of chemical pesticides used in management of crop pests. Natural products decompose quickly which makes them safer for use in the environment.

Pesticides from natural sources have very short re-entry intervals which guarantee safety for the applicant. Biopesticides are also used in decontamination of agricultural soils through introduction of important microbial species.

PEST MANAGEMENT WITH BIOPESTICIDES

Pesticides may be defined as chemicals (or mixtures of chemicals) used to restrict or repel pests such as insects, weeds, fungi, nematodes, and other organisms that adversely affect food production, ecosystem function or human health. Pesticides may also be toxic contaminants in our food supply and environment (air, water, soil), and responsible for illness or injury to people and wildlife. Biopesticides are attracting interest as alternatives to conventional pesticides, but without many of the non-target effects, by offering improved safety in pest control practices.

Current trends

Pesticide use in California and the US peaked in the early 1980s. Since then there has been a trend toward less pesticide use[1]. This trend reflects a combination of several factors: the banning or phase-out of high volume use synthetics, like toxaphene, chlordane, and methyl bromide; development of more efficient application technology which delivers more chemical to the target and allows less chemical loss by volatilization and wind erosion, to soil, and by surface runoff; and the introduction of transgenic modifications in some crops like cotton, corn, and soybeans so that resistance to pests or tolerance is carried by the crop without need for, or minimal use of, external chemical application for pest control.

An example is the technology underlying the use of transgenic Bacillus thuringiensis (Bt) toxin to control insects. Farmers are also using more integrated pest management tools such as intercropping, cover crops, biocontrol, and crop rotation, along with reduced risk chemicals such as synthetic pyrethroids, avermectins, and spinosads that are generally effective at lower application rates than conventional pesticides. These tools all work to reduce the amount of chemical applied to crops to obtain economically acceptable levels of pest control. They will also improve the safety of food production by eliminating or minimizing the use of costly chemicals that leave toxic residues in foods, or lead to illness in farm worker populations, and do so at lower costs in many cases.

Biopesticides

"Biopesticides" would be all of these things included in the definition of pesticides but with several modifications, in particular that they are naturally occurring or derived from natural products by straightforward chemical modification. The EPA definition is that biopesticides are natural compounds or mixtures that manage pests without a toxic mode of action, such as cholinesterase inhibition which is a characteristic of most organophosphate and carbamate insecticides. Third generation pest control agents, reduced risk pesticides, and biobased pesticides are other terms sometimes used in place of "biopesticides".

The common elements of biopesticides can include some or all of the following: naturally occurring, reduced toxicity to non-target organisms, low persistence in the environment or in ecological food chains, compatibility with organic farming methods, and low mammalian toxicity so that they are safe to handle, and not restricted in use by regulatory agencies. Few products will fit all of these criteria, but the intent is clearly to stimulate 'green' environmentally benign technologies for sustainable pest management and control.

Although the market is growing for biopesticides, none of the top-use pesticides in recent years in California or the US clearly meet the biopesticide definition. One could argue that sulfur is close, since it is naturally occurring and of low non-target toxicity, and useable in organic farming, but sulfur has been used for pest control for centuries and is not among the new generation of pest control agents.Furthermore, sulfur must be refined from its natural sources in the environment before it can be marketed and sold for use as a pesticide. Various mineral oils used for weed control, some plant essential oils (e.g., orange oil for termite control), and corn gluten for weed control might be considered within the realm of biopesticides as well.

Spinosads well represent the commercial possibilities for biopesticides, recently gaining a large market share for protection of apples, pears, strawberries and other high-value crops. In part this is because the residues left by spinosads are of low toxicity, and product is considered safe for consumers, including infants and children, when the product is applied in the manner specified on the label, and are approved in the US for organic production. Spinosyns - there are two active forms, A and D differing only by placement of a methyl group - are produced by soil-borne fungi (Saccharopolyspora spinosa) which can also be used in fermentation culture to produce the technical product. The use of fermentation to produce chemical control agents is another potential advantage for biopesticides since it can reduce or eliminate the need for extensive chemical processing facilities associated with pesticides based on petrochemicals.

A number of specialty pesticide companies, like Trécé, Certis USA, and Marrone Bio Innovations have marketed new biopesticides, and now larger companies like Monsanto, DuPont, Bayer, BASF, and Dow are developing and/or marketing biopesticides along with conventional (2nd generation) pest control chemicals. EPA has helped to move the biopesticide technology forward by offering a "fast-track" for registration of reduced toxicity pesticides.

The newer bioproducts include fungicides, insect repellants and attractants (semiochemicals), insecticides, nematicides, herbicides and products from genetic manipulation. In China, India, and several other developing economies, technologies for pest control are being developed based upon farmers experience in those countries. In the Philippines, for example, rice can be grown along with Tilapia in the paddies.

The tilapia fish consume insect pests of rice, so that no chemical control may be needed, and then become a nutritious co-product harvestable along with the rice crop.

Bioherbicides are a future target for development, although research in this area has been somewhat slow since the last new mode of action herbicides were introduced more than 20 years ago.The evolution of weeds resistant to the leading herbicide glyphosate may accelerate developments in this area. Non-synthetic chemical management of weeds in organic culture is a serious problem limiting wider use of organic farming methods.

The few bio- or green-products for weed control use high application rates or multiple applications, and even then are somewhat unpredictable in efficacy. Mechanical control methods include hoeing, mowing, burning, or solarization under plastic tarps, as well as use of grazing sheep and goats. Many of these methods are laborious, costly, and often unreliable, or accompanied by undesirable environmental side effects. The development of the new triketone herbicides is a case in point, where observation of herbicidal activity from an ornamental plant (bottlebrush), led to the isolation of the naturally occurring bioactive principle, leptospermone, and various synthetic analogs that are now commercial herbicides, such as mesotrione.

Mesotrione was brought to market by Syngenta in 2001. It is a synthetic analog of leptospermone which mimics the herbicidal effects of this natural product. It is a member of the class of HPPD inhibitors which work by inhibiting 4-hydroxyphenylpyruvate dioxygenase. HPPD is required by plants for carotenoid and plastoquinone biosynthesis; carotenoids protect chlorophyll from sunlight-induced degradation and plastoquinone is required for photosynthesis. When the HDDP inhibitor is present in plants, carotenoids are prevented from being made and photosynthesis is inhibited, causing chlorophyll to degrade, followed by plant death. Sales by Syngenta were more than 400 million USD per year in 2011, but expiration of patents beginning in 2012 has opened the market to other synthetic triketone herbicides.

Another future major market to be addressed is for bionematicides for soil application and use in stored products, given the mandated (Montreal Protocol) phaseout of methyl bromide, and off-target movement and exposure issues with present fumigants like methyl isothiocyanate (MITC), formed from synthetic metam products, and chloropicrin. Avermectins show some promise for nematode control, particularly in combination with other control measures, and farmers are adopting cultural methods (crop rotation, intercropping with Brassica species, solarization, etc) to address nematodes in the more susceptible crops like strawberries and carrots.

There is renewed interest in discovering botanical and related pesticides with novel structures and activity that can be used directly, or to inspire synthetic modification to form the pest control agents of the future. As researchers have pointed out, plants produce a bewildering array of "secondary metabolites" thought to play an ecological role in defending plants from attack by herbivores and pathogens, as well in chemically inhibiting competing plant species. These range from the familiar biopesticides of long standing interest, such as pyrethrins, rotenones, and alkaloids, to complex mixtures of terpenes, carbohydrates and proteins. Natural pesticide discovery is particularly pursued in Asia and Latin America as a way to overcome costly regulatory requirements in industrialized nations. The products are often complex unrefined mixtures of active ingredients and other components of presently unknown utility to the source plant or microbes.

Pesticide discovery has experienced renewed interest ranging from exploration of plants and microbes that produce chemicals and mixtures of potential utility to mainstream agriculture, to niche products that can be exploited by synthetic and biotechnological modifications in the future. There are plenty of areas yet to explore, by chemical synthesis, biotechnology, breeding, and by ecosystem scientists and engineers. The success of the current group of biopesticides will likely be expanded as more scientists and practitioners are attracted to the critically important field of pest control in sustainable food production.

Other alternatives to pesticides

The excitement over biopesticides, semiochemical communication cues, and other alternative controls, which was evident in the 1970s when the 3rd generation pest control movement was launched, is now regaining momentum. Over half of the new registrations for pesticides and pest control agents at EPA are for products associated with the features of biopesticides, and the market share is growing for these products.

High throughput screening methods, nano-based encapsulation methods, and further development of semiochemicals for both monitoring and population control of pests offer promise for further developments. Semiochemicals are already far along in crop protection applications. Pheromones or synthetic analogs are widely used to survey for pest populations, so that insecticide applications can be timed and positioned to be most effective. Mass trapping or confusion approaches have also been used with some success, using pheromones or synthetic or naturally occurring alternatives that disrupt pest insect populations.

An example is the pheromone and a naturally occurring alternative with pheromone-like attractant activity found in pear leaves that can aid in control of codling moth in apple, pear, walnut, almond, and other crops susceptible to economic damage by codling moth. Controlling this damaging pest, and other boring insects that affect cotton seed and peanuts, is a critical element in controlling invasion of Aspergillus fungi, which can affect the pome fruit, nut, or seed and produce aflatoxins—a group of carcinogenic fungal metabolites.

A combination of aflatoxin in foods and a hepatitis B-susceptible human population such as exists in many parts of Africa and Asia can lead to high levels of liver cancer and elevated mortality—a major public health and food safety concern. This is an example of how the use of effective pest control carries with it the added benefit of reducing the carcinogen load due to toxic natural products such as aflatoxins in the food supply—a topic of food safety.

GMO CROPS

Chemical control of pests is widely practiced, but more and more the use of crops and animals genetically improved to resist pests (insects, disease, nematodes, weeds) are being explored and developed to offset chemical usage while protecting valuable food sources, in wheat, rice, and with many other crop staples and in food animals. In some cases, the resistance genes are engineered into the crop, giving farmers new genetic resources for insect resistance (e.g., Bt toxin genes in corn and soybeans). In these improved varieties, little or no external chemical insecticide application may be needed.

Other crops (papaya and plums) have been made resistant to viral diseases by transgenically imparting production of viral coat proteins that stop virus reproduction. Gene-based technologies, such as RNA interference (RNAi), are underpinning new technologies in pest control. RNA interference is a natural process that affects the activity of genes. Research has successfully led to artificial RNAs that target genes in pest insects, slowing growth or killing them.

The development of GMO crops that make RNAi harmful to their pests is under active exploration. As with most new technologies, there are safety concerns that RNAi might also harm desirable species. Plant scientists are using CRISPR gene editing to make sustainable agriculture crops with higher precision than possible before and less potential for undesirable side effects. The first examples will be in corn, planned for release for commercial plantings in about 2020.

While these new technologies like gene editing will have a huge impact in agriculture and in design of new drugs, it is not clear whether they will be embraced by consumers, at least in the case of food products. Some environmental organizations have indicated that they will resist introduction of new crops and farm animals improved by gene modification.

Even though new GMO technologies offer the promise of safer and more abundant foods, many consumers and activists are also interested in preserving the natural qualities of foods, including taste, texture, color and growth related characteristics that influence availability and market choice.

Seemingly, for every technological advance in developing resistance, the target pest evolves a strategy for overcoming the protection, as happened so often with resistance in insect and fungal pests previously controlled with synthetic pesticides and with antibiotic use in farm animals.

This is possible with genetically modified crops and biopesticides, so requiring close monitoring of fields for early signs of resistance, and then applying an alternative pest control strategy from a "tool box approach" which may include conventional chemical pesticides, biopesticides, cultural methods, and other approaches is desirable to preserve these new technologies.

SMART APPLICATION SYSTEMS

Only small fraction of applied pesticides reach the intended pests, but rather bypasses the target and enters the soil, non-target vegetation, or are carried away by wind. Agricultural engineers and systems scientists have developed more effective spraying techniques using spray drift control technology, electrostatically charged spray droplets, controlled release technology, or smart systems that direct spray just to the optimal position for contacting the target.

These improvements save on the amount of pesticide needed for a particular situation, and also prevent inadvertent residues that can harm unintended crops, water way quality, or livestock and wild animals. But, they can also lead to more effective control of the target pest. Related to this, information is more accessible that clearly delineates the mere presence of a pest from the presence of a pest population of sufficient magnitude that can cause economic damage or impair safety in the harvested commodity.

Not spraying at all can sometimes be the best strategy, generating a cycle of pest control by beneficial natural enemies, a means of biocontrol that, once fostered, can lead to sustainable pest suppression for many years. This can have important ramifications in safety of foods, as lower chemical use may lead to safer foods as well as foods that are more widely available at relatively low costs.

Recent discussions with farming groups and food producers in China's Weifang Food Valley area of Shandong Province illustrate local efforts to reduce the use of conventional pesticides and fertilizers, including:

- Develop farming practices for organic produce and meat;
- Encourage use of biopesticides and biofertilizers;
- Control nematodes by non-chemical means, such as soil solarization;
- Develop novel ways of reducing development of resistance to pesticides;
- Introduce non-chemical insect and disease controls by biological or cultural means;
- Explore Traditional Chinese Medicine (TCM) approach to new antibiotics for use in farm animals and pest control;
- Assist in adoption and registration of biopesticides using an IR-4 type regional or national system such as in use in the US;

LIMITATIONS FACING USE OF BIOPESTICIDES

While biopesticides provide such advantages as safe environment and healthy food for human consumption, there are factors that limit their full adoption as pest and disease management options. High doses of the constituent compounds are needed for efficacy under field conditions. The concentration of the bioactive compounds in plants is dictated by the environment under which they grow.

The constituent active compounds are also dictated by the diversity of plants and their varieties resulting in differences in the responses to pathogens. The quality of botanical extracts is also dependednt on the method of extraction used. During formulation, it is sometimes challenging to get the right proportions of the active and inert ingredients needed. There are also no standard preparation methods and guidelines for efficacy testing especially under field conditions. While the in vitro tests produce excellent results, there are always inconsistencies at the field due to low shelf life and sometimes poor quality of source materials or preparation methods.

Adoption of biopesticides of predatory nature need a lot of consideration such as host crops and dispersal capability. Crop coverage and exposure time are essential and for a small acreage this could prove expensive since application may be manual.

Registration of the products requires data on chemistry, toxicity, packaging and formulation which is not always readily available. The cost of producing a new pesticide product is usually high and has a lot of resource limitations. Lack of a readily available market makes it hard to invest in biopesticides. There are insufficient facilities and capital for production of biopesticides especially in the slowly developing countries. The shelf life of natural products is dependent on many factors such as temeperatures and moisture which are sometimes difficult to control. Biopesticides also face high competition from synthetic pesticides and if the former were produced for a small agricultural activity, the costs may be relatively high and therefore not feasible.

There is insufficient awareness about biopesticides especially among the small-scale growers, stake holders and policy makers. In the case of microbial pesticides, there is usually no trust in the value and use chain between producers, buyers and users and considering the risk of importation, synthetic pesticides appear reliable.

CONCLUSIONS

The field of pest management, in agriculture, public health and home use, is undergoing change as sole reliance on synthetic chemicals gives way to a variety of approaches, some with little or no pesticide in the mix. Health concerns over exposures to pesticides among agricultural workers and consumers of residue-tainted food products, as well as wildlife and non-target species, have been partly responsible for this, as have demands by consumers, translated through the food supply chain to large commercial retail outlets, commodity organizations, and regulatory agencies, for pesticide free, organic, and "green" products. Still, the reality is that feeding a world population expected to top 9 billion by 2050 will require the use of chemical pesticides as a primary tool in combatting pests in the field and in stored products, as well as for public health, for years to come. It may in fact increase use of pesticides as emerging economies of China, India, Brazil, and other nations expand agricultural production. But, the options for pest control have expanded, also illustrated by biopesticides, and advances in the fields such as genetics, biotechnology, sustainability, and more targeted pesticide application technology. These factors may in the long run reduce the need for chemical pest management tools to a fraction of that used presently.

Other transformative research opportunities that will require mathematical and physical science research efforts to improve the sustainability of agriculture include:

- Ensuring a sustainable water supply for agriculture;
- Closing the loop for nutrient life cycles;
- Crop protection as noted in this manuscript;
- Innovations to prevent waste of food and energy;
- Sensors for food security and safety;
- Maximizing biomass conversion to fuels, chemicals, food and biomaterials.

6

BIOPESTICIDES AND BIOFERTILIZERS IN ORGANIC FARMING

INTRODUCTION

Organic growers have available a large array of biopesticides that may be applied for the management of plant diseases. This article will define what biopesticides are and provide general information on their potential benefits and limitations. While information discussed is applicable to both organic and conventional crop production, this chapter will address only those materials approved for use on organic farms at the time of writing.

Biopesticides are produced from toxins found in biological organisms that are harmful to the pests that invade plants. They do not affect the plant and in fact, reduce soil pollution and erosion. Hence, organic farming requires biopesticides for good crop production.

Following are the examples of biopesticides:

1. Bacillus thuringiensis that causes intestinal diseases of various lepidopteran, coleopteran, and dipteran species. The toxin from this bacteria has also been directly expressed in plants through genetic engineering, such as Bt cotton and Bt brinjal.
2. Trichoderma, a plant root fungus, acts as a biocontrol agent against several plant pathogens. Biopesticides can be classified as 1. Microbial pesticides consisting of bacteria, fungi, viruses, and sometimes nematodes.
3. Bio-derived chemicals that are various substances naturally occurring that control pest populations such as neem oil and other aromatic compounds.
4. Plant-incorporated protectants or Genetically Modified (GM) crops.
5. RNAi (RNA interference) pesticides. Some of them are applied topically on plants and are absorbed by them.

BIOPESTICIDES: DEFINITION AND SCOPE

Many biologically-based products are currently available, and these products are often referred to as "biorationals" or "biopesticides". Biorational is an undefined term used in broad reference to any biologically-based product used in agriculture that includes fertilizers, pesticides, herbicides, plant growth regulators, and various other products.

A biopesticide is defined by the U.S. Environmental Protection Agency (EPA) as a pesticide derived from natural materials. The EPA divides biopesticides into three categories: biochemical pesticides, microbial pesticides, and plants containing added genetic material.

Biochemical pesticides contain naturally occurring substances that control pests. Substances that control diseases in this category include potassium bicarbonate, hydrogen dioxide, phosphorous acids, plant extracts, and botanical oils. Microbial biopesticides contain microorganisms as the main active ingredient that function as biological control agents, affecting the pathogen directly or indirectly through the compounds they produce or by stimulating specific plant responses.

Plant-incorporated protectants (PIPs) are the least common type of biopesticide. These are pesticidal substances produced by plants that contain genetic material added to the plant often through genetic engineering. Biopesticides also exist for the management of weeds, insects, and nematodes.

While generally a logical fit for managing diseases in organic crops, there are some biopesticides that are NOT approved for organic production. Some biopesticides, like salts of phosphorous acid and all genetically-engineered PIPs, are not allowed for organic production. Note that some fungicides approved for use in organic production systems are not biopesticides, including mineral oils, copper, and sulfur. It is critical to verify that a biopesticide, like any other product, is approved for organic production and registered in your state prior to any application.

Biopesticides are considered as a naturally occurring organisms or bio-based formulations that control pests through different mechanisms of action. They are products or byproducts derived from animals (nematode; Heterorhabditis spp.), insects (Trichogramma spp.) plant parts or extracts (example, a finely ground flower of Chrysanthemum cinerariaefolium) and microorganisms (Bacillus thuringiensis, Verticillium lecanii, Neodiprion sertifer).

However, there is a deviation in the generally accepted definition, as it is that natural products should be regarded as chemical pesticides, if they can have an impact on the pest nervous system.

For example, rotenone, as a plant-based pesticide, has the potency of killing and controlling insect pests by exerting its toxic effect primarily on nerve and muscle cells. Similarly, nicotine (a fast-acting nerve toxin), sabadilla (affects nerve cell membrane action), pyrethrins (interrupts the normal transmission of nerve impulses), and fluoroacetate (causes depletion of glutamic acid thereby affecting the nervous system) are all effective pesticides that interfere with the insect nervous system. However, most literature classifies these compounds as botanical pesticides. A chemical homologue of a biopesticide can also be regarded as a biopesticide. Biopeticides intended for the control of herbivorous insects or pests can be regarded as phytosanitory biocontrol agent or bioproduct.

PRODUCTION, COMMERCIALIZATION AND MARKET PROSPECT OF BIOPESTICIDES

Biopesticide production starts with bioprospecting from the natural environment, screening (in vitro), in vivo experiments, purification, formulation and registration. Botanical pesticides are obtained from plants in the natural environment while microbial bioprospecting can be obtained from compost, manure or rhizosphere. Botanicals are first cleaned of impurities, extracted and the extracts are then screened in vitro against pest of interest using different methods.

From the screened plants, the most effective plant can be evaluated through field-trials, and from these plant extract the most active constituents can be identified and ascertained for optimum formulation. The active compounds are usually identified with either or in a combination of different spectrometric techniques including high-performance liquid chromatography (HPLC). The active ingredients are further combined with carriers, surfactants, emulsifiers and other components, followed by intensive in vitro and laboratory trials and optimization until such a product is ascertained to be efficacious.

Once the efficacy trial is confirmed to be successful, the product can be registered with concerned regulatory body or bodies. For microbial biopesticides, isolated pure cultures are maintained in agar slants. In vitro efficacy trials are conducted through culture and diffusion methods. Suitable substrate medium for the active microbial agents is prepared in the lab and mixed with other components (that will not inhibit microbial growth) such as enhancers, carrier materials and stabilizers.

The efficacy of the product is ascertained after repeated laboratory and field trials against the target pest(s). The stability and efficacy of active compounds of biopesticides lie in the formulation of the active compounds. Formulation of pesticides, in general, is a necessity for effective control of pests however for a particular application method.

Formulation of the active ingredients of bio-based pesticides are in principle the same as that of the formulation required for synthetic pesticides. Usually, the formulation consists of the active ingredients, adjuvants and carriers that will guarantee sustained bioactivity of the active agent, protection of the products from environmental conditions, storage stability, easier handling, and interaction with the target pests when applied in field settings. Based on the physical state, biopesticide commercial products are divided into dry and liquid formulations. Liquid formulations come as emulsions, oil dispersions, suspension concentrates.

Dry formulations come in the forms of dust, powder, granules, wettable powders, and water dispersible granules. Emulsions are liquid droplets dispersed in a different immiscible liquid with phase droplet size of between 0.1 and 10 μm. Emulsions are usually regarded as oil in water but can also be in the inverted form (water in oil as in essential oils). Suspension concentrate consists of an insoluble finely ground active compound in a liquid (water) phase. Their particle size distribution is between 1-10 μm. Oil dispersions are solid active ingredients dispersed in non-aqueous liquids, especially in plant oil. Biopesticide dusts are formulations made of an active ingredient and finely ground solid mineral powder with a particle size of between 50- 100 μm. Powder formulations are produced by mixing an active ingredient, carrier and adjuvant that will facilitate adherence of the product to seed coats.

Granules formulation are similar to dust formulations except that the granular particles are heavier and larger (100-1000 μm). Wettable powders are formulations made up of finely active ingredients (5 μm), surfactants, dispersing agents and inert fillers which are used after suspension in aqueous medium. Water dispersible granules are a powder formulation to be used when dispersed in water.

ROLE OF BIOPESTICIDES IN ORGANIC FARMING

Over the past couple of decades, people have become more conscious than ever of how the food we produce impacts our bodies and the world around us. More than ever, consumers are demanding that food and other products are produced in an ecologically-responsible way, but they're also not willing to compromise on quality.

This shift in awareness and attitudes has been the driving force behind the organic farming revolution. Some farmers all over the world have begun embracing new techniques for growing crops without toxic or otherwise hazardous substances. However, the problems faced by mass crop production that caused food producers to adopt non-organic practices in the first place still remain. If more farmers are going to make the switch while maintaining the size of their yields and the health of their crops, then the agriculture industry is going to need innovative organic solutions.

These innovations have come in a number of forms, but one of the key needs of the organic agriculture industry is options for non-toxic biopesticides to protect crops from disease and organisms that could destroy them.

Pest control, biopesticides and the organic foods market

The rising demand for organic foods has been more than enough for some food producers to make the switch. A report by Hexa Research projects the United States Organic Food Market to reach an incredible $70.4 billion in value by 2025 while the global organic food market will hit $323 billion by 2024 according to a 2018 report by Zion Market

Research. As of now, more and more consumers are choosing organic and the result is supply vastly outpacing demand. Many producers remain wary of the risks of organic production, worrying about smaller yields and the chance of losing crops to pests and disease. Organic production capacity in the US increased by 20 percent from 2011 to 2018, but still accounts for less than one percent of total US farmland.

As a result, US organic food companies have been looking to foreign suppliers for raw product. Organic disease and pest control have come a long way in recent decades, but concern from farmers about the efficacy of these products agricultural hazards is understandable. Making the switch to organic farming can be risky and complicated. Government and private funding and support for conventional farming still vastly outweighs that of organic farming, which requires very different practices in order to keep plants healthy. Transboundary plant pests and diseases have become more common in recent years and these crop-destroying factors can be potentially catastrophic for farmers' livelihoods.

In order for organic food production to finally meet the ample consumer demand, farmers require reasonable assurance that going organic will be a smart investment. This is why further development of non-toxic organic bio-pesticides will be crucial to the success of organic foods. Defined as naturally-occurring substances that control pests, common biopesticides include various plant extracts, botanical oils, potassium bicarbonate, hydrogen dioxide and phosphorous acids. Biopesticides that are recognized by the US National Organic Program (NOP) have advantages like leaving little to no toxic residue, carry little risk of pests and disease developing resistances and more. Downsides to today's common biopesticides include lower effectiveness and shorter persistence compared to conventional pesticides.

Common biopesticides may also be more complicated to use effectively than conventional pesticides, requiring a greater degree of expertise by the user. Despite current limitations, biopesticides are primed to see huge growth in the coming years as organic farming continues to make headway. A 2017 report by Markets and Markets projects the global biopesticides market to grow from $2.83 billion in 2016 to $6.60 billion in 2022 with a compound annual growth rate of 15.43 percent.

There's a number of factors driving that growth including the threat of pest resistance to conventional pesticides, faster regulatory approval for non-toxic substances, low costs for raw materials and, most importantly, the surging growth of the organic foods industry. Agri-food isn't the only agricultural industry that will be a factor in the growth of the organic biopesticides market.

The emerging legal cannabis industry is also driving demand for organic, non-toxic agricultural innovation. Like food, cannabis crops require robust pest control systems to prevent pests and disease. Like food, cannabis in legal jurisdictions like Canada is heavily regulated for consumer health.

Brassica as a pest control

The demand for reliable, inexpensive non-toxic biopesticides is spurring plenty of innovation in the space. Brassica, a genus within the mustard family, is one promising biopesticide that is now being seriously considered for the organic foods and cannabis spaces. Organic compounds contained within the brassica plant have been found to benefit agricultural production. Brassica has yet to be commercialized in the biopesticide space, but Canadian agricultural biotech company MustGrow Biologics (CSE:MGRO) has been exploring how the plant's natural mechanisms function against nematodes, fusarium, botrytis, powdery mildew, verticillium, pythium, and other fungal diseases and pests and whether it can be applied to mass agricultural production. The company has developed a proprietary method for isolating the plant's glucosinolate and enzymes for trials and for commercial production in the near future.

So far, the over-110 independent field trials that have been conducted on brassica MustGrow's technology and have shown strong potential for superior crop performance and enhancement of grower return on investment. One of the main upsides for brassica as a commercial biopesticide is its affordability and availability, particularly in Canada, where 28 percent of the global mustard crop is produced. This means that with brassica the Canadian North Amercian agriculture industries have easy access to an abundant, inexpensive and most importantly effective organic biopesticide.

Advantages of Biopesticides

Biopesticides exert their inhibitory effects through multiple mode of actions such as growth regulators, gut disruptors, metabolic poison, neuromuscular toxins, and non-specific multi-site inhibitors. These multiple modes of action against targeted pests obliterate the chances of developing resistance as is common with chemical pesticides. The intensive use of conventional pesticides in industrial-scale farming over a long period, especially in the Green Revolution era, also created challenges such as pesticide-related pollution, post-harvest chemical consumption through bioaccumulation, biodiversity losses, and insurgence of secondary pests and elimination of natural/beneficial enemies.

These negative consequences are not associated with the use of biopesticides. Thus, prohibitive restrictions are continually imposed on synthetic pesticides to reduce their numbers with time. For instance, there has been a reduction to 250 active ingredients of conventional pesticides in 2009 as opposed to more than 1,000 in 2001 while entrance of new conventional pesticides into the market reduced from 70 in 2000 to 28 in 2012. The direct result of the declining number of classical pesticides is the increased demand for biopesticides for some beneficial reasons.

These benefits include, but are not exclusive to, altering the course of pest resistance, low toxicity properties, complementary input to synthetic pesticides, eco-friendliness, specificity (thus have little or no negative impact on non-target organisms and humans), biodegradability, and little or no problem of post-harvest contamination, stability against abiotic stress, and compatibility in integrated pest management (IPM).

Prospects and Limitation of Biopesticides

In the nearest future, biopesticides could replace synthetic pesticides without significantly affecting productivity and yield, if their potentials are fully maximized. The limitation confronting the full adoption of biopesticides are the high cost of refined commercial products, inability to meet global market demand, differing standard method of preparations and guidelines, dose determination of active ingredients, the susceptibility of biopesticides to several environmental factors, ephemeral stability, and slow action among others.

While expecting these limitations to be addressed through research breakthroughs in the coming years, farmers, especially in rural areas, can take advantage of a crude plant extract in protecting plants for improved yield. The effectiveness of the application of pesticides, including biopesticides, is maximal when incorporated in IPM.

An IPM is a pest-control option, equivalent to an effective multi-faceted approach involving the combination of cultural practices and other suitable control tactics into one management program to achieve a long-lasting reduction of pest population and associated problems.

Positive externalities that go along with biopesticides besides significant pest population reduction are social acceptability, economic viability, and environmental stewardship, the three-domain concept of sustainable development.

Biopesticides as Drivers of Sustainable Agriculture and Sustainable Development

As with other natural resources, such as biostimulators and biofertilizers, biopesticide application in mechanized farming creates a balance amongst sociocultural relevance, economic productivity, and environmental protection that is considered pivotal to sustainable agriculture.

Integration of public policy into these four domains (technology inclusive) would yield a much higher concept known as sustainable development. The most current strategies aimed at achieving sustainable development are contained in the United Nations 2030 Agenda (17 SDGs).

Green Chemistry in Sustainable Agriculture and SDGs

GC most critical concept is that of design, and it aims to achieve sustainability at the very molecular level. As a result, GC covers a wide range of fields including agriculture. The direct connection of sustainable agriculture in the SDGs is dependent on the immediate positive impact of agriculture, judicious use and conservation of critical resources, application of green agro-products, sustainable technology and green energy, or the combination of any two or more of these. The use of green agro-products can influence SDG2 and SDG1, while SDG6, SDG12, SDG14, and SDG15 can be influenced by conservation principles, environmental sustainability, and judicious use of critical resources.

These goals are reflected in GC, which aims to mitigate harmful chemicals from production down to application, process safety, optimal energy consumption, and improved economy. The two priority mass-based metrics upon which the GC partly rest on are atomic economy (AE) and environmental impact factor (E factor).

The AE measures the content of reactants present in the final products while the E factor measures the waste amount in 1 kg mass of the target product, exempting water. In the context of GC, the higher the AE/lesser E factor, the better for the sustainability goals affiliated with agriculture.

The pursuit of ideal AE (100%) and E factor (0) in the production of a target chemical product demands 100% resource efficiency: mass of reactants = mass of target products. Beyond mass-based metrics is the life-cycle assessment, which comprehensively considers factors such as process input (reagents, solvents, metals) sustainability, environmental impact (carbon footprint, energy consumption, eco-toxicity, risk factor), implicated material/products, and analytical methods that work on the concept of minimized hazards and solvents, operator safety maximization, device and sample miniaturization, and greener solvent usage.

These metrics are covered under the 12 principles of GC. GC and its 12 principles can play a useful role in defining the concept of sustainability in six of the highlighted eight SDGs. Though the 12 principles are more oriented to the chemical process at the bench scale, they can still be linked to agriculture with efficient utilization of renewable resource and waste elimination as the most influential dimensions.

LIMITATIONS OF BIOPESTICIDES

Many biopesticides function by interfering with a pathogen's ability to infect a susceptible plant either directly through parasitism or indirectly through the production of secondary compounds. Therefore, timing and frequency of application are critical. Effective control requires that the application of materials begins prior to conditions favorable for disease development or immediately following the first symptoms of disease.

In foliar applications, these materials provide a protective barrier on plant surfaces that interfere with the pathogen. Repeated applications are required, often on a weekly interval, due to loss of product from degradation or rain/irrigation washing materials off foliage and fruit, or with the growth of new foliage.

Some biopesticides, notably Regalia and Companion, are thought to turn on plant defense responses. Such products should be applied before infection to be effective. Similar strategies are at play when biopesticides are applied to seeds, roots, or soil.

These materials can provide temporary protection to roots, but results can vary depending on the crop, pathogen, and environment. Subsequent root growth or rain/irrigation may wash biochemical materials away from the root zone. Many biological agents can grow within the root zone and can persist longer than chemicals.

However, because they are often unable to compete with native soil microorganisms, they tend to persist at levels ineffective for control. Many biopesticides are effective in the control of root rots associated with seed germination and seedling/transplant establishment.

Again, the efficacy of biopesticides and most fungicides is limited against established infections, so the sooner applications are started, either prior to or during a disease outbreak, the more likely control will be realized. The frequency of application of any product is related to its residual time, or the time required for a product to degrade in the environment.

Biopesticides in general have a short residual time, typically much shorter than compounds such as copper. On the other hand, this short residual time also limits most concerns about build-up in the soil that are associated with copper.

In addition, the activity of biopesticides is generally targeted to specific pests and closely related organisms, and they are usually inherently less toxic to non-target organisms. Therefore, they do not have the same potential to affect birds, beneficial insects, and mammals. Because of the short residual time, low toxicity, and reduced risk to nontarget organisms or the environment, the EPA generally requires less data to register a new biopesticide compared to a conventional pesticide.

However, sufficient data regarding the composition, toxicity, degradation, and other characteristics of the pesticide is still required to ensure that the product will not have any adverse effects on human health or the environment. The EPA, recognizing that biopesticides tend to pose fewer risks than conventional pesticides, has taken steps to promote development and adoption of new products.

The registration process is quicker for biopesticides than conventional products, often taking less than a year, compared with an average of more than 3 years for conventional pesticides. To facilitate their registration, the Biopesticides and Pollution Prevention Division was established in the Office of Pesticide Programs in 1994.

Some biopesticides are defined as minimum-risk pesticides through the FIFRA Section 25(b) rule because their active and inert ingredients are generally recognized as safe (GRAS). These consequently are exempted from the regulation requirements of FIFRA and thus can be used on any labeled crops for any target since they do not need to be registered as a pesticide. "Exempt from EPA registration" is stated on the label of these products. When selecting a biopesticide it is important to obtain information about the product's mode of action, residual time, and target disease(s).

Unfortunately, data on the efficacy of biopesticides is limited. Many products are broadly labeled with separate lists of registered crops and labeled diseases. Data proving efficacy is not required for the registration of pesticides in the USA. Most biopesticides are produced by small companies that lack funding to support replicated field trials needed to obtain sufficient efficacy data by experienced university and other independent researchers. To help fill this gap, the IR-4 Biopesticide and Organic Support Program funds grants to obtain efficacy information for biopesticides in development as well as those already registered. These funded projects help the program meet its objective, which is to further the development and registration of biopesticides for use in pest management systems for specialty crops (which include all vegetables) or for minor uses on major crops.

There are databases of labels and projects at the IR-4 web site. Several factors can affect performance of biopesticides, or any product used for disease control. These need to be considered when reviewing results from an efficacy experiment or on farm use. It is especially challenging to assess efficacy of a product used on-farm because there are no plants left untreated or treated with products known to be effective for comparison. A product may appear to be effective when actually conditions became unfavorable for the disease to develop or the pathogen was not present. Conversely, conditions may be so favorable for disease development that it is not possible to suppress the pathogen even with a conventional product demonstrated previously to be effective. The earlier in disease development that applications of a product are started, the more effective the product will be.

Disease spots (lesions) cannot be 'cured' and once a pathogen has infected a plant it cannot be killed, even with most conventional fungicides, in contrast with insect pests which remain on plant surfaces and are accessible to treatment. Product performance can also be affected by spray coverage and frequency of application. Treatment timing is a common potential explanation for poor control with a product that has been effective in other situations.

Sometimes in efficacy experiments the pathogen is introduced artificially rather than relying on natural inoculum. This may result in disease pressure that is greater or less than what would occur naturally Some products have continued to be developed and improved following registration, thus results obtained with an early formulation might not reflect the degree of control obtainable with the current formulation. Performance of some products can be improved by using an adjuvant, but on the other hand, it has been suggested that some products have been negatively affected by the adjuvant used. Laboratory testing provides an indication of product activity, but these results alone are not sufficient for predicting field efficacy because of the many environmental factors that can affect performance. Finally, most efficacy studies have not been conducted in organic systems, where healthy soils, crop rotations, and biodiverse agroecological environments may effect outcomes.

ROLE OF BIO-FERTILIZERS IN ORGANIC FARMING

"Biofertilizers are substances that contain microorganisms, which when added to the soil increase its fertility and promotes plant growth."

Organic farming has emerged as an important priority area globally in view of the growing demand for safe and healthy food and long term sustainability and concerns on environmental pollution associated with indiscriminate use of agrochemicals. Though the use of chemical inputs in agriculture is inevitable to meet the growing demand for food in world, there are opportunities in selected crops and niche areas where organic production can be encouraged to tape the domestic export market. Bio-fertilizers are being essential component of organic farming are the preparations containing live or latent cells of efficient strains of nitrogen fixing, phosphate solubilizing or cellulolytic micro-organisms used for application to seed, soil or composting areas with the objective of increasing number of such micro-organisms and accelerate those microbial processes which augment the availability of nutrients that can be easily assimilated by plants.

Biofertilizers play a very significant role in improving soil fertility by fixing atmospheric nitrogen, both, in association with plant roots and without it, solubilise insoluble soil phosphates and produces plant growth substances in the soil. They are in fact being promoted to harvest the naturally available, biological system of nutrient mobilization. The role and importance of bio-fertilizers in sustainable crop production has been reviewed by several authors. But the progress in the field of BF production technology remained always below satisfaction in Asia because of various constraints.

Indiscriminate use of synthetic fertilizers has led to the pollution and contamination of the soil, has polluted water basins, destroyed micro-organisms and friendly insects, making the crop more prone to diseases and reduced soil fertility. Depleting feedstock/fossil fuels (energy crisis) and increasing cost of fertilizers. This is becoming unaffordable by small and marginal farmers, depleting soil fertility due to widening gap between nutrient removal and supplies, growing concern about environmental hazards, increasing threat to sustainable agriculture.

Besides above facts, the long term use of biofertilizers is economical, eco-friendly, more efficient, productive and accessible to marginal and small farmers over chemical fertilizers. Biofertilizers are substance that contains microbes, which helps in promoting the growth of plants and trees by increasing the supply of essential nutrients to the plants. It comprises living organisms which include mycorrhizal fungi, blue-green algae, and bacteria. Mycorrhizal fungi preferentially withdraw minerals from organic matter for the plant whereas cyanobacteria are characterized by the property of nitrogen fixation.

Nitrogen fixation is defined as a process of converting di-nitrogen molecules into ammonia. For instance, some bacteria convert nitrogen to ammonia. As a result, nitrogen becomes available for plants.

Types of Biofertilizers

Following are the important types of biofertilizers:

Symbiotic Nitrogen-Fixing Bacteria

Rhizobium is one of the vital symbiotic nitrogen-fixing bacteria. Here bacteria seek shelter and obtain food from plants. In return, they help by providing fixed nitrogen to the plants.

Loose Association of Nitrogen-Fixing Bacteria

Azospirillum is a nitrogen-fixing bacteria that live around the roots of higher plants but do not develop an intimate relationship with plants. It is often termed as rhizosphere association as these bacteria collect plant exudate and the same is used as food by them. This process is termed associative mutualism.

Symbiotic Nitrogen-Fixing Cyanobacteria

Blue-Green algae or Cyanobacteria from the symbiotic association with several plants. Liverworts, cycad roots, fern, and lichens are some of the Nitrogen-fixing cyanobacteria. Anabaena is found at the leaf cavities of the fern. It is responsible for nitrogen fixation. The fern plants decay and release the same for utilization of the rice plants. Azolla pinnate is a fern that resides in rice fields but they do not regulate the growth of the plant.

Free-Living Nitrogen-Fixing Bacteria

They are free-living soil bacteria that perform nitrogen fixation. They are saprotrophic anaerobes such as Clostridium beijerinckii, Azotobacter, etc.

Among all the types of biofertilizers, Rhizobium and Azospirillum are most widely used.

Components of Biofertilizers

The components of biofertilizers include:

Bio Compost

It is one of the eco-friendly product composed of waste material released from sugar industries which are decomposed. It is magnified with human-friendly bacteria, fungi, and various plants.

Tricho-Card

It is an eco-friendly and nonpathogenic product used in a variety of crops as well as in horticultural and ornamental plants, such as paddy apple, sugar cane, brinjal, corn, cotton, vegetables, citrus, etc. It acts as a productive destroyer and antagonistic hyper parasitic against eggs of several bores, shoot, fruit, leaves, flower eaters and other pathogens in the field.

Azotobacter

It protects the roots from pathogens present in the soil and plays a crucial role in fixing atmospheric nitrogen. Nitrogen is a very important nutrient for the plant and about 78% of the total atmosphere comprises nitrogen.

Azolla biofertilizers

- Azolla is a water fern inside which grows the nitrogen fixing blue green algae Anabaena.
- It contains 2-3% nitrogen when wet and also produces organic matter in the soil.
- The Azolla-Anabaena combination type biofertilizer is used all over the world.
- The only constraint in Azolla is that it is an aquatic plant and water becomes limiting factor in growing it particularly in summer.

Phosphorus

Phosphorus is one of the essential nutrients for plants growth and development. Phosphate solubilizing microorganisms, hydrolyze insoluble phosphorus compounds to the soluble form for uptake by plants. Many fungi and bacteria are used for the purpose such as Penicillium, Aspergillus, Bacillus, Pseudomonas, etc.

Vermicompost

It is an Eco-friendly organic fertilizer that comprises vitamins, hormones, organic carbon, sulfur, antibiotics that help to increase the quantity and quality of yield. Vermicompost is one of the quick fixes to improve the fertility of the soil.

Rhizobium biofertilizer

- Rhizobium is a symbiotic bacterium forming root nodules in legume plants.
- These nodules act as miniature nitrogen production factories in the fields.
- The nodule bacteria fix more nitrogen (N2) than needed by legume plant and the bacteria.
- The surplus fixed nitrogen is then secreted and fertilizes the soil.
- Rhizobium is more efficient than-free living nitrogen-fixing bacteria.

Azospirillium biofertilizer

- These are aerobic free living nitrogen fixers which live in associative symbiosis.
- In this type of association bacteria live on the root surface of the host plant and do not form any nodule with roots of grasses.
- It increases crop yield and its inoculation benefits crop.
- They also benefit the host plants by supplying growth hormones and vitamins.

- These bacteria are commonly used for the preparation of commercial inoculants (vaccines, culture medium).

Blue green algae

- Blue green algae (BGA or cyanobacteria) like Nostoc and Anabaena are free living photosynthetic organisms also capable of fixing atmospheric nitrogen.

Mycorrhizal fungi biofertilizer

- Mycorrhizal fungi acts as biofertilizer and are known to occur naturally on roots of forest trees and crop plants.
- Mycorrhizal fungi resist disease in plants. The plants also show drought and salinity resistance. Plants can tolerate adverse soil, pH, high temperature and heavy metal toxicity.
- In soils low in available nutrients there is an increased absorption of nutrients by plants infected with Mycorrhiza.
- The fungus has the ability to dissolve and absorb phosphorus that plant roots cannot readily absorb.

Importance of Biofertilizers

Biofertilizers are important for the following reasons:

- Biofertilizers improve the soil texture and yield of plants.
- They do not allow pathogens to flourish.
- They are eco-friendly and cost-effective.
- Biofertilizers protect the environment from pollutants since they are natural fertilizers.
- They destroy many harmful substances present in the soil that can cause plant diseases.
- Biofertilizers are proved to be effective even under semi-arid conditions.

Applications of Biofertilizers

Following are the important applications of biofertilizers:

Seedling root dip

This method is applicable to rice crops. The seedlings are planted in the bed of water for 8-10 hours.

Seed Treatment

The seeds are dipped in a mixture of nitrogen and phosphorus fertilizers. These seeds are then dried and sown as soon as possible.

Soil Treatment

The biofertilizers along with the compost fertilizers are mixed and kept for one night. This mixture is then spread on the soil where the seeds have to be sown.

Potential role of bio-fertilizers in agriculture

The incorporation of bio-fertilizers (N-fixers) plays major role in improving soil fertility, yield attributing characters and thereby final yield has been reported by many workers. In addition, their application in soil improves soil biota and minimizes the sole use of chemical fertilizers.

Under temperate conditions, inoculation of Rhizobium improved number of pods plant-1, number of seed pod-1 and 1000-seed weight (g) and thereby yield over the control. The number of pods plant-1, number of seed pod-1 and 1000-seed weight. In rice under low land conditions, the application of BGA+ Azospirillum proved significantly beneficial in improving LAI and all yield attributing aspects.

It is an established fact that the efficiency of phosphate fertilizers is very low (15-20%) due to its fixation in acidic and alkaline soils and unfortunately both soil types are predominating in India accounting more than 34% acidity affected and more than seven million hectares of productive land salinity/alkaline affected.

Therefore, the inoculations with PSB and other useful microbial inoculants in these soils become mandatory to restore and maintain the effective microbial populations for solubilization of chemically fixed phosphorus and availability of other macro and micronutrients to harvest good sustainable yield of various crops.

BIOPESTICIDES: TO UNDERSTAND THEIR GROWTH IN THE GLOBAL MARKET

The global biopesticide market is projected to reach USD 6.6 billion by 2022, according to Markets and Markets forecast, which represents a 15% CAGR during the forecast period between 2016 and 2022. Other forecasts by the industry are even more optimistic and raise this figure to 17%.

To confirm the scope of these perspectives we just need to go back to the most recent past. Between 2005 and 2010, the world market for biopesticides grew at a double-digit rate (10%): in those 5 years, it grew from USD 670 million to USD 1 billion, according to figures from BPIA, the organization grouping the world's main biopesticides producers.

The reasons strongly supporting the growth prospects of the biopesticide market are to be found in the increased global demand, the growing adoption of organic farming practices and the changes in international legislation. Asia-Pacific appears to be the region with the greatest potential for growth, not only because of its population rate but also because of its agricultural land area.

The powerful social trend towards sustainability and natural food puts the companies developing this production strategy within the agricultural industry in a position of competitive advantage. These are the 5 reasons driving the growth of the biopesticide market globally.

Production needs of agriculture worldwide

According to the FAO, the world's population will grow by 39% over the next two decades, reaching 9.1 billion by 2050. To meet all this demand, it will be necessary to increase world food production by 60%. This scenario, in the short and medium term, is drawn in a context highly influenced by the negative consequences of climate change and its effects on agriculture in the form of water scarcity or temperature increase, among other factors.

Agriculture faces the dual challenge of becoming more productive (soon there will more people to feed) and, at the same time, reducing its impact on the environment. In this context, the use of biopesticides for pest and disease control in crops appears to be an even more necessary alternative to conventional chemical pesticides because of their harmful effects on nature and human health.

Market demand and changes in the production chain

Biopesticides are developed from substances of botanical and microbiological origin. They provide natural protection for crops when it comes to controlling pests and diseases. Biopesticides also allow food to be obtained without chemical residues. In addition, biopesticides are compatible with auxiliary fauna.

The use of biopesticides in agriculture is fully aligned with market trends that promote healthy eating without neglecting environmental conservation. Consumers are increasingly demanding residue-free food. The trend is becoming more and more powerful and the market has not been reluctant to understand this whole movement in a very clear way. There are figures confirming it: in 2016, for the first time in history, Spain surpassed the figure of two million hectares dedicated to organic farming, specifically 2,018,802 hectares. This is a 8.5% more than in 2015. Between 2012 and 2016, the area under organic farming grew by 18.7% in the EU, according to Eurostat data.

Going through crop by crop, we only have to stop at wine, a key sector in Spanish agriculture. According to data from Ecovalia, a total of 106,000 hectares are dedicated to organic grape cultivation in Spain. That's almost 11% more than the year before. a total of 106,000 hectares are dedicated to organic grape cultivation in Spain. That's almost 11% more than the year before.

The change is also visible in streets and in the prevalence that organic products have begun to take on the shelves of shops and supermarkets. The French distribution giant Carrefour opened last year its first exclusive shop for organic products in Madrid. Its arrival joins other chains specialized in organic food such as Oh! My Bio, Veritas or or Espacio Orgánico, which have recently opened new venues in Madrid too.

Biopesticides are within the strategies of the major players on the market

Large international players within the food industry, such as Tesco and Wal-Mart, have introduced changes within the production chain to be able to distribute products with zero residues on the shelves of their supermarkets. This is the requirement they have begun to introduce into the agreement with their suppliers. Last year, WalMart also announced the addition of the blockchain technology to improve traceability and transparency in its food supply network.

The use of biopesticides in agricultural production makes it possible to obtain food free of residues. Also, in integrated pest management programmes, the inclusion of rotating biopesticides with chemicals contributes to reducing residue levels in food.

International legislation is aligned with the use of biopesticides

In 2009, the European Union voted a directive to phase-out a number of chemical pesticides in agriculture The leading policies in this area are those of the governments of Sweden, Denmark and the Netherlands, which shortly afterwards announced measures to require a 50% reduction in on-farm chemical pesticide use.

France is another of the countries with the most active policies in this area in recent years. The main example is the so-called Plan Écophyto, an initiative to halve the consumption of pesticides in agriculture by 2025. Approved in 2007, Ecophyto initially set the time horizon of 2018 to achieve the goals, although this date was later extended to 2025.

International legislation is aligned with the use of biopesticides, although the registration of plant protection products continues to be a road full of barriers within the biopesticide industry, especially for those small and medium-sized companies that embark on this process. The registration process for biopesticides is the same as for conventional chemical pesticides. The investment, in terms of time and money, is very high, which slows down the arrival of new solutions on the market. In addition, there is no harmonisation of requirements at international level, so that the processes largely vary according to the country in which they are to be registered.

New technologies capable of creating new developments

The growth in the use of biopesticides in agriculture is also supported by the technological development of industry. At the end of the 1990s and the beginning of the new 21st century, important advances were made in microbiological and biochemical research, which later served to pave the way for the industry. The investment and advances in R&D allowed to develop new solutions with very high efficiency and to improve the application, conservation and shelf life of products.

Seipasa, a company specialized in the formulation, development and manufacture of biopesticides, biostimulants and fertilizers for agriculture, is an example of the application of research and innovation to the generation of new natural treatments in agriculture. Pirecris, the biopesticide for pest control developed from a unique formulation based on the perfect balance of PI and PII molecules; or Fungisei, the biological fungicide whose patented technology allows a liquid formulation based on microorganisms, are exponents of this new generation of products within the biopesticides market.

BIOPESTICIDES: AN ALTERNATIVE APPROACH FOR ORGANIC FARMING AND ENVIRONMENTAL SAFETY

Disparaging activities of many pests like plant foes (bacteria, fungi, nematodes etc.), insects and weed escorting the ascending destruction of field crops. Agriculture and plant territories contributed higher amount to maintain ecological, social, economical system, but over a decade ago by the application of conventional pesticides insects become resist to them and encompasses many chemical substances which would biomagnified time respectively. Insects are small in size but in large quantity they became detrimental and contributed major insesticidal effect. Greater than 60% of population directly dependant on agricultural activities and remain on indirectly involved in easily pest affected food industries, forestry, poultry, dairy and many other agribusiness.

From past decades farmers and many food product manufacturing companies, used synthetic chemicals to produce, process and preserve the food & crops to enhancing the crop production. Local farmers exploited conventional pesticides in crop fields to get higher field products and these insecticides are responsible to acute and chronic toxicity. These conventional pesticides became the key factors for lot of diseases like lung, prostate, skin, throat, breast cancers, leukemia, lung diseases, respiratory problems, reproductive problems, hypertension and skin disorders.

By the overconsumption of the pesticides and their adverse effect on human, animal and environmental health they become a global concern; to inhibit their lethal properties there is a need of sustainable, healthy and immunoprotective biopesticides which will of low cost, pest specific, non toxic, ecologically safe and reducible. Biopesticides are any naturally occurring pesticides or derived from natural material (plants, animals, bacteria, fungi and minerals etc.) to destroy or kill the host specific insect-pest.

They are biological controlling agents of weeds, insects, crop pest, fungi, bacteria, virus, rodents, microbes and derived from living organism as plants (Azadirachta, Datura, Eucalyptus, Calotropis etc.) and their products (phytochemicals, microbial products) or byproducts (semiochemicals) to inhibit the target organism by non toxic methods.

AGRICULTURAL OUTPUT AND ENVIRONMENTAL SAFETY

Agriculture sector strengthen the Indian economic value through crops export and also by agribusiness to improving and enhancing the agricultural food by genetically modified advance technologies; From 1960s, commonest way to control the pest has been the excessive use of conventional pesticides and in 1940s such pesticides were implemented as DDT (dichloro-diphenyl-trichloroethane) with carbamate pesticides & organophosphate.

Green revolution technology in which water, biofertilizers, synthetic chemicals were used to enhances agricultural output, but with the increasing food and feed the severe use of these pesticides and their residues in field contaminated the water, air, soil, and other natural resources; beside this bioaacumulation of chemicals inherits carcinogens. The crop productivity increases with the increasing use of agrochemicals but with that they also express adverse health and environmental effect.

Unsystematic use of these agrochemicals affect soil health and quality, water quality, air and also affect the health of human and warm blooded animals by gene variation, gene erosion, deformities, cancer, teratogenesis, lower birth rate, high mortality etc.

With the ecofriendly, safe and environmentally suitable approach biopesticides application play major role in agriculture, forestry, food and environment also. Many research are already in practicise to controlling crop pest & insects; because through vector born diseases they reflected agriculture output, food chain and non targeted organism.

From past decades in India biopesticides become the major source of healthy, organic, immunoenhancing and defensive key for food and environment. There are some pesticides which are registered and applied to insect management-

Table 1: Biopesticides registered as Insecticides Act, 1968.

S.NO.	BIOPESTICIDES
1	Bacillus sphaericus
2	Bacillus thuringiensis var.galleria
3	Bacillus thuringiensis var. kurstak
4	Bacillus thuringiensis var. israelensis
5	Trichoderma harzaianum
6	Trichoderma viride
7	Pseudomonas fluoresens
8	Neem based pesticides
9	Beauveria bassiana
10	NPV of Helicoverpa armigera
11	NPV of Spodoptera litura
12	Cymbopogan

GLOBAL MARKET OF PESTICIDES

Globally biopesticides or plant protectants (bioinsecticides, bionematicides, biofungicies) are the key driver of the insects control and they become field crop improving agents. 700 products are available in the market and 175 registered biopesticides active component already accessible in the market. The North America biopesticide market was valued at

USD 1159.66 million in 2017, and is expected to reach USD 2304.2 million by 2023.

The US has the largest share with 80% of the market share. North America and Mexico is the fastest growing market with 16.4% CAGR during 2018-2023. Indian biopesticides market produced returns of $102 million in 2016 and is expected to provide $778 million by 2025, growing at a CAGR of 25.4%. China at the position of fastest growing biopesticides using country with CAGRs of 18.5% for volume consumption and 19.4% for value demand over the 2014-2020 analysis periods.

In India biopesticides contributed only 4.2% of overall plant protectants with the less residual property, no harmful effect, higher efficacy it is anticipated to articulate ascending growth of their use 10% accordingly in next coming years. Now only 12 biopesticide have been registered for insect pest management according to Act 1968 (as on 2008), including Azadirachta indica (neem), Bt, Trichoderma Nuclear Polyhedrosis Virus, they all are originated and used in India.

Globally many bacteria, virus, nematodes and plants contributed the major percentage of plant protectants as Bacillus thuringiensis (Bacteria), Metarhizium anisopliae (fungi), Baculovirus (Virus), Nematodes Steinernema (Rhabditia), Nosema (Protozoa), botanical insecticides.

BENEFITS OF BIOPESTICIDES

The increasing potency of biopesticides use is based upon the advantages which relied with this are-

1. Biopesticides has zero/less harmful property.
2. They are insect specific or act against only targeted organism without effecting beneficial pests.
3. They are ecologically safe and environmentally acceptable.
4. With the lesser cost and higher efficacy. They cause zero residual effect.
5. Biopesticides are not altering the gene information.
6. They are not changing the nutritive value of crops, fruits, nuts or vegetables.

LIMITATIONS

Now, in a 21st century when everyone believes on fast working and readily available product our farmers are also become smart, advance and used to of the conventional pesticides for higher agriculture production in a limited season or time . Our farmers are well known to hazardous, noxious ecological, social and environmental effects of these synthetic pesticides but for the financial need and higher crop yield they become avoider. So it is necessary to provide readymade, fully packed, cheap and processed biopesticides for direct applications in crop fields.

Other Environmental Benefits

- Most carbon in the soil is lost as greenhouse gas (carbon dioxide, CO_2) into the atmosphere if natural ecosystems are converted to agricultural land.
- Soils contain 3.3 times more carbon than the atmosphere.
- This makes soils an important source of greenhouse gases but also a potential sink if right management is applied.
- The use of crop residues for bio-energy production reduces the carbon stocks in cropland.
- Further the dedication of cropland to bio-fuel production increases the area of cultivated land and thus carbon loss from soils and vegetation.
- Bio char remains stable for millennia, providing a simple means to sequester carbon emissions.
- If bio char is returned to agricultural land it can increase the soil's carbon content permanently and would establish a carbon sink for atmospheric CO_2.

7

BIOPESTICIDE USE IN HORTICULTURAL CROPS

INTRODUCTION

The modern agricultural practices such as increased use of fertilizers, pesticides, high yielding varieties etc., coupled with intensive and extensive cultivation of crops have contributed for increased agricultural production in the country during past 3 decades, but at the same time a number of new problems have emerged especially in the field of plant protection.

Over dependence on pesticides has led to problems such as development of resistance in pest species, resurgence in pest populations, emergence of new crop pests, environmental pollution and disruption in biodiversity. Moreover, with increased cost of crop protection, the net income per unit area in case of crops like cotton, paddy and certain vegetable has come down sharply, making production of these crops uneconomical.

The concept of Biopesticides was born out of the discontent and crisis brought on by the unrestricted use of pesticides. In this regard biopesticides play important role in the management of insect pests of horticultural crops. These are safe to non target organisms including humans.

According to WTO 3 million agriculture workers experience severe poisoning from pesticides. Among them about 18,000 die each year.

Major problems with Pesticides

- Pest species (plant and animal) develop resistance to pesticides
- Pesticides that don't degrade easily can cause problems for humans and other organisms
- Bioaccumulation - build up of pesticides in body
- Biomagnification - organisms higher up in the food web have higher concentrations of pesticides in their body.

What are biopesticides?

The pesticidal compounds derived from living organisms are biopesticides. The biopesticides presently being used in plant protection are mainly derived from plants and micro-organisms.

EPA Definition

Biopesticides are certain types of pesticides derived from such natural materials as animals, plants, bacteria, and certain minerals. For example, neem cake and pongamia cake have pesticidal applications and are considered biopesticides.

TYPES OF BIOPESTICIDES

- Botanicals
- Bacterial antagonists
- Entomopathogenic Bacteria
- Entomopathogenic Fungi
- Entomopathogenic Nematode
- Entomopathogenic Virus
- Fungal Antagonists
- Parasitoids
- Predators
- Weed Feeder

BOTANICAL INSECTICIDES

Insect pests, nematodes and plant diseases of horticultural crops are generally controlled by frequent applications of chemical pesticides with an objective to increase crop productivity and obtain greater profit in conventional farming. With consumers' awareness and perception, horticultural crops without residue of chemicals are being preferred in local and export markets. For this purpose, plant-derived crude products or formulated pesticides can be eco-friendly, effective and economical for an average producer.

Neem (Azadirachta indica A. Juss.) is being widely used in the form of water extracts of leaves or seeds and neem oil owing to its effects as an oviposition deterrent, sterilant, antifeedant, repellent and pesticide against insects, and as an inhibitor of spore germination and development of pathogens and affecting penetration by nematodes into roots. Other indigenous plants also show potential in pest management.

Further research needs to be focused on the synthesis and commercial formulations of botanicals for control of pests and pathogens. Plants, during their long evolution, have synthesized a diverse array of chemicals to prevent their colonization by insects and other herbivores. They produce terpenoids, alkaloids, flavonoids, phenolic compounds, unusual amino acids etc.

These chemicals exhibit repellent, feeding and oviposition deterrent and insecticidal properties besides interfering with their normal growth, development and reproduction. Several plant based insecticides such as nicotinoids, natural pyrethrins, rotenoids, neem products, etc have been used in the past for suppression of pest species.

Among them, neem based products are ideally suited for practical utilization because of their greater effectiveness against diverse group of insect pests under field conditions, ready availability and simple methods of preparation of insecticidal products.

Neem has been evaluated against more than 500 species of insects at global level and most of these insects have been found to be susceptible. Neem products give good control of caterpillars such as diamondback moth on Cole crops and gram pod borer on red gram, cotton, tomato etc.; grass hoppers, leaf hoppers, plant hoppers, leaf miners, whiteflies, scales and mealy bugs. Neem kernel powder at 1 to 2 per cent gives good control of pulse beetle and other storage pests.

Soil application of neem and pongamia cakes suppress nematodes. Soil application of neem cake (250 kg/ha) at planting and repeated at flowering reduces leaf miner and fruit borer in tomato, shoot and fruit borer in brinjal, leaf miner and fruit fly in cucurbits, pod borers in vegetable pigeon pea, dolichos lab lab, etc. Coconut mite infestation can be reduced by tying cloth bags containing neem cake in crown region. Cake should be replaced at 45 days interval.

Neem and Pongamia oil soaps

The potassium soaps (Soft soaps) of neem and pongamia oils at 1 per cent are effective against diamond back moth, leafhopper, aphids, leaf miners, fruit borers and fruit fly in vegetable crops. They are also effective against mango leaf hopper and citrus leaf miner. Efficacy may be reduced at high temperatures.

In recent years, commercial neem products with 300, 1500, 10,000, 30,000 and 50, 000 ppm of azadirachtin have been introduced. Field and laboratory studies have shown that the aqueous neem seed kernel extract (NSKE at 4-5%) is more effective and economical to use compared to commercial neem based products.

Whole neem seed powder extract can also be efficiently used in place of neem seed kernel extract. Similarly, neem oil emulsions (2-4%) prepared by using soap can be effectively used against scales and mealy bugs on fruit trees such as guava, custard apple and sapota.

Preparation of Neem seed kernel extract (NSKE)

Select good quality seeds free from fungal infection. Remove seed coat by pounding and grind the kernels into fine powder. Use 4 kg of kernel powder to prepare 100 litres of 4 per cent extract. Soak kernel powder in 20-25 litres of water overnight. Pass the soaked material through thin cloth and collect the filterate. Make up the volume to 100 litres by adding water. Add liquid soap (1 ml/litres) before spraying the extract.

Preparation of Neem seed powder extract

As often farmers find it difficult to prepare NSKE, they can also use pulverized whole neem seed powder extract (PNSPE). To prepare 4 per cent extract, 4 kg of pulverized neem seed powder is soaked in 20-25 litres of water overnight. After filtering, the volume is made up to 100 litres and used for spraying. The neem seed powder can be stored up to 3 months without losing its insecticidal property.

Preparation of neem and Pongamia oil emulsions

Oil emulsions are normally used at 1-2 per cent on vegetables and 3-4 per cent on fruit trees like guava, custard apple and sapota. To prepare 2 per cent oil emulsion, oil (2 litres) and liquid soap (500 ml) should be mixed thoroughly and then added to 100 litres of water to get stable emulsion. The emulsion should be agitated constantly while spraying. However, oils may be toxic at high doses and high temperatures.

Tobacco leaf extract

Tobacco leaf extract is effective against soft-bodied insects like aphids, thrips, etc. 250 g of tobacco dust is soaked in 4.5 litres of water overnight, filtered, made up to 25 litres and sprayed by adding soap solution (0.5 ml /l).

Microbial Insecticides

The insecticidal substances obtained from the microorganisms are microbial insecticides. The insecticidal crystal proteins (endotoxins) produced by the bacterium, Bacillus thuringiensis kurstaki are effective against lepidopteran pest species. These toxins are very specific in their action, easily bio-degradable and being stomach poisons, safer to non-target organisms.

Presently, many commercial products are available in the market for use against caterpillar pests. However, some pests like diamondback moth on Cole crops have developed resistance to certain Bt. endotoxins.

Other microbial products that have been recently introduced for pest management are the insecticidal toxins produced by soil borne Actinomycetes organisms. The Avermectins (abamectin) derived from Streptomyces avermitilis possess potent anthelminthic, acaricidal and insecticidal properties at very low doses. Like wise, Spinosyns (Spinosad), toxins obtained from another Actinomycetes organism, Saccharopolyspora spinosa exhibit very good insecticidal property against lepidopteran pests like diamondback moth, gram pod borer; etc

Bacterial antagonists

Bacterial antagonists are those bacterial species, which act against the harmful microorganisms and their products. Pseudomonas fluorescens is common non-pathogenic saprophyte that colonises in soil, water and on plant surfaces. It produces a soluble greenish fluorescent pigment.

Pseudomonas florescens

The foliar application of this bacteria is supposed to reduce many insect pests like leaf miner, rice leaf folder etc. and diseases like bacterial wilt, fungal diseases and nematodes. Regular sprays are advocated to minimize the disease incidence in main field and these are believed to induce resistance in plants. Treating the nursery @50 g/sq. meter with this antagonistic bacteria induces disease resistance against bacterial wilt in tomato. Pseudomonas fluorescens suppress plant diseases by protecting the seeds and roots from fungal infections. This bacterium is mass-produced using fermentation technology. It is most effective against soft rot causing bacteria. Erwinia carotovora (carrot) Induced systemic resistance through foliar spray and seed treatment of Pseudomonas florescens against H. armigera . Pseudomonas florescens influences the growth and development of insectsat all stages of their growth. Pseudomonas maltophi affects the growth of larval stage of Helicoverpa zea, leading to reduced adult emergence.

Among the variousdoses of Pseudomonas florescens used as foliar spray and seedtreatment the doses of 10 g/lit of water and 30 g/kg of seed recorded higher amount of biochemical compounds like phenol, peroxidase, and PAL which increased gradually up to 20 days after spraying.

Entomopathogenic Bacteria

Bacillus thuringiensis (commonly known as 'Bt') is a spore forming gram positive bacterium. It is commercially produced worldwide using fermentation technology. The commercial Bt products are produced as dust, wettable powder and emulsifiable concentrates. The toxin genes have also been genetically engineered into several crop plants.

A field study was carried out on the efficacy of different sequential application of some novel insecticides viz., novaluron 10 E.C., indoxacarb 14.5 S.C., bifenthrin 10 E.C., lambda cyhalothrin 5 E.C., and biopesticides viz., nucleopolyhedrovirus (NPV) of Helicoverpa armigera, Bacillus thuringiensis var. kurstaki and neemarin, against H. armigera in comparison with sequential application ofconventional insecticide i.e. endosulfan 35 E.C. and untreated control on tomato hybrid Pusa Ruby.

Results showed that among different sequential application of insecticides, indoxacarb with lowest fruit infestation of2.53 and 2.83 and highest yield of39.45 & 38.85 q/ha were recorded' during both the seasons, respectively. While among the biopesticides, neemarin followed by Bt and NPV with mean fruit yield of 30.27 and 29.60, 28.17 and 27.61 and 26.70 and 26.11 q/ha were optained in two seasons, respectively.

Relatively higher numbers of predatory coccinelids (Coccinella septumpunctata.) were recorded in endosulfan insecticides as well as biopesticidesapplied plots as compared to treated plots.

The treatments of Bt and NPV ranked in moderately effective group for the control of fruit borer and find support from the work of Bhosle et al. (2007). They reported that dipel (Bt) was effective next to carbaryl in controlling the fruit borer. Similarly researchers found biolep (Bt) @ 0.20% effective next to insecticides used.

Entomopathogenic fungi

Fungal species can act as a parasite of insects and kill or seriously disables them. These fungi usually attach to the external body surface of insects in the form of microscopic bodies (usually asexual, mitosporic spores also called conidia).

- Verticillium lecanii ('white – halo' fungus): It is effective against coffee green bug, thrips,nematodes and certain other homopterans.
- Beauveria bassiana : It is naturally occurring in soil throughout the world. It is useful against Coffee berry borer, Helicoverpa armigera, Diamond backmoth, Thrips, Grasshoppers, Whiteflies, Aphid, Codling moth etc.
- Metarrhizium anisopliae : is a widely distributed soil-inhabiting fungus. The spores of M. anisopliae can be formulated as dust and sprayable formulation. It is used to control termites, mosquitoes, leafhopper, beetles etc.

Entomopathogenic nematodes

Steinernematids (Steinernema carpocapsae) exclusively soil organisms, are entomopathogenic on a variety of insect pests. It is particularly effective against lepidopteron larvae, including various webworms, cutworms, armyworms, girdlers and woodborers.

Entomopathogenic Virus

Entomopathogenic viruses are used as biological control agent as it infects or kills the pests by causing pathogenicity to the insects. Commonly used viruses are Nuclear Polyhedrosis Virus (NPV), Granulosis virus (GV). NPV are host specific. Larvae infected with GV and NPV usually die within 5-12 days after infection. These viruses are produced on the host insects and are formulated both as liquid and dust formulation . The safety aspects of botanicals and biopesticides to non-target organisms had been already studied by several workers. HaNPV was not found pathogenic to T. chilonis.

Sources of NPV formulations

- Pest control India Ltd., Bangalore- HaNPV, SINPV
- Biopest Management, Bangalore – HaNPV, SINPV
- TNAU, Coimbtore

- Agricultural University, Rahuri
- IARI, New Delhi

TRADE NAME	TARGET PEST	CROP	MANUFACTURER
Halt	DBM	Cabbage	Wockhardt Ltd.,
Biolep, bioasp, Biotox	*H.armigera, Earias spp,*	Cotton	Biotech International
	Pectinophora spp		
Delfin	*Spodoptera litura,&*	Cole crops	Rallies India
	Plutella xylostella		
Biobit	*lepidopterans*	Many crops	TAC
Spicturin	*Helicoverpa armigera*	Cotton	TAC
Dipel	DBM	Cole crops	Lupin Agro
Dipel WG	DBM	Cole crops	Margo

Fungal antagonists

These are those fungus species, which act against the harmful microorganisms and their products.

- Trichoderma spp. i.e., Trichoderma viride, T. harzianum and T. virens are most extensively used fungal antagonists. They are used for the management of soil-borne pathogens by seed treatment and soil application.
- Aspergillus niger : Aspergillus niger , an ubiquitous fungus presents in all types of soil with no specific moisture and pH requirements. It controls a number of devastating soil borne pathogens e.g. Fusarium oxysporum, Macrophomina phaseolina, Pythium aphanidermatum, Rhizoctonia solani , and Sclerotinia sclerotiorum
- Myrothecium verrucaria : It is used to control root-infecting fungus, Sclerotium rolfsii on groundnut.

Light and pheromone traps: can be used to monitor and suppress certain pest species. Yellow sticky traps can be used to monitor aphids and whitefly. In cabbage and cauliflower, light raps are useful in reducing diamond back moth incidence. A bulb (100 watts) is hung in front of a thin gunny bag or polyethylene sheet (3'x4') smeared with oil or grease.

A bucket with water can also be kept below the bulb. About 6-8 light traps are to be set up per hectare. In Bangladesh, brinjal shoot and fruit borer was effectively managed by errecting 32 pheromone traps per acre and clean cultivation by destroying infested shoots and fruits. Lures like methyl eugenol, cue-lure can be used to suppress fruit flies. Red palm weevil on coconut can be controlled by using sex pheromone traps.

Growing crops under nylon net

Shadenets are constructed using various materials like plastics, wood, GI pipes, angle irons to name a few house is a framed structure made of materials such as GI pipes, angle iron, wood or bamboo. This framed structure is covered mostly with plastic net which are mostly polyethylene based materials. It modifies the microclimate of crop by altering the percentage of shade which is the result of reduction in intensity of light and heating parameters. As different crops have different photoperiodic requirement, this can be partially done by the shade net structures which is able to control the abiotic parameters partly. Before construction of the shade house structure care should be taken regarding the type of crop to be grown, local climatic conditions and locally available materials. There should be enough facilities of subsequent expansion which is the key to future plans.

Predators

An entomophagous species that generally consumes more than one prey individual to complete its development is called a predator.

- Lady bird beetle is a voracious feeder of mealybugs in both nymphal and adult stage. It also feed on cabbage aphid.
- Chrysoperla carnea : The first instar larvae are released in the field @ 1.25 lakh/ha to control sucking pests and eggs and first instar larvae of lepidopteran insects.
- Weed Feeder Those organisms, which feed on the weeds depleting or eliminating them, are called Weed Feeder.
- Zygogramma bicolorata . This is Chrysomelid beetle used as a biocontrol agent for the control of the carrot weed, Parthenium hysterophorus. Larvae and adult beetles feed on the leaves of this weed.

ADVANTAGES OF BIOPESTICIDES

- Do not leave harmful residues
- Substantially reduced impact on non-target species
- When locally produced, may be cheaper than chemical pesticides
- In the long-term may be more effective than chemical pesticides (e.g. as demonstrated by the LUBILOSA Programme)
- When used as a component of Integrated Pest Management (IPM) programs, biopesticides can greatly decrease the use of conventional pesticides, while crop yields remain high.
- These products are more selective, eco-friendly and leave no toxic residues in the environment unlike synthetic insecticides. Disadvantages of Biological Controls

- Some farmers are unwilling due to the ease of using chemical pesticides
- Biological controls are slow-acting whereas chemical pesticides are fast acting
- Less avalibility. Conclusion
- With renewed interest on organic farming, use of biopesticides gaining further momentum in order to avoid the pesticides residues on the produces.
- Biopesticides are safer to non-target organisms including humans.
- But farmers usually prefer chemical pesticides.

MICROBIAL CONTROL OF ARTHROPOD PESTS OF TROPICAL TREE FRUITS

Entomopathogens

Viruses

Many viruses have been identified from hundreds of arthropod species. For most of them, those with particles occluded in protein bodies (OBs) [Baculoviridae, Entomopoxviridae, Reoviridae (Cypoviruses)] have been used successfully in microbial control programs. The Baculoviridae (nucleopoly-hedroviruses and granuloviruses) are the most studied and used as microbial control agents. They are normally transmitted per os and gain access to host tissues via the midgut where the OBs that surround the virus rods are dissolved. The currently unclassified virus of Oryctes rhinoceros L. is the most successfully used non-occluded virus. Viruses comprise some of the most host-specific entomopathogens but their main drawbacks are the requirement for in vivo production and their sensitivity to ultra-violet degradation.

Bacteria

Although several species of bacteria have been used as microbial control agents of a variety of insects, only Bacillus thuringiensis Berliner (Bt) has been used for practical pest control. It is the most widely used inundatively applied microbial control agent. Several isolates of Bt are commercially produced with activity against Lepidoptera, Coleoptera, and Diptera. The safety of Bt for applicators and vertebrate and invertebrate non-target organisms is well documented. Its insecticidal activity is associated with delta-endotoxins located in parasporal inclusion bodies (or parasporal crystals) that are produced at sporulation and must be ingested by the target organism in order to be active. A limiting factor of Bt is its fairly narrow host range.

Fungi

The majority of fungi that naturally regulate insect and mite populations is in the order Hypocreales (Ascomycetes, including the vast majority of conidial entomopathogens in more than two dozen genera formerly classified among either the Hyphomycetes or Fungi Imperfecti) and Entomophthorales. The latter are difficult or impossible to mass-produce and, hence, they have not been commercially produced or applied inundatively on a large scale. On the other hand, several asexually reproducing species in the Hypocreales are amenable to mass production and commercialization. The most studied for control of insects and mites belong to the genera Beauveria, Metarhizium, Paecilomyces, Aschersonia, Hirsutella, and Lecanicillium (formerly Verticillium). Because the normal route of invasion is through the cuticle, fungi are especially suitable microbial control agents for sucking insects (Hemiptera).

Phylogenetic studies now confirm that the Microsporidia are highly derived organisms correctly placed among the lower fungi rather than extremely ancient and simple organisms classified with the Protozoa. Although many microsporidians are common pathogens of arthropods, few have been included in microbial control programs because certain fundamental characteristics (complex life cycles, obligate parasitism, and chronic rather than acute effects) inhibit their use.

Entomopathogenic nematodes (EPNs)

Nematodes in the families Steinernematidae and Heterorhabditidae are effective control agents of dozens of insect species in soil and cryptic habitats. These nematodes are associated with symbiotic bacteria (Xenorhabdus spp. and Photorhabdus spp.), which are housed in the intestine of the infective juvenile (IJ) nematode (also referred to as the Dauer stage). The IJ, the only free-living stage, occurs in the soil and searches for an insect host. Upon finding a host, it enters through the mouth, anus or spiracles and penetrates into the body cavity. In the case of heterorhabditids, IJ penetration can occur directly through soft cuticles. In the body cavity, the IJ releases mutualistic bacterial cells, which multiply rapidly and kill its insect host, usually within 48h. In addition, the bacterial cells digest host tissues and produce antibiotics that protect the host cadaver from saprophytes and scavengers, and allow the nematodes to develop and reproduce.

The nematodes feed on the mutualistic bacterial cells and on degraded host tissues. Depending on host size, there may be one to three nematode generations in the host cadaver. When host nutrients are depleted, the pre-IJs sequester the mutualistic bacterial cells in their intestines. The resulting IJs leave the host and search for new hosts. In the absence of a host, IJs can persist for months in moist soil. However, the IJs have their own natural enemies (i.e. nematophagous fungi, predatory mites and other soil predators) and must also contend with abiotic factors such as temperature extremes, low soil moisture, and ultraviolet radiation that affect their survival. Several EPN species are commercially produced and available for large-scale application. For small-scale experimental testing, EPNs can be produced in vivo and on artificial media.

RESEARCH AND APPLICATION OF ENTOMOPATHOGENS FOR CONTROL OF ARTHROPOD PESTS OF TROPICAL FRUITS

The literature on microbial control agents of tropical fruit pests has, for the most part, concentrated on key pests of a few major crops (e.g. citrus and banana). In this review we bring information on the use of microbial control of arthropod pests of citrus, banana, coconut, mango, guava, papaya and pineapple.

Citrus

Because of the diversity of cultivars and climates in which they are grown, citrus is perhaps the most widely distributed tree fruit crop ranging from tropical and subtropical climates to temperate habitats around the world. Consequently a huge range of arthropod pests is reported from citrus varieties. Successful microbial control of several pests has been reported using fungi, bacteria, viruses and EPNs.

Citrus rust mite (CRM), Phyllocoptruta oleivora (Ashmead) (Acari: Eriophyidae). This is a major pest of citrus in several countries, including Brazil and the USA. Studies on the use of fungi for its control are limited number. Researchers assessed the pathogenicity of five concentrations of Beauveria bassiana (Bals.) Vuill. ranging from 10^6 to 10^8 condia/ml under laboratory conditions (25°C, 12h photophase, 98% RH). Mortality was time and dosage dependent and ranged from 24% to 91% with an LC_{50} of 4.23 x 10^6 conidia/ml five days after treatment. LT_{50} at the highest concentration was 2.74 days.

Hirsutella thompsonii Fisher (Entomophthorales: Moliniaceae) is infectious for CRM and several other mite pests of greenhouse crops, coconut and turf (Samson et al. 1980, McCoy 1996). In the early 1980's it was mass-produced and formulated by Abbott Laboratories (Chicago, IL, USA) and registered for CRM control in the USA (McCoy & Couch 1982). Under optimal conditions, H. thompsonii can control CRM within one to two weeks.

Field applications of mycelia led to production of conidia within 48h and provided suppression of CRM for up to 14 weeks. McCoy et al. (2007) provided protocols for the field evaluation of H. thompsonii and other fungi intended for CRM control. Unfortunately, commercial development of the fungus was discontinued by Abbott Laboratories.

Citrus red mite, Panonychus citri (McGregor) (Acari: Tetranychidae). This mite can be a serious pest of citrus in certain locations. In southern China, researchers evaluated the efficacy of four rates of B. bassiana (ranging from 1.2 x 10^{12} to 3.0 x 10^{13} conidia/ha) and a combination of the fungus with a low rate of pyridaben for control of P. citri in orange groves. All of the B. bassiana application rates produced significant mortality in P. citri and the combinations with pyridaben led to better control.

Two applications of > 1.5 x 10^{13} conidia/ha plus low rate pyridaben with a 15 day spray interval resulted in good control of P. citri for 35 days with mite density declines of 74-91%.

Where P. citri is a pest in arid regions, it is not recommendable to use fungi against this pest. Researchers summarized research and results of applications of a non-occluded virus found in P. citri. The virus has apparently been responsible for decimating epizootics in P. citri populations in Arizona and California. Researchers reported control of the mite after application of triturated mites infected with the virus. A major limitation of this virus is the need for its mass production in P. citri.

Broad mite (BM) or white mite, Polyphagotarsonemus latus (Banks) (Acari: Tarsonemidae). This is a cosmopolitan pest of a variety of plants including citrus, papaya, and mango. It is abundant during warm and humid condition and thus an ideal candidate for control with fungi. However, very little research on microbial control has been conducted on this pest. Researchers reported natural infection of BM with H. thompsonii. Peña et al. (1996) assessed the infectivity of B. bassiana, H. thompsonii, and Paecilomyces fumosoroseus (Wize) Brown and Smith conidia in laboratory bioassays on bean leaves.

The LC_{50} values for the fungi were 1.16 x 10^6, 2.39 x 10^3, and 1.29 x 10^5 conidia/ml, respectively. Mortality due to B. bassiana was most rapid in mite densities between 65 and 125 mites/leaf. The efficacy of B. bassiana and P. fumosoroseus and other agents were also evaluated against BM in a greenhouse test on potted bean plants. Treatments with B. bassiana were the most efficacious and persistent, and resulted in 88% mortality.

False spider mite (FSM), Brevipalpus phoenicis (Geijskes) (Acari: Tenuipalpidae). FSM, also known as the red and black flat mite, is a polyphagous widely distributed tropical-subtropical species that has been reported from several hundred plant hosts including citrus, banana, macadamia, orchid, papaya, passion fruit, coffee and tea. FSM is an important citrus pest as it is a vector of the citrus leprosis virus.

Researchers assessed the activity of 52 isolates of fungi including B. bassiana, Metarhizium anisopliae (Metsch.) Sorokin, Paecilomyces spp., H. thompsonii, Lecanicillium spp. and others. The most active isolates were all H. thompsonii, causing 90-100% mortality six days after treatment. All other species of fungi produced less than 30% mortality six days after treatment. The authors observed conidiogenesis of H. thompsonii with development of mycelium and condiophores emerging from the posterior and anterior parts of mites 120h after spraying with conidia.

Whiteflies and blackflies (Hemiptera: Aleyrodidae). About 30 species of whiteflies and blackflies have been reported attacking citrus worldwide, six of which are considered major pests.

Most of research on microbial control of whiteflies in citrus has focused on Aschersonia spp., which has produced spectacular epizootics in conditions of high humidity and rainfall. In Florida, USA, Fawcett (1944) reported epizootics in Dialeurodes citri (Ashmead) and D. citrifolii (Morgan) in citrus groves during the summer when high humidity promoted conidial sporulation and host infection, and frequent rains enabled effective dispersal of the conidia. Researchers observed A. aleyrodis Ashby infecting

D. citri in Texas. Researchers also published on the distribution of the citrus backfly, Aleurocanthus woglumi Ashby, and its natural enemies in four localities in Costa Rica. In addition to parasitoids and predators, mortality due to A. aleyrodis was significant. The same agent was unable to control A. woglumi following its introduction into and spread through El Salvador. Outbreaks of disease caused by Aschersonia spp. among populations of Bemisia giffardi (Kotinsky) on citrus in Taiwan were observed.

In Brazil, the occurrence of Aschersonia spp. on whiteflies is very common in all areas where citrus is grown and coincides with the periods of greatest rainfall. In addition to natural occurrence, Aschersonia spp. has been successfully applied against whiteflies in citrus groves in the Georgia, China and Japan. In field tests in Japan Aschersonia sp. was pathogenic to nymphs, pupae and eggs of D. citri, and the mortality increased with the concentration of conidia (Uchida 1970). Researchers introduced several isolates of Aschersonia from six countries into orange groves near Adzharia, Georgia, being A. placenta Berk & Br. from Vietnam and China the most effective leading to up to 90% parasitism in favorable weather.

Scales (Hemiptera: Coccidae and Diaspididae)

These are regarded as the most abundant and injurious citrus pests. Researchers observed up to 49% infection of the diaspidid Parlatoria ziziphi (Lucas) by A. aleyrodis and A. goldiana Sacc. & Ellis in Cuba. Researchers noted that the main control agent of a diaspidid, Selenaspidus articulatus (Morgan), in a grove near São Paulo, Brazil, was A. aleyrodis. They observed fungal infections caused by Podonectria coccicola (Ellis & Everhart) Petch, Pseudomicrocera henningsii (Koord.) Petch, and Sphaerostilbe aurantiicola (B. & Br.) Petch in the coccids Chrysomphalus aonidum (L.), P. ziziphus and Lepidosaphes beckii (Newm.) in citrus groves in Taiwan. In South Africa, Researchers reported several fungi attacking four scale pests of citrus. No publications on the applied use of fungi for control of scale in citrus were found in the literature.

Aphids (Hemiptera: Aphididae)

Aphids, particularly those in the genus Toxoptera, are important pests of citrus, especially due to their roles as vectors of diseases such as citrus tristeza. Fungi are important natural enemies of aphids under warm and humid conditions but studies on their use in citrus groves have been limited. Researchers demonstrated good potential for control of the brown citrus aphid, Toxoptera citricida (Kirkaldy), with B. bassiana. Application of the fungus in field trials at 2.5×10^{13} and 5.0×10^{13} conidia/ha resulted in 79.8% and 94.4% control, respectively.

Diaprepes root weevil, Diaprepes abbreviatus (L.) (Coleoptera: Curculionidae)

This weevil is native to the Caribbean and has become a major pest in Florida citrus since its introduction was first reported in 1964.

Notching along the margins of young leaves is a typical sign of feeding by the adult. Eggs are laid on older leaves and, after hatching, larvae drop to the ground, enter the soil, and feed on roots for most of the year. Injury caused by the weevil appears to be cumulative; root damage impedes the plant to take up water and nutrients, and can result in tree mortality. In addition, this injury provides an avenue for fungal root rot infections by Phythophora spp. A single larva can kill young trees, whereas several larvae can cause decline of older, established trees.

Since larvae are below ground, it is difficult to detect them before decline of above ground portions of the host plant is observed. EPNs (Steinernematidae and Heterorhabditidae) are effective in soil habitats. Because D. abbreviatus larvae are vulnerable while entering the soil, numerous studies have been conducted on the use of EPNs for their control. In Florida, EPNs have been marketed for weevil control for over 15 years. Currently, two commercially available species, Steinernema riobrave Cabanillas, Poinar & Raulston and Heterorhabditis indica Poinar, Karunakar & David are used for control of the weevil. These nematodes appear to be most effective at high temperatures (27 ± 2°C) in coarse sandy soils. Larval mortality of over 90% has been reported for field trials with S. riobrave when applied at 1.2 x 10^{10} infective juveniles (IJs)/ha. The use of irrigation systems for application of EPNs has been effective in delivering IJs into the zone below trees where larvae enter the soil. Indigenous entomopathogenic fungi infect adults and larvae of D. abbreviatus and other weevils in the soil.

Research demonstrated that a commercial oil formulation of B. bassiana (Mycotrol) or a combination of the fungus and a sublethal concentration of imidacloprid (a chloronicotinyl insecticide) provided effective control of neonate larvae and teneral adults when applied as a soil barrier. Researchers pointed out that the efficacy of B. bassiana as a weevil control agent was limited by its poor persistence in soil. Researchers demonstrated larvicidal and sublethal activity of elevated concentrations B. thuringiensis subsp. tenebrionis for D. abbreviatus in artificial diet and potted citrus tests.

Citrus root weevils, Pachnaeus spp. (Coleoptera: Curculionidae)

Pachnaeus litus (Germar) and P. opalus (Oliver) are native to Florida and normally are considered minor pests, although they can damage young citrus plants. The adults feed on tender foliage of citrus and eggs are laid on mature leaves, often on the same trees on which the adults have been feeding. Neonate larvae drop to the ground and quickly burrow into the soil where they feed on roots. Field applications of 5 x10^6H. bacteriophora IJs/tree resulted in significant reduction of P. opalus adults (76%) as compared to controls. In another experiment, two applications of S. riobrave or S. carpocapsae at 2 x 10^6 IJs/tree provided an overall 64% and 53% reduction of P. litus, respectively.

Fuller rose beetle (FRB), Asynonychus godmani Crotch (=Pantomorus cervinus (Boheman) (Coleoptera: Curculionidae)

Adults feed on citrus foliage and cause leaf notching. Larvae develop for 6-10 months in the soil where they feed on the roots. The beetle does not usually cause economic damage but the presence of eggs on exported fruit requires fumigation. Researchers reported the results of field application of rather high rates of S. carpocapsae IJs against late-instar larvae under Valencia orange trees. A single application of the Kapow or All strain of S. carpocapsae, each applied at 50, 150, and 500 IJs/cm^2, reduced the number of emerging adult FRB a combined 55% and 38%, respectively, the year following treatment and 79% and 82%, respectively, the 2nd year. Based on EPN recovery six months after application and continued reduction of FRB emergence in the second year, the authors concluded that the EPNs persisted and recycled in the environment.

Citrus fruit borer, Ecdytolopha auratiana (Lima) (Lepidoptera: Tortricidae)

In the late 1980's, E. auratiana became a key pest of citrus in São Paulo, Brazil. Females lay eggs on the fruit surface where neonate larvae enter the fruit and feed for approximately 20 days, rendering them worthless for consumption and processing. When fully grown, larvae leave the fruits and pupate in the soil. Laboratory tests with EPNs against sixth-instar larvae in pots containing sandy soil showed that H. indica applied at 1.6 IJ/cm^2 resulted in 92% mortality.

False codling moth (FCM), Cryptophlebia leucotreta Merge (Lepidoptera: Tortricidae)

This is an important pest of citrus in Africa and outlying islands. A granulovirus of FCM has shown promise in field trials. Researchers conducted a small-scale field trial of the virus formulated with skimmed milk powder and a wetting agent and applied at 10^8 and 10^9 granules/ml on FCM-infested citrus in Cape Verde, resulting in a 77% reduction of FCM population. Researchers provided a synopsis of research conducted on this virus in South Africa and concluded that it plays a natural regulatory role in FCM populations. Application of the granulovirus at rates of 10^{14} to 10^{15} OBs/ha provided up to 60% reduction of FCM infestations in navel oranges.

Subsequently, 17 field trials have been conducted on citrus in three different provinces in South Africa. Where spray coverage was thorough and FCM pressure was moderate, infestation was reduced by 70% following a virus treatment three weeks before harvest. The virus has been registered for use and is now commercially produced in South Africa.

Old world bollworm, Helicoverpa armigera (Hübner) (Lepidoptera: Noctuidae)

This insect attacks a wide variety of important crops in portions of Africa, Asia, Australia, and Europe. In South Africa and parts of Asia, it is a serious pest of citrus. B. thuringiensis subsp. kurstaki (Btk) has been applied for control of H. armigera in citrus, but growers have reported dissatisfaction with its efficacy. Field trials of the nucleopolyhedrovirus of H. armigera conducted by researchers demonstrated that it was superior to Btk for suppression of the bollworm. Application of 7.26 x 10^5 and 1.15 x 10^6 OBs/ml resulted in 100% reduction in bollworms within 14 days of application. Damage to fruit was reduced by up to 75-84% and rejection for export was reduced by 62-96%.

Other lepidopteran pests

The citrus leafminer, Phyllocnistis citrella Stainton (Phyllocnistidae) is a significant pest of citrus worldwide. Field application of Bt against the leafminer in the Azores resulted in significant larval mortality 48h after treatment. Researchers observed that leaf damage and number of P. citrella larvae were significantly reduced after 21 days by treatments of Bt plus a wetting agent. Researchers demonstrated only limited potential for control of P. citrella with S. carpocapsae. Researchers reported successful control of the citrus butterfly, Papilio demoleus L. (Papillionidae), on sweet orange following applications of Bt at two locations in India. The citrus leafroller, Cacoecia occidentalis Walsingham (Tortricidae) is a minor pest of South African citrus.

Fruit flies (Diptera: Tephritidae)

Some species, such as the Mediterranean fruit fly, Ceratitis capitata Weidemann, are key or major pests of citrus in tropical production areas. Several authors have reported the results of research for control of C. capitata and other tephritid species using EPNs and fungi in laboratory studies and field trials in other agroecosystems.

Banana

Several varieties of banana and plantain (Musa spp.) are grown throughout the tropics and into the sub-tropics. In addition to providing indigenous populations with rich sources of carbohydrate and other nutrients, banana and plantain are valuable export crops. A wide variety of insects and mites attack banana and include species that bore into the trunk, pseudostem, rhizomes, corm and roots, and species that attack flowers, fruits and foliage.

Banana weevil (BW), Cosmopolites sordidus (Germar) (Coleoptera: Curculionidae)

This weevil is reported as the most important insect pest of banana and plantain (Gold et al. 2001). Oviposition takes place at the base of the plant and neonate larvae bore into the corm. Heavy infestations can result in crop failure in newly established stands and reduced yield and shortened life span of plants in established stands.

Strains of B. bassiana have shown good potential to control adult BW

Mortality in adult weevils of up to 60% was reported. Godonou et al. (2000) evaluated two formulations of B. bassiana (oil palm kernel cake-based formulation of conidia [OPKC] and conidial powder) applied to the planting holes and suckers of banana. Both formulations resulted in 75% mortality in artificially released weevils. Under natural infestation conditions, the OPKC performed better (42% mortality) than the conidial powder (6%). Researchers reported up to 61% reduction of BW adults after treatment with B. bassiana.

A major constraint to the use of B. bassiana is the lack of an economic means of effectively applying the fungus. In Brazil, baits were made of sections of banana pseudostem treated in a suspension of B. bassiana conidia, mycelia and medium or with a paste that resulted in application of 5 x 10^9 conidia/bait. Fifty baits/ha were recommended being baits replaced 15 days, until less than 5 BW are captured/bait. Researchers improved targeted delivery of the fungus in and near traps baited with an aggregation pheromone that is attractive to both sexes. B. bassiana is transmitted horizontally among BW individuals, and this greatly improves the dissemination of the fungus within populations. Also, as most mycosed adult cadavers are found in the leaf sheath at the base of plants, the likelihood that ovipositing females and mating pairs to come into contact with recently produced conidia is increased.

Researchers reviewed the research of several authors on the potential of endophytic fungi (Fusarium spp., Acremonium spp., Geotrichum spp.) for BW suppression. The most effective species in this role was Fusarium oxysporum Schltdl.

Laboratory, greenhouse and field assays demonstrated the activity of EPNs against BW

Researchers evaluated exotic and native EPNs in Venezuela and found some native isolates of Heterorhabditis with good potential for BW control. Researchers evaluated S. carpocapsae, S. glaseri (Steiner) and S. feltiae (Filipjev) against BW in greenhouse tests in Puerto Rico.

The nematodes significantly reduced the number of tunnels made by larvae in plantain corms at 400,4,000 and 40,000 IJs/four-month-old plant. At the two higher rates, 100% larval mortality was achieved.

Researchers assayed 32 strains and species of Steinernema and Heterorhabditis against larvae and adults of BW and reported the greatest activity of the BW strain of S. carpocapsae against adults. They also described a method for introducing S. carpocapsae IJs into banana corms that involves removing cones (50 mm diam. by 150 mm long) from residual corms with a desuckering gouge and adding 2.5 x 10^5 IJs/cavity. The cone is then reinserted to produce a protected cavity that is attractive to adult weevils.

Researchers also reported significant mortality (43- 68%) of BW larvae in banana rhizomes after applying S. carpocapsae with a water thickener into cuts or holes made in residual rhizomes. Mortality of adult BW attracted to the application sites on treated rhizomes was also observed. In contrast, bimonthly treatments with Heterorhabditis zealandica Poinar and S. carpocapsae applied in a thickened aqueous solution into 200 mm deep incisions in the residual rhizomes of harvested plants from November to May failed to produce adequate control.

The researchers speculated that the treatments were not effective possibly because of early nematode mortality caused by free water in the spike holes and/or because of the need for more frequent applications. They demonstrated synergy between the insecticide chlordecone and S. carpocapsae for control of BW.

West Indian sugarcane borer, Metamasius hemipterus sericeus (Olivier) (Coleoptera: Curculionidae)

This pest is also known as the silky cane weevil and rotten stalk borer of sugar cane. It can be an important pest of banana in certain areas of the Americas. As with BW, fungi and nematodes have potential to control this pest. Researchers reported infection of low density populations of the insect by B. bassiana in a three-year old banana field. They observed up to 70% infection when more than ten weevils were captured/trap. Unlike BW, M. hemipterus sericeusis are strong fliers and could disperse fungi from sources of inoculum (i.e. through attractant traps) into neighboring populations.

Banana moth, Opogona sacchari (Bojer) (Lepidoptera: Tineidae)

This insect is only a minor pest of banana, but a serious pest of certain types of palms.

Opsiphanes tamarindi Felder (Lepidoptera: Brassolidae)

This is a major defoliator of plantains during the dry season in the region south of Lake Maracaibo, Venezuela. Broad-spectrum insecticides have been ineffective for its control. Researchers described an IPM system that combined cultural practices, application of Bt against early larval stages, and relying on natural enemies (parasitoids and predators) to control late larval and pupal stages. The seasonal application of Bt helped to eliminate first instars without affecting natural enemies.

Bagworm, Oiketicus kirbyi Guilding (Lepidoptera: Psychidae)

This is a defoliating pest of banana in Costa Rica and Colombia. Researchers noted the occurrence of B. bassiana and a Nosema species in bagworm larvae in Costa Rican bananas.

Coconut

Coconut is significant and sometimes predominant sources of income for several tropical countries. Although a multitude of insects and mites exert varying degrees of economic impact on it, microbial control agents have been used on relatively few of them.

Palm rhinoceros beetles, Oryctes spp. (Coleoptera: Scarabaeidae: Dynastinae)

These beetles are serious pests of coconut throughout the old world tropics. Oryctes rhinoceros (L.) is one of the most important pests of coconut in Southeast Asia and several South Pacific islands. Adults attack the heart of plants and feeding can reduce yield and kill trees. Larvae develop mainly in rotting palm trunks.

One of the most successful uses of an entomopathogen for classical biological control is reported for O. rhinoceros. Researchers described a non-occluded virus of O. rhinoceros from Malaysia that demonstrated potential for long-term control. Adults become chronically infected via oral contact with the virus and subsequently serve as reservoirs and disseminators. The midgut epithelial cells of adults become heavily infected; individual beetles may produce and excrete up to 0.3 mg of virus per day. Viral transmission among adults occurs during mating, when they feed in palms contaminated with feces containing the virus, or in larval breeding sites.

There are no external symptoms of the disease in adults and it is not immediately fatal. However, it shortens lifespan and reduces fecundity of infected adults. Infected beetles stop feeding, fly less frequently, and the males mate less often. Transmission to larvae occurs when virus-infected adults defecate in breeding sites. Viral infection in larvae is always lethal. Studies with other Oryctes species and Strategus aloeus (L.) revealed that the virus is cross-infective to certain other Dynastinae (Lomer 1987). Control of O. monoceros (Olivier), a serious coconut pest in Africa, with the Oryctes virus was reported.

The preferred method for disseminating the virus in coconut plantations has been the infection and release of O. rhinoceros adults, which resulted in establishment of the virus within larval and adult habitats in several locations in Asia, Africa, and the South Pacific where it was previously absent. As few as ten infected beetles can successfully establish the virus on an island. Introduction of virus into artificial and natural larval habitats has also been used successfully to inoculate beetle populations. Since 1967, the introduction of virus into coconut plantations in several South Pacific islands and other locations has resulted in significant control of O. rhinoceros. Integrated measures that include removal or covering of old palm logs that serve as breeding sites along with inoculative releases of the virus have reduced the density of Oryctes populations to below economic thresholds in many locations.

Despite the successes of introducing the virus into previously virus-free islands, O. rhinoceros remains a serious threat in the coconut and oil palm plantations of Southeast Asia and the Pacific. The prohibition of burning palm logs in the 1990's as a method for maintaining plantation hygiene has dramatically aggravated the problem.

In such situations, O. rhinoceros outbreaks are not caused by the absence of the virus disease, but by ecological disturbances in the transmission cycle of the virus, like the availability of large number of coconut palms for insect breeding. Under these conditions there is little contact between virus-infected and healthy individuals, whereas a low density of dead-standing palms creates good conditions for spread of the virus. Ecological methods for promoting the spread of the virus are to: 1) hide trunks of felled palms from the beetles by promoting the growth of cover crops over them, rather than attempting to burn the trunks, and 2) leave about five dead-standing coconut palms/ha.

DNA analysis in Malaysia revealed several distinct viral genotypes with different virulences. The most virulent for larvae and adults (type B) was produced in vivo and released into healthy populations. Examination of beetles from the release site and vicinity demonstrated the spread and persistence of type B with concomitant reduction in palm damage.

Decreased control has been reported from other earlier release sites. Researchers reported considerable genetic variation in the virus that suggests its rapid evolution. They recommended a renewed coordinated effort for the selection and distribution of virulent viral strains. Earlier work showed some distinct differences in virulence among strains of the Oryctes virus. Researchers demonstrated that re-release of the virus could result in an increase of the infection rate with a reduction of palm damage.

The combined use of the fungus M. anisopliae in larval breeding sites and release of virus has also been proposed, but the fungus does not spread well between breeding sites. The Oryctes virus has remained the most important biological control agent, however O. rhinoceros populations in Java and the southern parts of Sulawesi, Indonesia, are suspected of having developed resistance to the virus.

Root weevils (Coleoptera: Curculionidae)

The rotten sugar cane borer and the Diaprepes root weevil are reported as pests of several palm species. Research on the use of EPNs and fungi for control of these pests has been conducted in banana and citrus, respectively. Experiments with S. carpocapsae applied at 8 x 10^6 IJs/palm conducted by researchers on M. hemipterus sericeus-infested Canary Island date palms (Phoenix canariensis Hort. ex Chabaud) resulted in 51% mortality in weevil larvae. Because of high weevil production/palm, researchers recommended that EPNs should be applied frequently and over a long period of time for effective management. The precise interval of application and duration of treatments will depend on environmental conditions and population density of the weevil.

Mango

Over 260 species of insects and mites have been recorded as pests of mango worldwide. Key pests that require regular control measures include fruit flies, seed weevils, tree borers and various Hemiptera.

Fruit flies (Diptera: Tephritidae)

Fruit flies are among the most serious pests of tropical fruit and are regarded as the principal pests of mango. Anastrepha ludens (Loew) develops in a variety of fruit crops, but is especially damaging in mango and citrus. It is widely distributed in Mexico, most of Central America and southern United States. The Caribbean fruit fly, Anastrepha suspensa Loew, is a pest of mango and several other tropical fruits and is distributed within the Greater Antilles, Bahamas and Florida. Anastrepha obliqua (Maquart) is a significant pest of mango in Brazil and most of the new world tropics, but it has not yet been tested for susceptibility to entomopathogens. C. capitata and other Ceratitis spp. have been reported from mango and many other fruit worldwide. The oriental fruit fly, Bactrocera dorsalis (Hendel), is a pest of a wide range of fruit in Asia and is a key pest of mango in India. Fruit fly females oviposit in ripening fruit, and larvae burrow into the pulp. Fully grown larvae exit the fruit, usually after it has fallen to the ground, and pupate in the soil.

EPNs and fungi have been evaluated as alternatives to conventional insecticides for control of some important fruit fly pests of mango. Although A. ludens is susceptible to a variety of EPN species under laboratory conditions, extremely high rates are required for control in the field (2.5 x 10^2 IJs of H. bacteriophora/cm^2). Similarly, laboratory and field research conducted on the effectiveness of EPNs against C. capitata, revealed susceptibility of larvae to several nematode species, but high application rates are required for control in the field (5 to 50 x 10^2 IJs of S. carpocapsae/cm^2. Research on the susceptibility of other fruit flies of mango has been limited to the laboratory. Researchers reported on the susceptibility of B. dorsalis to S. carpocapsae and reported on the evaluation of A. suspensa susceptibility to several steinernematids and heterorhabditids.

Testing of fungi on fruit flies of mango has been predominantly on M. anisopliae and B. bassiana. Laboratory and field research on M. anisopliae against Ceratitis spp. Elucidated the effect of various factors on the activity of the fungus that included temperature, moisture, gender, life stage and fly species.

Researchers investigated the effect of fungicides, acaricides, insecticides and herbicides on M. anisopliae activity against C. capitata in laboratory exposures in field-collected soil. Significant pupal and adult mortality occurred in soil treated with the fungus with and without pesticides. No larval mortality was observed. Pesticides affected fungal activity slightly with the most significant effect due to the fungicides chlorothalonyl and tebuconazole.

Laboratory and field studies on M. anisopliae activity against A. ludens were reported by researchers. When M. anisopliae was applied in field cages at 2 x 105 conidia/cm2, adult emergence was reduced by up to 43% in loam soil. Researchers studied the activity of strains of M. anisopliae and P. fumosoroseus against C. capitata adults and reported LD50 values of 5.1 x 103 and 6.1 x 103 conidia/fly, respectively, for the two most active strains. They also noted a sublethal effect of the fungi on fecundity. Researchers reported on the laboratory evaluation of B. bassiana and B. brongniartii against adults of C. capitata and Researchers studied the effects of B. bassiana on A. ludens. In both studies, adult flies were very susceptible to infection by conidia, but researchers reported negligible effects on larvae and pupae.

Mango seed weevil, Sternochetus mangiferae (F.) (Coleoptera: Curculionidae)

This is a widespread pest of mango. Eggs are usually laid on green fruit and larvae tunnel to the seed where they develop. Researchers reported laboratory and field tests with two strains of B. bassiana, but neither strain had an effect on S. mangiferae. Researchers described the isolation of a virus infecting larvae of S. mangiferae in India. It caused reduction in feeding, sluggishness, browning of the integument and milkiness of the haemolymph. The authors discussed the similarity of the virus to that reported in O. rhinoceros.

Rhytidodera bowringii White (Coleoptera: Cerambicidae)

Researchers reported on the evaluation of B. bassiana against Rhytidodera bowringii White. The fungus was isolated from dead R. bowringii adults and subsequently used to produce inoculum for field experiments in China. Treatment of two older mango orchards with B. bassiana resulted in 84% mortality of the beetle.

Mango mealy bug, Drosicha mangiferae Green (Hemiptera: Margarodidae)

Researchers found B. bassiana infecting nymphs of D. mangiferae in mango orchards in five localities in Lucknow, India. In field trials on infested mango panicles, spray application of a suspension of 4.8 x 10^6 conidia/ml reduced populations of D. mangiferae by 33-100% in ten days. Researchers demonstrated the dose-mortality relationship of B. bassiana against D. mangiferae first-instar nymphs in laboratory assays. Researchers showed insecticidal activity for Photorhabdus luminescens (Akhurst) (the symbiotic bacterium isolated from H. indica). Application of a formulation of 1.4 x 10^6 cells/ml of P. luminescens on D. mangiferae-infested mango twigs resulted in 92.5% mortality of second-instar nymphs after 48h.

Mangohopper, Amritodus atkinsoni Lethierry (Hemiptera: Cicadellidae)

Researchers reported that a 75 min exposure to 10^9M. anisopliae conidia/g of inert dust caused 100% mortality of A. atkinsoni after 96h. Concentrated aqueous suspensions of the fungus (10^9 conidia/ml) were considerably less effective. Researchers reported natural infection by the fungi Lecanicillium lecanii (Zimm.) and B. bassiana in populations of the leafhopper Idioscopus clypealis (Lethierry) on mangoes in Uttar Pradesh, India.

Guava

Psidium guajava L. is native to the American tropics but is currently grown in more than 50 subtropical and tropical countries. Brazil is the principal red guava producer followed by Mexico whereas India is the major producer of white guava. Different pests attack fruits, leaves and trunk, causing more or less damage depending on the region or country. Main pests of fruits are the guava weevil (Conotrachelus psidii Marshall) and fruit flies (C. capitata and Anastrepha spp.). On the leaves, the main pest is a psyllid (Triozoida sp.) that causes damage mainly after pruning when new leaves start growing.

Fruit flies (Diptera: Tephritidae)

Fruit flies are very important pests in guava because the adults lay eggs in the fruit, and resulting damage by larvae lowers its quality. In Brazil, the main species in guava are

Anastrepha fraterculus (Wied.), A. obliqua, A. sororcula (Zucchi), A. zenildae (Zucchi) and C. capitata. Most of the research on microbial control of these pests has been conducted in other tropical fruit crops, most notably in mango (see mango section).

Guava weevil, Conotrachelus psidii Marshall (Coleopera: Curculionidae)

This is a major pest of guava in certain areas in Brazil. Females lay eggs in immature fruit (3-4 cm diameter) and larvae progress through four instars as the fruit develops. Infestation leads to acceleration in fruit maturation and fruit drop when ripe. At this moment, larvae crawl into the soil where they develop into prepupae. Individuals may remain in this stage for up to six months before pupation and development into the adult.

Control methods involve weekly applications of insecticides to suppress adults, but most of those currently in use for guava weevil control will be discontinued soon. Without chemical control, the percentage of damaged fruit in heavily infested orchards can reach 100%. The amount of fruit attacked has been increasing over the past three years possibly due to the development of insecticide resistance.

Poorly timed chemical applications and the tendency for adult weevils to hide in the litter around trees and avoid contact with the chemicals could also be involved.

The virulence of four species/strains of EPNs to fourth-instar larvae was assessed in the laboratory. In petri dish assays with sterile sand at 100 IJs/larva, larval mortality ranged from 33.5% to 84.5%, with the heterorhabditids being the most virulent. In sand column assays with H. baujardi LPP7, H. indica Hom1, and S. riobrave 355 at 100, 200 and 500 IJs/larva, significant mortality was observed only for H. baujardi (62.7%) and H. indica (68.3%) at the highest dose. For H. baujardi LPP7, the LT_{50} and LT_{90} for 100 IJs were 6.3 and 9.9 days, whereas the LC_{50} and LC_{90} over seven days were 52 and 122.2 IJs. In a greenhouse study with guava trees in 20-L pots (ten weevil larvae/pot), and doses of 500, 1000 or 2000 IJs/pot, H. baujardi LPP7 caused 30% and 58% mortality at the two highest doses.

In small farms in Cachoeiras de Macacu, RJ, Brazil, a combination of different control methods against the guava weevil is being implemented. H. baujardi LPP7 is applied in orchards as infected cadavers. Initial results indicate a 40% to 70% decrease in adult weevils by applying 20 cadavers/tree. In addition, removal of all damaged fruit from their orchards helps to reduce pest population in the following year. Another alternative is the weekly application of neem oil against adults and neem cake applied to the soil for control of larvae. By eliminating pesticides, these strategies have effectively reduced production costs by 40%. Recently, fruit quality and yield are being assessed.

Leaf-footed bug, Leptoglossus zonatus Dallas (Hemiptera: Coreidae)

This is usually a secondary pest on fruits and flowers. Few attempts to control this pest with microbials were published. Three B. bassiana isolates and one isolate of M. anisopliae were assessed in the laboratory against adults, being M. anisopliae NB the most efficient.

In a field trial, mineral oil-based ultra low volume controlled droplet applications of M. anisopliae NB at 10^{10} conidia/tree caused 94% adult mortality. When B. bassiana was applied, there was a 28% increase in fruit yield.

Papaya

Carica papaya L. originated in southern Mexico, Central America and northern South America and is cultivated in most tropical countries. A total of 134 species of arthropods are reported to attack papaya, some of which are important vectors of major pathogens of papaya. Since most fruit production is for exportation and the presence of pesticide residues is not tolerated, the use of alternatives to chemicals, including cultural methods and microbial control, are being increasingly employed.

Fruit flies (Diptera: Tephritidae)

Fruit flies comprise the most important pests of papaya in most producing regions. Research on the use of microbial control agents (EPNs and fungi) against several species that attack papaya (Anastrepha spp., Ceratitis spp., B. dorsalis) has been conducted in other tropical fruit crops, most notably in mango.

Twospotted spider mite, Tetranychus urticae Koch (Acari: Tetranychidae)

This is the most important pest in papaya in Brazil, and is responsible for a major portion of the production costs. Because of the high temperatures and humidity in areas where papaya is grown, fungi have potential as microbial control agents of mites. Researchers reported use of B. bassiana on over 1000 ha of commercial papaya production in Brazil for control of T. urticae.

Most of the research on B. bassiana and T. urticae has been conducted in other crops. Laboratory tests with B. bassiana against eggs, deutonymphs, protonynphs, larvae and adult stages of T. urticae on green bean showed positive results. The LC_{50} for juvenile stages and adults was 3184 and 1949 conidia/ml, respectively. No significant differences in mortality were observed among egg age classes (24-96h-old eggs) at the tested concentrations (1400-22,800 viable conidia/ml).

Natural epizootics of fungi in the genus Neozygites have been responsible for spectacular declines in T. urticae populations in other cropping systems.

Broad mite, P. latus

This is also a significant pest of papaya in some areas. Researchers investigated the use of B. bassiana and other fungi against this pest (see citrus section).

Aphids (Hemiptera: Aphididae)

Although aphids do not colonize papaya, they are considered important potential vectors of papaya diseases. Many fungi (several Entomophthorales and L. lecanii) have resulted in massive natural epizootics in other cropping systems in some of the same aphids that attack papaya (e.g., Aphis gossypii Glover, Myzus persicae (Sulz.). Their potential to control aphids and other hemipteran papaya pests warrants further attention.

Other hemipteran pests

Papaya scale, Philephedra tuberculosa Nakahara and Gill (Hemiptera: Coccidae) attacks papaya and annona fruits. It is naturally infected by the fungus L. lecanii, which can lead to 90% mortality during summer. A variety of mealybugs, whiteflies, and leafhoppers attack papaya. Although literature on microbial control of these pests on papaya is scant, several fungi are reported to attack these insects in other agroecosystems, including tropical fruit. Research on the use of fungi for control of these species in papaya is warranted.

Pineapple

Ananas comosus (L.) Merril is native to Brazil, Bolivia, Peru, and Paraguay, and is currently cultivated in most tropical countries. It is the third largest fruit crop (after bananas and mango) harvested in the tropics. These authors provided a summary of arthropod pests of pineapple worldwide and considered pink pineapple mealybug, Dysmicoccus brevipes (Cockerell) as the key pest worldwide.

Mealy bugs, Dysmicoccus brevipes and D. neobrevipes (Hemiptera: Pseudococcidae)

Mealy bugs cause wilting due to the toxic effect of their feeding. There is potential for using EPNs against this pest based on work done with a closely related species, D. vaccinii Miller & Polavarapu.

Large moth, Thecla basalides (Lepidoptera: Lycaenidae)

This insect is an important pest of pineapple in Brazil, mainly in the cultivar 'Pérola'. Females lay eggs from the beginning of flowering until fruit formation. Larvae penetrate the flowers and complete development in 13 to 16 days. In Northern Brazil, the dose of 600 g/ha of Bt is recommended for control. In Southern Brazil, researchers reported on the natural enemies of T. basalides and effectiveness of insecticides. Application of Bt resulted in effective control.

White grubs (Coleoptera: Scarabaeidae)

The larval stages of 23 species of scarabs in three sub-families were reported to attack the subterranean organs of pineapple plants in several locations worldwide, with the most serious pests reported from Australia and South Africa. Although no specific studies on the use of microbial control agents have been reported for white grubs in pineapple, pathogens have been successfully applied for their control in other crops. Candidate control agents include fungi [M. anisopliae, Beauveria brongniartii (Sacc.) Petch], bacteria (Bt, Paenibacillus spp., Serratia entomophila Bizio), and EPNs (Heterorhabditis spp., Steinernema spp.).

THE ROLE OF MICROBIAL CONTROL IN INTEGRATED PEST MANAGEMENT (IPM) IN TROPICAL FRUITS

Integrated Pest Management plays a significant role in crop protection being an important aspect of sustainable agriculture that attempts to minimize the negative environmental impacts and other deleterious effects due to the use of chemicals. Individual components of IPM are often evaluated as stand-alone tactics without consideration of their interactions with other components of the agroecosystem.

An integrated approach that is based on pest densities and their relation to economic injury thresholds will ultimately be required for each cropping system and location before agriculture will be truly sustainable.

When selective insecticides are used, the negative impact on beneficial insects is reduced. Biopesticides provide an alternative means of control that further minimizes impacts on beneficials and other non-target organisms (NTOs). This is due to the specific nature of many microbial control agents.

Safety testing data for entomopathogens indicate that they are generally safe for most NTOs, especially vertebrates.

However, it will be necessary to determine their effects on the beneficial organisms under the specific conditions in each agroecosystem.

The way in which entomopathogens are utilized, i.e. augmentation, inoculative introduction (classical biological control) or conservation, will depend on the characteristics of the pest and the fruit crop in which it causes damage or yield loss. Fruit crops are stable agroecosystems where any of the above strategies for pathogen use could be considered. In addition to the use of commercially available biopesticides, it may be useful to consider employing native entomopathogens. Surveys should be undertaken in different agro-ecological zones to identify prevailing environmental conditions and the presence of native pathogens and natural enemies that may be better suited for the targeted location than an exotic species or strain.

On the other hand, an exotic pest may require importation of natural enemies from its native range. In classical biological control, natural enemies, including entomopathogens, are sought in the region of origin of the invasive pest, imported and established in an area where they do not naturally occur. Typically this is a geographic area where the pest has invaded without its natural enemies and there are no effective native natural enemies in the invaded region.

When microbial control agents are formulated as biopesticides, they are predominantly used for inundative applications and often treated much like chemicals, with the expectations that they will perform at the same standards. In general, this has not always been possible. On the other hand, there are biological control agents capable of doing what chemicals are not able to do, i.e. EPNs that have a capacity to find their pest host, kill it and reproduce in it. In fact, many entomopathogens have the capacity to reproduce in the host and hence produce secondary inoculum able to attack and kill other individuals in the pest population. This numerical increase response, of which chemicals are incapable, needs to be better exploited in tropical conditions.

Several other advantages of entomopathogens over chemicals are presented by Alves. The cost of producing natural enemies must be judged in terms of the value of the crop protected by using the agent and in comparison to the cost of competing control options such as chemicals.

In Europe the costs of biological control agents used in protected crops and horticulture have proven to be economic and comparable to chemicals. In Florida, the use of EPNs is an integral part of IPM in citrus indicating that the benefit/cost relationship is positive. There are crops that have few or no registered pesticides and consumers who prefer to buy pesticide-free produce. In those cases, microbial control is strongly supported.

Interaction of Entomopathogens and other Biological Control Agents

Parasitoids and predators can interact synergistically/additively (e.g., enhanced transmission and dispersal of insect pathogens) or antagonistically (e.g., parasitism/ infection, predation and competition) with entomopathogens. In most studies examining the interaction between entomopathogens and other natural enemies, the pathogen almost always dictates the population dynamics of other guild members. Most studies indicate the positive nature of these interactions with respect to the control of insect populations.

Various studies have shown the capacity of parasitoids to identify and avoid oviposition in hosts infected by the different entomopathogen groups. This rejection is usually due to visual changes that occur in hosts and/or chemical cues associated with biochemical changes in hosts late during the disease development. As for predators, this does not appear to be true. The combination of EPNs with other nematode species, fungi and viruses often results in additive effects on pest mortality, whereas nematode-bacteria interactions range from antagonistic to synergistic.

Interaction of Entomopathogens and Chemical Pesticides

Unlike parasitoid, entomopathogens are generally compatible with chemical pesticides. Exceptions include use of several fungicides and most nematicides with entomopathogenic fungi and EPNs, respectively. Certain combinations of entomopathogens with chemical pesticides can be synergistic, such as reported for sublethal concentrations of imidacloprid and fungi or EPNs, respectively. The economic feasibility of such combinations will depend on how much of each component can be reduced compared with their recommended application rates when they are applied individually. The compatibility of chemical pesticides with arthropod natural enemies will be another consideration when integrating pesticides and entomopathogens into a pest management program.

Ecological Engineering and Manipulation of Environment to Enhance Activity and Persistence of Entomopathogens

Ecological engineering in the context of biological control and IPM is the manipulation of agricultural habitats to be less favorable for arthropod pests and more attractive to beneficial organisms. Under optimal environmental conditions many entomopathogens have the natural ability to cause disease at epizootic levels due to their persistence in the environment and efficient transmission. hen an insect pathogen is capable of becoming established in an environment it also has the potential to confer long-term regulation of a pest population. In order for a disease to become epizootic in an arthropod population three factors are required: presence of the host; presence of the pathogen; and proper environmental conditions. Habitat manipulation techniques have been employed to optimize environmental conditions to increase entomopathogen (endemic, inoculative release, or inundatively applied) activity and facilitate their persistence.

These include augmenting moisture in the habitat and creating habitat refuges such as mulches, hedgerows or grass banks, which maintain soil humidity and temperature favorable to microbial activity, propagation, and survival. Manipulated habitats have also been used to provide host plants for alternate arthropod hosts and nectar sources for parasitoids and other natural enemies. Microbial agents are susceptible to ultra-violet light, heat, and desiccation, but the effects will vary with the microbial species or strain and habitat.

Some are capable of remaining viable for just hours (e.g., non-occluded viruses) or days (e.g. fungal conidia, occluded viruses). Others may persist for months (e.g., EPNs, fungal resting spores) or years (e.g., Paenibacillus spp., fungal resting spores). In addition to habitat manipulation, formulation of pathogens with humectants, nutrient sources, and UV protectants has been used to enhance persistence under field conditions.

AVAILABILITY OF ENTOMOPATHOGENS AS A LIMITING FACTOR

Factors that favor accelerated growth and use of biopesticides are their improved performance and cost competitiveness in the face of increasing insect resistance to chemical insecticides, environmental hazards and lack of selective chemical pesticides. Recently, major agrochemical companies have taken a greater interest in microbial pesticides. In all, 281 biopesticides were available on the market in 1993, with active ingredients of bacteria, EPNs, fungi, and viruses.

Although the market for microbial insecticides is growing, it represents less than 1-1.5% of the total crop protection market. Many microbial control agents can be produced on artificial media using fairly simple methods and there are potentially ample markets for them. However, the bottleneck in the use of microbial control in many countries is their local production and availability. Although there are successful examples of important pests being controlled with biopesticides, many microbial control agents are not universally available to growers.

CONCLUSIONS

Sustainable agriculture will rely increasingly on alternatives to conventional chemical insecticides for pest management that are environmentally friendly and reduce the amount of human contact with hazardous pesticides. Microbial control of arthropod pests of tropical fruits, in conjunction with other IPM components, can provide effective control. The challenge will be to find successful combinations of entomopathogens, predators, and parasitoids along with other interventions. Aspects that warrant further study and attention are improved formulation, storage, marketing, and transfer of technology to growers.

8

NANO BIOPESTICIDE: AN EMERGING TECHNOLOGY FOR PEST MANAGEMENT FOR FUTURE PERSPECTIVES

Pest affects plants in nature worldwide, leading to excess use of mineral fertilizers and toxic pesticides, effecting environments, plants, animal and causes serious health problems to farmers. The indiscriminate use of bio pesticide has led pest resistance, reduce soil fertility and finally there is an emergence of novel product to intercept the pest situation. Nanoscience a new discipline have great deal of application in various fields and may also be useful in agrochemical and plant protection area to control pest to considerable extent. Until now, nanoparticles were used in formulation of nano based pesticides and insecticides, encapsulated nanoparticles, nanoparticle-mediated gene or DNA transfer in plants and bio sensors for remote sensing for precision farming.

The nanopesticides of biological origin named as bio- nanopesticide could be fabricated using any metal such as Ag, Cu, SiO_2, ZnO with broad-spectrum pest protection efficiency. However, extensive research and spectrum of various field studies are specially required to develop understanding of interaction between nanoparticles, microorganisms, soil, plants and humans.

The plant such as weed plant that having no importance in agriculture could be dissects to identify many useful secondary metabolites that can be effective against pests. There are some plants with inherent capacity to deal pest, their extract shows different insect resistant compounds such as for example camphor, limonene and b-caryophyllene from Ocimum kilimandscharicum.

The neem extract is pioneered for characteristic of larvicidal and pesticidals activity. The weed plant Argimone maxicana, Calotropis procera, Dathura, Catharanthus roseus contents phytochemicals and secondary metabolites essential to minimize insect population. The need to formulate the biopesticide with efficient insect killing property is attempted by mixing different secondary metabolites in various ratios. However the constant use of the best sale biopesticide has put itself in danger for resistance in pest. It has allowed scientists and researchers all over the world to assemble the efficient secondary metabolites against insect-pest. The other alternative hidden in nature against pest needs to open up with the existing technologies. The greener method to avoid pollution and safer use with minimum toxicity for animals and plants is the best alternative. The Green chemistry topic that uses the green natural material as starting source to formulate the nano size particle capped with the pest resistant secondary metabolites could be possible solution to defect pest in farms.

One of the main current challenges facing global agriculture is the need to control and reduce the intensive use of agrochemicals such as herbicides, fungicides, and insecticides. It is well known that these substances can have adverse environmental and social impacts, as well as cause resistance in target organisms. Furthermore, large amounts of pesticides are lost during application, due to volatilization, degradation, and photolysis, with less than 0.1% of applied pesticides effectively acting against the target organisms. Nanotechnology seems to offer a way to mitigate the harmful effects of pesticides on the environment and human health, since it can provide systems enabling the controlled release of active compounds, thus increasing the efficiency and safety of products, while reducing the quantities required in field applications.

Controlled release systems are emerging technologies that have attracted global commercial and scientific interest in recent years. Such systems are used in areas including medicine, cosmetics, engineering, food, and agriculture.

Considering their use in the agricultural and environmental sector, a survey using the Scopus database (considering the period from 2009 to 2019) showed that only 6% of research concerning nanoformulations was related to the agroindustry field, of which 77% concerned controlled release nanoformulations, while only 23% involved stimuli-responsive materials (Fig. 1).

Considering stimuli-responsive materials, pH was the most widely studied stimulus (37%), followed by photo and thermal stimuli (27% and 17%, respectively). Enzymes, redox, and other types of stimuli together accounted for 20% of the studies (Fig. 1).

The advantages of controlled release for the delivery of agrochemicals include reductions of phytotoxicity, leaching losses, volatilization, drift, and soil degradation, as well as increased safety during application [4]. However, the release of active agents from loaded particles mainly occurs by means of passive diffusion, capsule erosion, or osmotic pressure, resulting in poor control of pesticide release. Hence, stimuli-responsive systems are promising candidates for improving controlled release properties, promoting site-specific and smart release of pesticides in response to biotic or abiotic stimuli (Fig. 2).

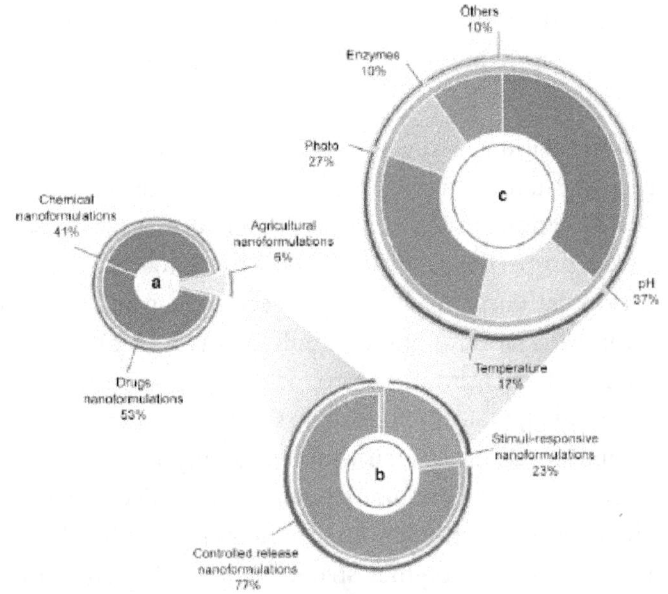

Fig. 1 Survey of published research available in the Scopus database for the period 2009–2019. The search was performed using the following keywords: a nanoformulations; b controlled release, stimuli-responsive nanoformulations; and c pH, temperature, photo, enzymes, redox, magnetic field-responsive

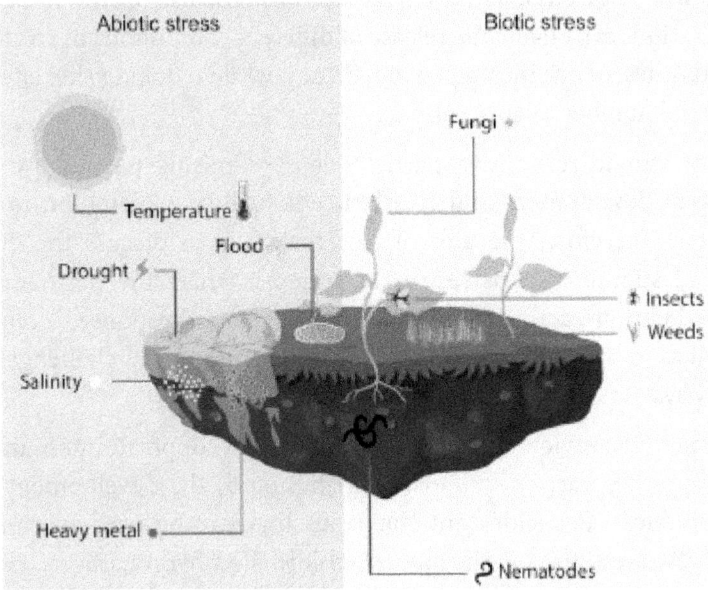

Fig. 2 Abiotic and biotic factors enabling the use of stimuli-responsive nanomaterials for site-specific release of active substances, in order to increase plant resistance

Biotic (plant pathogens, insects, and weeds) and abiotic (temperature, drought, flooding, and salinity) stressors are responsible for reducing global crop productivity by about 50%. Therefore, prospecting new responsive carriers that respond to environmental stimuli such as pH, temperature, redox conditions, light, and enzymes could be an important strategy for improving the resistance of crops to the stresses caused by these environmental factors.

The design of formulations using stimuli-sensitive carriers is a strategy to improve the controlled release system, where the macromolecules present an active response to small signals or modifications in the surrounding environment, resulting in changes in their physicochemical properties favoring the release of loaded substances. When used in agrochemical formulations, these sensitive carriers can allow the release of active compounds in response to abiotic or biotic stimuli such as light, pH, temperature, magnetic field, enzymes, among others. Stimuli-controlled delivery systems can be prepared using natural polymers (such as alginate, cyclodextrin, starch, chitosan, carboxymethylcellulose, and ethylcellulose) and silicas due to their wide availability and biodegradability. Recently, carriers based on pillararenes and curcubit[n]urils were studied in pesticide and herbicide delivery due to their responsiveness to environmental stimuli. Many physicochemical modifications can occur in the polymer structure, including changes in physical state, shape, conductivity, solubility, hydrophilic/lipophilic balance, and sol-gel transition.

These changes can be reversible, such that when the stimulus is withdrawn, the polymer returns to its initial structure. The release systems of different stimuli-responsive nanoparticles will be discussed in the next sections. The use of stimuli-sensitive carriers in nanoformulations can enable the release of different compounds in an effective manner and at defined times, hence increasing bioefficacy, while reducing side effects, the dosage required, and the number of applications.

The use of stimuli-responsive particles is a promising technology that is still in an early stage of development and clearly needs further exploitation to extend its use in agroindustry. Therefore, the aim of this review is to discuss the current state of development of stimuli-responsive carrier systems, where the mechanisms of release of active ingredients are activated by different triggers, in response to biotic and abiotic factors, as well as to highlight the perspectives and commercial challenges for the use of smart delivery systems in the agricultural sector.

In addition, this review indicates potential new opportunities and applications for stimuli-responsive carrier systems, contributing to the development of sustainable agricultural practices. Pesticides are chemicals that can be used to control pests and prevent plant diseases. Organophosphates, chlorinated hydrocarbons, carbamates, and carbamide derivatives are the most common active ingredients (AIs) in these substances. These compounds are commonly used in veterinary medicine, agriculture, and health.

They can harm humans and animals via skin absorption, inhalation, and dietary intake, as well as contaminating the environment. Improper use of pesticides in agriculture leads to changes in the level of antioxidant and oxidant enzymes in the human body, and thus results in the development of diseases caused by oxidative stress. Biopesticides, often known as biological pesticides, are insecticides derived from microorganisms or natural substances.

Biopesticides are divided into three categories: microbial biopesticides, botanical biopesticides, and plant-incorporated protectants. As an alternative to conventional insecticidal methods, biopesticides have recently gained much attention due to their potential target specificity, fewer harmful side effects, capacity to disintegrate fast, and high efficacy. Several substances have been investigated as biopesticides in recent years, including Clitoria ternatea extract, oxymatrine (an alkaloid component), stilbenes in grape cane, Talaromyces flavus strains (SAY-Y-94-01), and olive mill oil.

The usage of biopesticides, which represent less of a hazard to the environment and human health than synthetic pesticides, should be done with caution. Certain products have been licensed for usage as biopesticides, although they still pose health concerns. However, in this regard, nanotechnology has gained special attention and has become a novel field during the last couple of years due to its multidisciplinary applications in pharmacology, agriculture, pest management, and parasitology fields. Nano-particles are the most rapidly expanding area of nanotechnology, which provides the solution to many environmental problems due to their eco-friendly behavior and cost-effectiveness, small size (1-100 nm), and large surface area.

According to emerging evidence, nanotechnology has been proven to be an effective tool for the formulation of new nanocomposites against pests and improving crop varieties. The nano-biopesticides have superiority over the biopesticides and conventional techniques for many reasons, including environmentally friendly behavior, desired results within a few hours after applications, biodegradability, easy delivery to plants, and release slowly from the vector.

Furthermore, their small size makes them an effective carrier when combined with pesticides that can easily enter the plants. Another advantage of nano-biopesticides is that they did not have an adverse effect on soil microorganisms and phototoxicity of Ag-based nano-particles was suppressed by nano-coating them with biocompatible polyvinyl pyrrole compounds. The nano-biopesticides can be synthesized by following two ways: either by extracting the biological active pesticidal compound (APC) from plants and blended it with nano-particles and inserted it into a suitable polymer that acts as a supporting material, or APC secrete the metallic salt with bind with nano-particles (NPs) that hemolyze and merge into an appropriate polymer. The APC integrated with NPs and merged into a compatible vector including micelles, liposomes, nanosphere, polymer, and nanofiber. These ingredients were used as a spray to kill the insect pests for food protection. The accumulative data revealed that nano-biopesticides contain secondary plant metabolites and their mediated metal oxide nanomaterials.

It was found that biopesticides have gained importance over chemical pesticides during the past few decades due to their eco-friendly behavior, high efficiency, and fewer side effects. The evidence reported that recently much research had been carried out on nano-biopesticides; either pests are attaining chemical pesticide resistance, or a small number of insecticides have expired due to severe environmental and human concerns. However, this situation demands novel plant-based pesticides on the nanoscale to formulate the nano-biopesticides for pest management.

Moreover, the present chapter explains the mechanism of action, delivery, and environmental sustainability of nano-biopesticides.

Recently, different biopesticides have been reported against different pests such as Bacillus firmus and Bacillus sphaericus, which are used against diamondback moths, while Trichoderma harzianum and Trichoderma viride are used against root rots and wilts, Beauveria bassiana against mango hoppers, neem-based nano-biopesticides against whitefly, and Bacillus thuringiensis against israelensiss but they have not shown an effective impression on pests.

However, future studies are needed to overcome and improve nano-particle release rates and delivery. Therefore, the present chapter demonstrated the importance of nano-biopesticides for pest insect management on different crops for agricultural interest to overtake these situations.

PESTICIDES

Agricultural output has been increased dramatically in the early 20th century, especially in the United States, to keep up with the rapidly growing human population. During the last century, the world's population has tripled from 1.5 billion in 1900 to 6.1 billion in 2000. The world's population has grown by one billion people in the past decade, and the UN predicts it could reach 9.4 to 10 billion by 2050 if current growth rates continue.

A lateral increase in food production was needed during the 20th century to keep up with the increase in the world population. This was accomplished via the use of fertilizers and other agricultural inputs throughout the twentieth century. Bio-fertilizers (such as guano) were first used in the late 1800s; inorganic phosphate fertilizers (such as urea) were first used in the early 1900s and have steadily gained in favor ever since.

Phosphates helps to increase the crop diversity and yields and aided in the unprecedented "green revolution" for agricultural productivity. This caused a tenfold increase in grain production per unit surface area of agricultural land, leading to a global food surplus. An increasing global population and increased phosphate production were shown to be positively linked throughout the 20th century, with an R2 of 0.97 for the period 1900–1988.

Synthetic crop preservation agents were introduced to the market in the 1940s, which increased food output. It grew from 0.2 million tons in 1950 to over 5 million tons in 2000, up from 0.2 million tons the year before. Between 1950 and 2000, pesticide production grew by about 11%, from 0.2 million tons to over 5 million tons.

Crop preservation chemicals, sometimes known as pesticides, are composed of various composites, including growth regulators, neonicotinoids, organochlorines, pyrethroids, organophosphates, carbamates, and more recently, biopesticides. This wide spectrum of chemicals and insecticides has been developed throughout history to protect crops from pests and diseases. All sorts of pesticide sales grew; however, herbicides were the group that extended the most, accompanied by fungicides and insecticides.

Pesticide application has suffered owing to lack of global uniformity, high cost of chemicals, human resources, and the vast diversity of pests present in each climatic or geographical area. Using FAO data, it was found that the mean pesticide application rates per hectare of arable land ranged from 6.5 to 60 kg/ha, with the greatest mean values occurring in Asia and a few South American nations. Unlike Western Europe and North America, Asia has not witnessed a rise in the usage of herbicides in both urban and agricultural areas.

Compared to the widespread use of insecticides, herbicide usage in Asia has remained relatively low in recent years, according to World Bank and International Food Policy Research Institute data. Throughout the twentieth century, ancient synthetic pesticides intended for agricultural pest control, such as DDT, were often used to treat human parasites and animal ticks.

So, for example, DDT was designed to be used in agricultural pest control. Despite of being prohibited, it is nevertheless extensively used as a food prophylactic for various fish in South Asia, and to control home pests and malaria vectors globally, albeit seldom illegally.

Pesticides have been administered in agricultural settings for decades, employing techniques ranging from truck and aircraft spraying to old-fashioned field worker spraying. Studies on the effects and toxicity of manufactured chemicals on human health and well-being have shown that individuals report euphoria after pesticide application.

This research included peasants, farmworkers, and their families following pesticide treatments. They previously discovered that unintentional poisoning affects about 355,000 individuals annually and is linked to high susceptibility and poor chemical management. They also discovered that increased sensitivity and poor management of hazardous substances are closely linked to such toxicities.

The research was conducted to assess the number of pesticides in the environment that killed various animals. Among the animals that resided there were fish, birds, bees, amphibians, and tiny mammals. It was also noted how much they were killed and how they were slaughtered. Following the introduction of synthetic chemicals into the environment, it was only a short amount of time before it was thought that crop protection pesticides were causing disease both locally and internationally.

Many believe that sprayed on-crop DDT is deported into water bodies, quickly converted into DDE, and bio-accumulated in aquatic food systems before being reintroduced into the environment and ultimately reaching people. To manufacture endosulfan at this time, a rigorous and scientific decision-making procedure is undertaken. Additionally, this strategy includes scientific research to enhance food production, food safety, and environmental security in addition to the other objectives listed above. Common use of synthetic pesticides inhibits the development of plant pathogen strains resistant to these chemicals, causing the reemergence of illnesses in the environment. Pesticides are being used more often by farmers, which is good.

TYPES OF PESTICIDES

The purpose of pesticide production is to remove pests and weeds. Farmers use pesticides to protect and enhance the quality of their crops. Pesticides are divided into five groups based on their structure: fungicides, herbicides, insecticides, rodenticides, and bactericides.

Fungicides

Fungicides are pesticides that protect plants against fungal diseases by killing or incapacitating fungi. From 1940 to 1970, fungicides with new chemical structures were introduced, including dithiocarbamates and phthalimides (organic chemicals), which were more active, less phytotoxic, and easier to prepare than the previously known inorganic fungicides. The fungicides mancozeb and chlorothalonil, known as protective fungicides, were also made during this period. Systemic methods as well as foliar and seed treatment were developed in this decade. Experiments were switched from in vitro to in vivo, and new products were tested in vivo.

Fungicides have evolved since the 1970s. During this period, more attention has been paid to the environment and human health, and in 1993, the European Union passed a law, according to which the produced fungicides had to be certified safe by the Union.

During 2005–2021, attempts were made to produce fungicides that reduce the negative effects of artificial factors such as waste, resistance, and environmental pollution. In this regard, fungicides that was made may be selective effective, biodegradable, and less environmentally toxic. Several classes of fungicides introduced in this period. For instance, multi-site inorganic salts, organic compounds with protectant action, and single-site systemic fungicides with curative activity. However, the latest discovery is succinate dehydrogenase inhibitor (SDHI) fungicides. SDHI are active components that used in fungicides to control certain fungi. SDHIs inhibit their growth by blocking the enzyme involved in cellular respiration. In the next few years, biopesticides will probably completely replace current pesticides, because they are completely harmless to the environment and humans.

Herbicides

Herbicides are used to eliminate weeds. A weed is a plant that interferes with the growth of crop plants and causes significant problems for farmers. In the early days when humans discovered agriculture (10,000 BC), weeds were removed by hand. Over time, in about 6000 BC, primary hand tools were used to remove weeds. In 1000 BC, oxen and horses that pulled harrows were employed for this purpose. Since 1920, with the development of the agricultural industry, cultivators have been used, and in 1930 and 1947, biological and chemical herbicides were developed, respectively.

A mode of action is a functional or anatomical change that occurs when a living organism is exposed to a substance. According to some articles, the mode of action is limited to the cellular level. A mechanism of action, on the other hand, explains such changes at the molecular level. Herbicides are divided into the following groups based on their mechanism of action:

Acetyl-coenzyme A carboxylase (ACCase) inhibitors

These herbicides inhibit the activity of the enzyme ACCase and are used to control weeds during the cultivation of broadleaf plants. Aryloxyphenoxy-propionates (FOP), cyclohexanedione (DIM), and phenyl pyrazoline (DEN) are a few examples of these herbicides.

Acetolactate synthase (ALS) inhibitors

ALS is an enzyme that is involved in the synthesis of branched-chain amino acids (leucine, isoleucine, and valine) and the suppression of this enzyme causes the plant to wither and eventually die.

Root growth inhibitors

These herbicides inhibit cell division, thus preventing root growth and development. The purpose of these herbicides is to block the polymerization of microtubules, which disrupts the process of chromosome segregation during mitotic division, resulting in cell death.

Plant growth regulators

This group includes hormonal herbicides (synthetic auxins), which are used to protect wheat and corn against broadleaf grasses. This group of herbicides targets the ATP protein pump. By activating this pump, the metabolism of nucleic acids and the integrity of the cell wall are compromised. These herbicides are also called indole acetic acid (IAA) imitators.

Photosynthesis/photosystem II (PSII) inhibitors

The target of these herbicides is the photosynthetic pathways (especially the photosystem II pathway). As a result of the loss of chlorophyll and other pigments such as carotenoids from the cell membrane, the production of $NADPH_2$ and ATP, which are essential for plant growth, is disrupted, eventually leading to plant death.

Shoot-growth inhibitors

These herbicides are applied to the soil prior to the emergence of the weeds. The functional site of these herbicides is on very long-chain fatty acids (VLCFA).

Aromatic amino acid inhibitors

This group of herbicides inhibits the synthesis of amino acids. This action is specific to glyphosate. These herbicides destroy any plant they come in contact with, although they are only licensed for corn, cotton, canola, and soybeans. Glyphosate inhibits the enzyme essential for the production of 5-enolpyruvylshikimate 3-phosphate (EPSP), resulting in a decrease in aromatic amino acids, i.e., tryptophan, tyrosine, and phenylalanine, and eventually cell death.

Glutamine synthetase inhibitors

These herbicides suppress the activity of glutamine synthetase, which is involved in the conversion of glutamate and ammonia to glutamine, thus increasing the level of ammonia and lowering pH on both sides of the cell membrane. This activity disrupts various cellular functions, especially the uncoupling of photophosphorylation. One example of these herbicides is glufosinate.

Pigment synthesis inhibitors

These herbicides destroy chlorophyll, which is essential for photosynthesis in plants. As a result, they cause the plant tissues to turn white, which eventually causes the weeds to die. Some examples include isoxazolidinone and isoxazole.

Protoporphyrinogen oxidase (PPO) inhibitors

These herbicides damage cell membranes by inhibiting the enzyme PPO. This enzyme is implicated in chlorophyll and heme biosynthesis. Lipid peroxidation occurs following the inhibition of this enzyme, which destroys cell membranes and leads to plant death.

Insecticides

Insecticides are substances used to kill insects and are widely used in medicine, agriculture, and industry. Before 1900, only a few chemicals were used as pesticides; however, in the early 1940s, as the Second World War raged on, the insecticide DDT was produced. These insecticides were game-changers. Since then, farmers have been using an ever-expanding list of insecticides to protect their crops.

Their use has undoubtedly resulted in substantial increases in agricultural output and a decrease in disease transmission, but it has also resulted in major challenges, including health concerns for pesticide consumers and the general public, the rise of pesticide resistance, and environmental issues. Insecticides mainly attack three targets in the nervous system: acetylcholinesterase, which is a vital enzyme in nerve impulse transmission (organophosphates and carbamates), voltage-gated sodium channels through the nerve membrane (pyrethroids and DDT), and the acetylcholine receptor (neonicotinoids).

Selective insecticides, such as juvenile hormone mimics (fenoxycarb and pyriproxyfen), ecdysone agonists, and chitin synthesis inhibitors (diflubenzuron), impede insect reproduction and growth by acting on insect-specific targets. Azadirachtin, a feeding deterrent and insect growth regulator, derived from the Indian neem tree, suppresses fecundity, pupation, molting, and adult emergence. Modern insecticides used in pest control include imidacloprid, acetamiprid, and thiamethoxam, which selectively act on the insect's nicotinic acetylcholine receptor (neonicotinoids).

Avermectins are microbial insecticides that target GABA-gated chloride channels, while the ryanodine receptor is targeted by diamide insecticides [38]. Spinosyns are a new generation of insecticides derived from actinomycetes, which exhibit high specificity and low mammalian toxicity while also being environmentally friendly.

Biopesticides

Biopesticides are pest-control agents made from microorganisms, plants, and animals. Synthetic pesticides continue to strongly protect agricultural products, but their long-term use has negative effects such as carcinogenicity, long-term stability in the environment, and other similar consequences. To address these issues, the production of new pesticides became necessary. Biopesticides were considered, since they are safe and eco-friendly.

Fungal spores were first used as a biopesticide in the late 1800s to control insect pests. In 1835, Agostine Bassi demonstrated that spores of the white muscardine fungus (Beauveria bassiana) could protect silkworms from diseases. This was one of the first documented cases of biopesticide use. Since then, the application of biopesticides has been continuous throughout modern agricultural history, although it has remained minimal compared to conventional crop protection. The primary distinction between biopesticides and synthetic pesticides is their mode of action. While most synthetic insecticides, if not all, are neurotoxic to pests, many biopesticides have other modes of action, such as anti-feeding, mating disturbance, desiccation, and suffocation. Biochemical, microbial, and plant-incorporated protectants (PIPs) are the three types of biopesticides identified by the US Environmental Protection Agency (EPA).

Microbial pesticides

Microorganisms such as viruses, bacteria, fungi, protozoa, and yeasts are used to produce biopesticides. Microbial pesticides are a more effective alternative to chemical insecticides. Their pathogenicity to the target pest varies depending on the species. The effect of microbial pathogens is induced by the invasion of the pathogen through the skin or the gut of the insect, which results in pathogen multiplication and the death of the host, i.e., the insect. Insecticidal toxins produced by pathogens are critical in their pathogenesis.

The majority of toxins produced by microbial pathogens are recognized as peptides, but their structure and toxicity vary considerably. Humans and other non-target species can benefit from the effectiveness and safety of the pesticides. These pathogens leave their meals with little or no residues. Microbial pesticides are ecologically safe and remove threats from other natural pests, resulting in increased biodiversity in managed ecosystems. As a result, microbial agents are highly specific against target pests, allowing beneficial insects to thrive in treated crops. This is the reason why, over the last three decades, microbial insecticides have been introduced as biological control agents.

Different pests can be controlled using different microorganisms as the AIs, such as bacteria, fungi, viruses, or protozoa; however, each active ingredient is relatively specific for its target pest. Some fungi, for example, control certain weeds, while others kill particular insects. As another example, Bacillus thuringiensis toxin may be more effective against Aedes aegypti, while the sphaericus strain is effective against a variety of insects, including Culex quinquefasciatus.

Biochemical pesticides

Biochemical pesticides are substances naturally occurring in the environment, which control pests via a non-toxic mechanism. The mechanism of action of biochemical pesticides is different from that of traditional pesticides.

Traditional pesticides directly affect and destroy their target, but biochemical pesticides act indirectly. For example, they disrupt the sexual function of their targets. Natural plant-derived products such as terpenoids, alkaloids, phenolics, and other secondary chemicals may be used as biopesticides.

Pesticidal properties have also been discovered in certain vegetable oils, such as canola oil.

PIPs

PIPs are pest-control substances made by plants and the genetic material needed for the plant to make the substance. The majority of pesticides in use are synthetic organic compounds with low molecular weights (LMWs, less than 500 g/mol or 0.5 kDa). These pesticides' environmental and analytical chemistry, as well as their potential effect on human and environmental health, have been thoroughly investigated. While LMW-synthetic pesticides have dominated the market, biopesticides are becoming an increasingly important part of the overall pesticide market.

The global market for all biopesticides is currently estimated to be worth $34 billion, accounting for approximately 6% of the total pesticide market. PIPs are biopesticides that are expressed directly in the tissue of genetically modified (GM) crops to protect them from pests such as insects and viruses. Because of the widespread use of insecticidal PIPs around the world, as well as the recent emergence of new PIPs targeting insect pests, this section focuses on PIPs developed against insects.

Insect pests consume PIPs when feeding on the transgenic crop tissue. Cry protein and double-stranded ribonucleic acid (dsRNA) PIPs in the insect gut affect insect development or cause insect mortality in various ways. Cry protein PIPs interact with specific receptors on epithelial cells in the insect midgut, insert into the cell membrane, and eventually form transmembrane pores, which lead to cell lysis and pest death. There are several types of Cry proteins, each with a specific structure and toxicity that is unique to certain insect orders.

Lepidoptera (e.g., the corn borer) is poisoned by Cry1 proteins, while Coleoptera (e.g., the corn rootworm) is poisoned by Cry3 proteins. Cry proteins were the first-generation insecticidal PIPs. The next-generation dsRNA PIPs have been recently approved. Following ingestion, the dsRNA PIPs in the pest insect are transported into a target cell. dsRNA is cleaved into small interfering RNA molecules (siRNA, 20 nucleotides) within the cell, which guides the insect's endogenous RNA interference (RNAi) machinery to degrade the complementary mRNA. The degradation of the targeted mRNA prevents it from being translated into basic pest insect proteins, causing sublethal effects (e.g., decreased growth) or pest mortality (Fig. 3). The first dsRNA PIP to be approved by the FDA targets the corn rootworm (Diabrotica virgifera virgifera) by interfering with the synthesis of the Snf7 protein, which is an important vacuolar sorting protein.

NANOTECHNOLOGY

From 1959 to 1960, developments in nanotechnology and nanoscience have been made to explore the synthesis and role of nano-particles prior to using them for different biomedical applications. Norio Taniguchi, a professor at Tokyo University of Science, made several successful attempts to synthesize nanometer-sized semiconductors in 1974. Later, it laid the foundation for research to perform experimentation on different types of nano-particles and nanocomposites. Nano-particles are found naturally in plants such as algae in the form of superoxide nano-particles and insects in the form of nanostructures.

Fig. 3. A naturally occurring bacterium that produces a protein toxic to certain types of insects. The gene inside the bacteria, which is responsible for producing that toxin, i.e., the Bt gene, can be transferred to crops, making them more resistant to the or responding insect

Nano-particles can be synthesized through physical, chemical, and biological methods. Nano-particles fabricated via physical, chemical, and biological methods are classified by their chemical composition, Nanoparticles in the form of metals such as Cu, Fe, Zinc, Au and in the form of oxides such as ZnO, CuO, AlO, in the form of semiconductors such as ZnS, CdS, ZnSe, carbon-based nano-particles in the form of graphene, diamond, fullerenes, in the form of silicates such as nano clays, in the form of nano-particles based on dendrite with long chains of fibers.

Different nano-particles are divided into different dimensions on the basis of their application in different biomaterials. The one-dimensional object possesses thin layers and fine surfaces. Second-dimensional possesses the wires with excellent flexibility and long tubes. Third-dimensional materials can be synthesized from metal oxides through physical and biological methods. These dimensions of the nano-particles have different applications in the fields of agriculture, medical, pharmaceuticals, pest management, and different industrial sectors.

Nano-pesticides

Engineered nanoparticles are now being used, or have the potential to be applied, as novel carriers for pesticide delivery. A range of formulation types has been suggested, including nano-emulsions, nano-encapsulations, nano-vesicles, nano-gels, nano-fibers, etc., which can be used to improve the efficacy of existing pesticide AIs or to enhance their environmental safety profiles, or both. The most common nanoparticles are illustrated in Fig. 4.

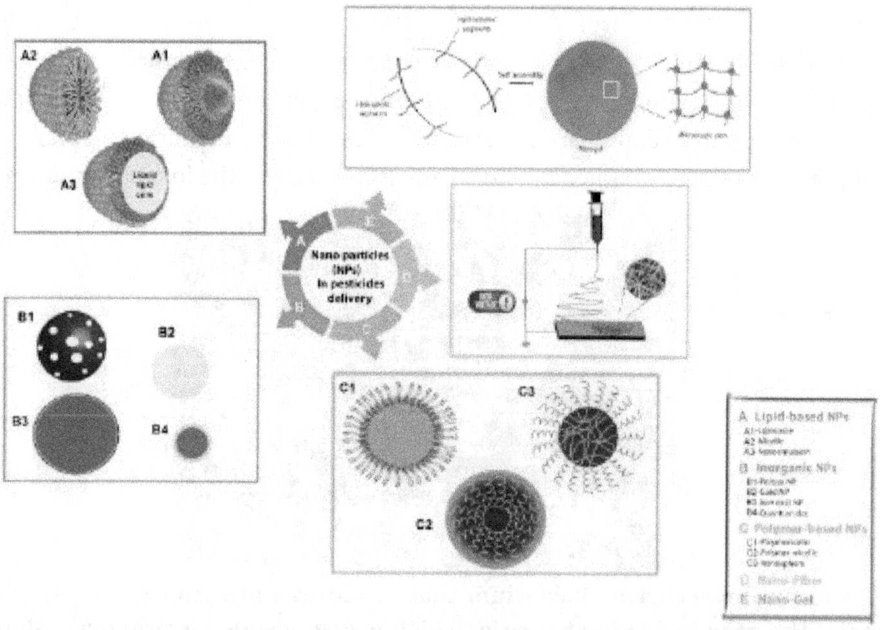

Fig. 4. Nanoparticles in pesticide delivery: A Lipid-based nanoparticles. B Inorganic nanoparticles. C Polymer-based nanoparticles. D Nanofibers. E Nano-gels

Nano-emulsion

Nano-emulsion is a biphasic dispersion system formed by mixing surfactants; AIs dissolved in the oil phase and the water phase. The nanometer-sized droplets (~ 20–200 nm) make this system kinetically stable and give it a transparent or translucent appearance. Nano-emulsion of pesticides is an oil-in-water (O/W) dispersion, which can dissolve poorly water-soluble pesticides into small oil droplets and greatly improve their bioavailability and efficacy.

In addition, nano-emulsion significantly reduces the use of organic solvents and surfactants compared to traditional pesticide formulations, and has attracted considerable attention from researchers in recent years [83, 84]. For example, Jiang et al. presented a nano-emulsion system that encompassed environmentally friendly surfactants, esterified vegetable oils, and 41% (w/w) herbicide glyphosate isopropylamine. This nano-emulsion had a small particle size (< 200 nm) and lower surface tension than the commercial cationic surfactant system (Roundup®), which would cause the droplets to be deposited uniformly on leaves with lower contact angle and increase the wetting, spreading, and permeation.

Visual injury assessment indicated that the nano-emulsion formulations showed significantly lower effective dose 50 (ED_{50}) than Roundup®, suggesting that they exhibited higher biological efficacy. In another study, researchers developed an abamectin (Abm)-loaded nano-emulsion containing 2% Abm, 5% castor oil polyoxyethylene, and 7.5% hydrocarbon solvent, which conformed to the quality indicators of the Food and Agriculture Organization (FAO). Compared with commercial oil/water emulsions (EW) and microemulsions (ME), the nano-emulsion exhibited various advantages such as a smaller dynamic contact angle on cabbage leaves, higher insecticidal activity, and lower cytotoxicity.

Although a nano-emulsion of pesticides with remarkable physical stability can be readily obtained, the most common approach is high-energy emulsification, and the oil phase and emulsifiers may still be toxic. To overcome these challenges, green nano-emulsions and low-energy methods (e.g., self-emulsification, phase transition, phase inversion temperature methods, etc.) are promising strategies. Researchers developed a green nano-emulsion of β–cypermethrin using renewable fatty acid methyl ester (methyl laurate) as the oil phase, "green surfactant" alkyl polyglycoside, and the non-ionic surfactant polyoxyethylene 3-lauryl ether ($C_{12}E_3$) as the mixed surfactant.

The nano-emulsion incorporating β–cypermethrin had a nearly monodisperse droplet size distribution (polydispersity index (PDI) < 0.2 Mw/Mn), which indicated excellent wetting and spreading properties on the hydrophobic surface compared with the commercial β–cypermethrin nano-emulsions, suggesting a biocompatible strategy for pesticide delivery.

Nano-encapsulation

Nano-encapsulation of pesticides is a delivery method in which AIs are encapsulated in various nanomaterials and released in a controlled way. Encapsulation by nanomaterials can protect AIs from premature degradation (e.g., photolysis, hydrolysis, biodegradation, etc.) and unnecessary losses by leaching and volatilization, and it is more effective in practical applications than traditional pesticide formulations. After the rational design, nanomaterials can also enable the encapsulation system to exhibit sustained release behavior to prolong the control period or give it stimulus-responsive release properties to achieve precision control.

Various nanomaterials including organic materials (e.g., synthetic and natural polymers, lipids, plant-derived nanoparticles, etc.), and inorganic materials (e.g., silica-, carbon-, calcium-, and clay-based nanoparticles, etc.) have been applied for nano-encapsulation of pesticides. AIs are often encapsulated into nanocarriers via incorporation, electrostatic complexation interactions, or covalent bonding, which will improve uptake, dispersibility, mobility, adhesion, and controlled or target release, thus leading to increased bioavailability and a longer effective lifetime of AIs.

Due to their excellent biocompatibility and biodegradability, as well as their abundant functional groups, various forms of polymer-based nano-encapsulated pesticides, including nano-capsules, nano-spheres, nano-micelles, nano-gels, and nano-fibers, have been extensively investigated. Lipid-based nano-encapsulation of pesticides not only can lead to effective encapsulation of hydrophobic or hydrophilic AIs without the need to use organic solvents but also can enhance their penetration and absorption in insect epidermis, resulting in an increased insecticidal activity. More recently, several published articles have demonstrated that the nano-encapsulation of the pesticides based on plant viral nanoparticles (red clover necrotic mosaic virus, tobacco mild green mosaic virus) could increase the mobility or distribution of Abm, a nematicide, within the soil compared to the free Abm, leading to enhanced crop protection against plant–parasitic nematodes in the soil.

Although nano-encapsulation of pesticides based on low-cost organic materials has been widely investigated, many challenges still need to be addressed, such as physicochemical instability and acid monomers formed by polymer degradation leading to the decomposition of AIs. Deteriorative reactions occur when polymers are subjected to heat, oxygen, and mechanical stress, and during the useful life of the materials when oxygen and sunlight are the most important degradative agencies. Moreover, degradation may be induced by high-energy radiation, ozone, atmospheric pollutants, mechanical stress, biological action, hydrolysis, and many other influences.

Porous inorganic materials such as silica and calcium carbonate are ideal carriers for pesticides because of their large surface area, tunable pore size, high loading capacity, and good biocompatibility. The controlled release properties and anti-photolysis ability of AIs are significantly enhanced after encapsulation in those materials. Some articles have reported that the release behavior of encapsulated AIs could be regulated through controlling the pore structures (e.g., 2D hexagonal channel, 3D open network structure, single large pore, etc.), thus further improving the efficacy in practice. In addition, some inorganic materials with specific shapes or structures have been reported for controlling the encapsulated AI loss.

For example, the Wu group reported a high-energy electron beam (HEEB)-modified natural nano-clay that could effectively encapsulate chlorpyrifos and increase its adhesion on crop leaves through the 3D network structure. In another study, Sharma et al. used graphene oxide decorated with copper selenide nanoparticles for the encapsulation of chlorpyrifos, which led to enhanced adhesion to cauliflower leaf due to the resistance of graphene oxide to aqueous runoff, the ability of carbon to bind to the organic surface, and the piercing effect by a sharp sheet on the plant leaf.

In general, nano-encapsulated pesticides have two typical release behaviors: (i) sustained (slow) release and (ii) stimuli-responsive release. For the sustained release of nano-encapsulated pesticides, AI concentrations remain within an effective control window against pests for a long period, avoiding frequent pesticide application. To achieve optimal control efficacy, developing stimuli-responsive nano-encapsulated pesticides that can precisely deliver AIs to the target is crucial. So far, various stimuli-responsive (e.g., pH-, enzyme-, temperature-, redox-, light-, ionic strength-, and humidity-responsive) nano-encapsulated pesticides have been reported for promoting the smart release of AIs in response to biotic (plant pathogens, insects, and weeds) or abiotic (temperature, sunlight, drought, soil texture, flooding, and salinity) stimuli, and have been discussed in detail in the previous reviews. These controlled release properties indicate the potential of nano-encapsulated pesticides for improving pesticide-use efficiency.

Vesicles such as liposomes and niosomes are versatile drug vehicles that are composed of various phospholipids or non-ionic surfactants. In general, they could carry both lipophilic and hydrophilic drugs anchored into their bilayer or encapsulated in their cavity, respectively. These vesicles have the potential to be used as carriers for pesticide delivery. Some studies have utilized vesicles for pesticide delivery. Researchers examined the impact of the nanoform of pyrifluquinazon on the destruction of the green peach aphid, Myzus persicae. The nanotype of pyrifluquinazon was constructed using liposomes and was subsequently covered with chitosan to make the unstable core materials more durable.

Moreover, to decrease the toxicity of paraquat, which is a widely used herbicide, photo-responsive and user-friendly vesicles loaded with paraquat were utilized. In this formulation, paraquat could only be released upon UV or sunlight irradiation. Researchers used liposomes to deliver the dsRNA of the P0 gene to the tick Rhipicephalus haemaphysaloides and evaluate the anti-tick characteristics of this system. Their findings revealed that P0 dsRNAs are virtually silenced. Some of the studies that focused on nucleic acid delivery using vesicles in the context of crop protection are presented in Table 1.

Table 1: Nano-vesicles employed for the delivery of nucleic acids (dsRNA) in the context of crop protection

Nano-vesicle composition	Target species
Lipofectamine	*Euschistus heros*
Lipofectamine	*Spodoptera frugiperda*
GenJe plus	*Blattella germanica*
Lipofectamine	*Drosophila suzukii*
Lipofectamine (Cellfectin,	*Drosophila melanogaster, Drosophila sechellia, Drosophila yakuba,*
DMRIE-C, Transfectin also used in one assay)	*Drosophila pseudoobscura*
Cellfectin	*Spodoptera frugiperda* Sf9 cell line
Effectene (Qiagen)	*Aedes aegypti*
Effectene (Qiagen)	*Aedes aegypti*

Nanoparticles

Nanoparticle-based pesticides discussed herein mainly include two types, nanoparticles directly used as pesticides and pesticide nanoparticles. Organic/inorganic nanoparticles are widely investigated as pesticides for the management of insect pests and bacterial and fungal diseases of plants.

For example, some articles have reported the insecticidal activity of silica nanoparticles, nickel nanoparticles, and titanium dioxide nanoparticles against weevils (Sitophilus oryzae), cattle tick (Rhipicephalus microplus, Hyalomma anatolicum anatolicum), mosquito (Anopheles subpictus, Culex quinquefasciatus, and Culex gelidu), and Haemaphysalis bispinosa. Many metalloids (e.g., gold, silver, copper, platinum, etc.) and metal oxides (e.g., silver oxide, zinc oxide, copper oxide, titanium dioxide, etc.), as well as nonmetal nanoparticles (e.g., sulfur, reduced graphene oxide, and chitosan, etc.), can be directly used as nano-pesticides owing to their antifungal or antibacterial activity.

In recent years, the use of bottom–up or top-down nanotechnology to formulate pesticide AIs as nanocrystalline particles has shown significant promise for improving the solubility and dispersion of poorly water-soluble pesticides [93]. After nanosizing pesticides, their solubility, permeability, dispersion, penetration, and bioactivity could be greatly improved. Compared with commercial emulsions in water, nanosuspensions showed increased retention on leaf surfaces, improved anti-photolysis, and enhanced bioavailability. As a result, the solid nanodispersion showed a greater than fourfold increase in fungicidal activity compared to the conventional water dispersible granule. Owing to their ease of large-scale production and wide applicability for almost all poorly water-soluble pesticides, the use of pesticide nanoparticles is emerging as the key innovative strategy for reducing pesticide consumption.

Nano-gels

Nano-gels have gained considerable interest from researchers as nanoscopic drug carriers and are mainly used for delivering bioactive mediators in a time-controlled or site-specific manner. Using van der Waals forces or covalent bonds, hydrophilic polymers are cross-linked and can absorb high volumes of water. Extensive studies have been conducted on nano-gels as the carriers of essential oils and pheromones.

The potential of nano-gel systems for the delivery of AIs in a constant, targetable, and controlled manner has been proven.

Pheromones are environmentally friendly and highly specific biological control agents which require to be protected against decomposition in ambient conditions and released slowly. As evidence has revealed, the evaporation of pheromones is significantly decreased in nano-gel carriers compared with the pure AIs. This extends their efficacy up to 33 weeks compared with that of the pure AIs, which is only 3 weeks.

The Institute of Pesticide Formulation Technology (Gurgaon) has recently presented a permethrin nano-gel formulation (a synthetic pyrethroid pesticide) for enduring the impregnation of this pesticide in clothes. This formulation protects individuals against mosquito bites while working in forest areas. Another research has developed chitosan nano-capsules containing lemongrass oil in the form of gel for long-lasting anti-mosquito effects.

Electrospun nano-fibers

Electrospinning has recently emerged as useful nanotechnology for the production of nanofiber materials that can be applied in different biomedical areas. Electrospinning technology can scale up and meet industrial manufacturing requirements. Recently, researchers have studied the nano-fibers generated by electrospinning for plant protection purposes. Although they are still in the early stages of agricultural application, electrospun nano-fibers have the benefit of evading release bursts, resulting in the facilitation of the field application of essential oils and pheromones.

Researchers presented an analogous nanofiber network made of cellulose and poly (lactic acid) nanocrystals. The thiamethoxam-loaded fibers were effective against whitefly over 9 days in a glasshouse experiment at 50% of the proposed dose for the pure AIs. Pesticides were delivered in a localized manner by coating seeds with electrospun cellulose diacetate (CDA) nano-fibers containing fluopyram or Abm, and it was found that nanofibrous seed coating could be potentially used as an alternate approach for controlling plant pathogens, such as fungi and nematodes. In another study, the authors produced the thiram/hydroxypropyl-β–cyclodextrin inclusion complex nanofiber (thiram/HPβCD-IC-NF) using electrospinning for efficient delivery of the pesticides

NANO-BIOPESTICIDES

Nano-biopesticides are attractive due to their tiny size, high surface-area-to-volume ratio, stability, enhanced efficacy, better solubility, mobility, and decreased toxicity. Nano-biopesticides are also suggested because of their low toxicity. Chemical pesticides are directly applied to plants can possess toxins released by air into the food chain and cause environmental issues. To control these issues, pesticides with formulations of nano-particles such as micelles and nanocomposites reduce the chances of both environmental and health issues. Similarly, clay-based nanotubes deliver pesticides to control pests.

Like nano-fertilizers, nano-biopesticides are contained in carriers that enable for regulated release of active ingredients to accomplish desired effects in a given environment. Stiffness and penetrability are two properties enhanced by adding nano-biopesticides to biopolymers. Crystallinity, thermal stability, solubility, and biodegradability are also enhanced.

When nanomaterials were applied to the soils, nano-biopesticides containing nanomaterials resulted in the growth of mutualistic microorganisms that promote the pants' activities. Sometimes, toxicity can be induced by coatings of silver-based nano-particles that could be reversed by biocompatible coatings, thus increasing the chances of seed germination in plants. Recently, nano-emulsions, nano-encapsulates, nanocontainers, and nano-cages have been reported as some nano-pesticide delivery techniques with different functionalities for plant protection.

Further research shows that cationic polymers may bind to polyanionic surfaces of bacteria, disrupt cell membranes, and kill pests. In agriculture, plants may be treated with biopesticides, such as nano-biopesticides, which can decrease microbial resistance, whereas chemicals applied directly to plants are unable to suppress a wide range of bacterial growth. Tertiary ammonium groups may be found in nano-particles as lengthy amino acid chains. Depending on their structure, these groups may attack various pests and illnesses, including bacteria. Because of their high activity in a wide variety of environmental and chemical conditions, polymers with quaternary ammonium groups in their chains are widely used. Many polymers with this characteristic have been found and researched throughout time. For example, amphiphilic copolymers, functionalized cationic polycarbonates, poly(amidoamines), polyethylenimine, poly(methyl methacrylates), amino celluloses and chlorinated cellulose acetates are now available.

Essential oils (EOs) are highly volatile secondary metabolites found in many higher plants and flowers and certain fruits and vegetables. In addition to their traditional uses in medicine and cosmetics, a new study indicates they represent a major natural source of ecologically friendly pesticides. Essential oils are often used to treat gardening pants to keep insects and bees out of the garden. Invertebrates become neurotoxic when their nervous systems are suppressed of GABA and acetylcholine esterase (ACE).

This 2007 research evaluated the anti-pest effects of plant extracts, essential oils, their purified components, and plant-based nano-formulations, as well as their modes of action. Temperature, light, and oxygen supply all have an impact on the EO's integrity. Researchers found that encapsulating flaxseed in gelatin and Arabic capsules may improve effectiveness by up to 84 percent, preventing the production of certain oxidants that stimulate the growth of some insects. Researchers summarized many techniques for encapsulating vegetable oils, including associated production processes, antimicrobial applications, insecticide/pesticide/pest repellant formulations, and antimicrobial applications. Purslane mustard, according to the manufacturer, is efficient against Sitophilus granaries and other grain-feeding insects. Using nanotechnology to control weevils and other pests improved their efficiency by about 7%, 21%, and 98%, according to prior research.

A variety of plants with nano-emulsions of ECs can be used to control the larval infections of different insects. These plants are Tagetes minuta, Ageratum conyzoides and Achillea fragrantissima used to control the growth of Callosobruchus maculatus.

These nano-emulsions can be applied to kill or inhibit the growth of eggs and larvae in the form of fumes with treatment ranges from 16.1–40.5 µL/L air and 4.5–243 µL/L air, respectively. The encapsulation of EOs can be performed through the use of nano-particles composed of liposomes and solid lipids. Encapsulation of EOs can be carried out through inverse gelation and oil emulsion. Encapsulation of EOs through liposomes is helpful against microbes by protecting the cell membrane from the effects of EOs.

However, different types of nano-particles are used for pest management. Microparticles are also used to stabilize the effects of EOs. The encapsulation of carvacrol was performed with a diameter of 0.5 µm with the cell wall of Saccharomyces cerevisiae to control the larvae of Rhipicephalus microplus and LC50 formulations of about 0.71 mg/mL. The cell wall of Saccharomyces cerevisiae appears more helpful in maintaining the low volatility of encapsulated carvacrol that maintains the acaricidal activity up to 60 hours.

In vitro nano-biopesticides bioassay

Nano-biopesticides can be tested against a specific pest in order to check their efficiency before applying them in different crops. Nano-biopesticides can be synthesized through the active pesticidal compounds and combinations of different nanomaterials such as zinc oxides, silver oxides, and aluminum oxides. The toxicity of nano-biopesticides can be measured through the minimum inhibitory concentration that employs the agar well diffusion method. Filter paper is usually coated with the outer surface of nano-biopesticides, and oral feeding directly applies to the target pest. The concentration of dead and alive pests can be precisely measured after 40 days of feeding.

Pupicidal activity

The pupicidal activity of nano-biopesticides is helpful in preventing the attack of pupae of different insect groups. It can be measured after applying the nano-biopesticides applied to the pupae of the target insects. This activity strongly measures the mortality rate after one day, which depends on the concentrations of nano-biopesticides. The work shows that nano-particles from the extraction of Leucas aspera show pupicidal activity in the concentration values (LC 50 and LC 90) for killing the larvae of mosquito vectors such as Aedes aegypti.

Larvicidal activity

The larvicidal activity of nano-biopesticides can be measured by the leaf disc method by introducing them into the leaf, and concentrations of larvae can be determined after 96 hours. Some plants show larvicidal activity, such as leaf extract of Ambrosia arborescens which is helpful for larvae of some pests in concentrations such as LC50 = 0.28 ppm and LC90 = 0.43 ppm.

Anti-feeding activity

The anti-feeding activity of nano-biopesticides can be measured by applying them to the leaf disc of pest food. The one-third-instar larva is introduced to the leaf, and the condemnations of leaf eaten by larvae can be measured every 24 hours. Anti-feeding activity has been observed about 92.4% in Helicoverpa armigera by applying silver-based nano-biopesticides.

In vivo nano-biopesticide treatment

Nano-biopesticides can be applied to plants in the right concentration in order to protect them from seasonal diseases. These concentrations (LC 50 and LC 90) aid in the identification of specific larvae, insets, and bee attacks. Nano-biopesticides are also applied in changing environments such as temperatures, humidity, and environmental stresses. In these conditions, nano-biopesticides are directly applied in the form of sprays to protect the plants from pest attacks. Therefore, the use of nano-biopesticides has become the most effective method in controlling the attack of animal vectors and disease-transmitting pests.

Activity against stored grain pests

Pests of stored grains are among the most difficult to manage in an agricultural system because of their large size. Recently, it has been shown that alumina, silica, SiO_2, zinc, and silver nano-particles have a substantial anti-pest effect against a range of pests when combined with other chemicals. According to the researchers, when sprayed on plants or crops, nano-emulsions have been shown to be efficient in deterring the attack of attack insects that cause harm to grains that have been stored for extended periods of time. The researchers discovered that nano-biopesticide emulsions effectively prevented the spread of the Tribolium castaneum fungus in one of the case studies they conducted.

In this study, oil/water emulsions of P. anisum essential oil (14 percent, v/v of the total coarse emulsion), ethanol (3 percent, v/v), and Tween 80 (3 percent, v/v), which together represented 20 percent (v/v) of the total coarse emulsion, as well as various components, were used. The emulsions were properly mixed and kept at 86°C for 1 hour. Separating the mixture from the water was accomplished by centrifuging it at 10,000 g for 15 minutes (which made up 80 percent of the mixture).

A technique known as photon correlation spectroscopy was used to evaluate a variety of characteristics such as conductivity, zeta potential of the emulsion, and polydispersity index. Contact and digestion techniques on the insects in the medium were used to assess the insecticidal activity of the P. anisum emulsified with nano-particles. During the testing against the insects in the medium, the researchers discovered that the P. anisum emulsion incorporating nano-particles proved to be very effective. Different concentrations of nano-emulsions can be prepared by acetone in 50 mL bottles containing the 20 g mash grains.

The positive and negative controls as emulsions of P. anisum. Different groups of 20 beetles were properly made in the sex ratio of 1:1 and then transferred to 50 mL bottles containing the treated and control grains, and each concentration was recorded as three replicates. The mortality rate was carefully recorded after treatments with 3-day intervals till 12 days, while on the other hand, the mortality rate with essential oil containing the beetles was recorded at intervals of 48, 72, and 96 hours. Those insets that were survived and were attached to grain were removed. F_1 progeny were recorded 60 days after insect infestation to avoid generation overlaps. Standard procedures were used to determine the LC 50 values that were used to control the attack of battles and to understand the morphological changes associated with the nano-emulsions.

USE OF NANOPESTICIDES FOR INSECT PEST MANAGEMENT IN TEA CROP

Nanopesticides refer to the utilization of nanotechnology for plant protection, an emerging research field, offers new methods to design active ingredients amid nanoscale dimensions, in addition to their formulation and delivery. This field comprises broad research aspects including the study of basic understanding of the interface between insects and nanoscale materials, formulation of AI into nano-emulsions and dispersions with accessible pesticides, preparation of new nano-pesticide using nanomaterials as active pesticide agents or nanocarriers for their delivery. Extensive research on nanopesticides is expected to deal with the restrictions of the accessible strategies used for pest control and endowed with nano-based novel formulations that enter into the target (pest), remain steady and active in the environment, without impact on non-target organisms, using cost-effective formulation.

Researchers defined nanoparticle (NP) as ultrafine particles comprising dimensions from 1 to 100 nm in size and have characteristics that are not shared by non-nano scale particles with a similar chemical composition. The key parameters for the NP for use as nanopesticides include some specific properties viz., chemical composition (organic, inorganic, metallic, polymeric, and carbon), shape (rods, spherical, irregular, and tube), the surface to volume ratio, size, and crystal phase (amorphous and crystalline) besides the toxicity.

NP-based pesticides tend to show several benefits in comparison to conventional pesticides viz. increased solubility in water and the stability of the formulation, its ability to eliminate toxic organic solvents and release the AI at a slower rate, its enhanced mobility, and its insecticidal activity.

Recent studies indicated that to formulate NP, varieties of materials were synthesized in various forms of chemical compositions including carbon, silicates, polymers, metal oxides, lipids, semiconductor quantum dots, ceramics, emulsions, proteins, and dendrimers.

NPs to Improve Pesticide Formulations to Control Insect Pest

Recent studies also showed different plant-synthesized NPs and their efficacy against a wide range of economically important insect pests, including mosquitoes, beetles, moths, hard ticks, lice, louse flies, etc. Hence, most of the formulation of pesticides are available in oil-in-water (O/W) emulsions or emulsifiable concentrates (ECs) and are sparingly water-soluble. Usually, EC formulations contain very expensive organic solvents that are more toxic and flammable or a mixture of surfactant emulsifiers. On the other hand, the O/W emulsions are based on a blend of non-ionic and polymeric surfactants, which requires high-energy input for emulsification. To address the disadvantages of these conventional pesticides, new formulations are being introduced based on micro and nanoemulsions.

For the management of a wide range of pests and diseases in different agroecosystems, several formulations based on micro-emulsion are available, including systemic fungicides and plant growth regulators. Micro-emulsions are steadier than the nano-emulsions (nano-emulsions require high energy input for scale-up for commercial production and are difficult for onsite production by the handlers) and advantageous over the conventional formulations (they have improved stability and tank mix compatibility, reduce toxicity to the handler and flammability, and have increased efficacy due to better uptake and penetration from surfactants having high solubilizing power).

Conversely, micro-emulsions showed various disadvantages viz., higher surfactant concentrations, with very low AI contents and the limited number of suitable surfactants. These limitations could be partly sorted out by different formulation concepts based on nano-suspension or nano-dispersion, where AI of nanocrystals creates nano-dispersions containing similar properties of solutions, and the development of such suitable release systems could enhance the pesticide's efficiency and performance, reducing adverse environmental effects.

In general, NPs are customized to carry a specific biomolecule to the organ, tissue, or even cell, and using a nanocarrier can easily penetrate the insect or plant cell and accurately deliver the products. For delivery and tracking purposes, various inorganic nanomaterials with distinctive chemical and physical properties, viz., silica, metal oxides, metals, semiconductors, and carbon-based materials, have been prepared. In addition to protecting crop plants from pests and diseases, nano-delivery vehicles can enhance plant growth, yield, seed vigor, and could even be used for manipulation of genes.

By creating new pores either through ion channels or by binding to a carrier protein, the NPs enter into the plant cells and can cause significant changes to these systems. Despite the remarkable capability of the plant cell wall during the entry of NPs, some unlikely impact of NP application has been raised, and studies are being continued focusing on phytotoxic effects of NPs and their influence on plant development.

Electron microscopic studies of the effect of NPs on insects and plants also revealed that they can penetrate the cell, mitochondria, or nucleus of insects and plant easily, and this suggests that NPs can be used as a pesticide for target delivery.

PESTICIDE DELIVERY SYSTEM USING NANOPARTICLES AS NANOCARRIERS

The concept "pesticide delivery system" (PDS) was developed based on the drug delivery concept used in medicine where NPs are used to deliver the therapeutics to the target organs. PDS is designed to make available of the AI to a specific target at specified durations and concentrations to achieve the projected biological efficacy and by minimizing the harmful effects on non-target organisms. For the optimized release of sufficient and necessary amounts of pesticides within a specific time, controlled delivery plays an important role. Possession of specific characteristics of NPs, viz., high effective loading capacity, larger surface area, fast mass transfer to the insect's body (target), and the ability for easy attachment of various pesticide molecules, encouraged the use of NPs as nanocarriers. Pesticide molecules after encapsulations, show more gradual release over time, hence it requires fewer applications; at the same time, due to degradation, NPs hinder the loss ineffectiveness of the pesticides. For loading of pesticide molecules on NPs, different concepts, viz., encapsulation, adsorption, entrapment inside the NP, and covalent attachment reconciled by various ligands, are adopted (Fig. 5).

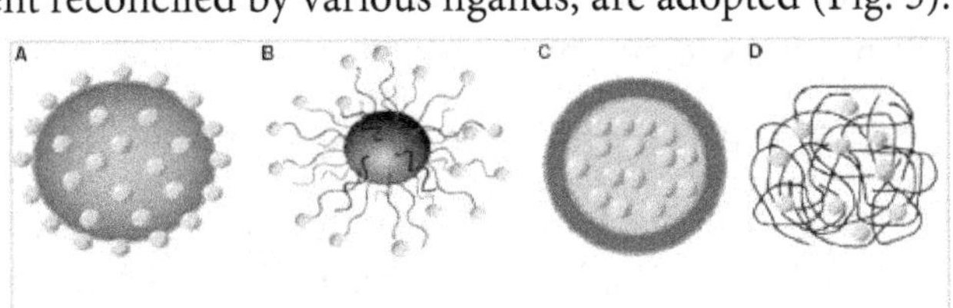

Fig. 5. Schematic representation of NPs for delivery of pesticides, (A) adsorption on NP, (B) attachment on NP by different linkers; (C) encapsulation inside polymeric hydrophobic or hydrophilic core (polymer micelles); and (D) entrapment inside polymeric nanoparticle

Based on the bonding of the AI to the material, degradation properties of the nanocarrier, and the environmental conditions, control and slow-release characteristics of the molecules could be attained. For the delivery of pesticides, nanocarriers like synthetic silica, polymer, titania, silver, alumina, and copper are mostly used. Utilization of nanoencapsulation technique indicates safer usage of pesticides and less exposure to the ecosystem. Among all the NPs available, the solid and mesoporous silica NPs have shown the most potential delivery agent of agrochemicals due to their structural flexibilities for the formation of NPs of different shapes and sizes and ability to form pores during loading the biomolecules.

POLYMER NANOPARTICLES AS NANOCARRIERS

Polymer NPs that include polymeric nanocapsules and nanospheres are being used as polymer nanocarriers due to their flexibility to design a complex pesticide delivery system with various modes of actions, biocompatibility, scalable preparation, and biodegradation. In polymer nanospheres, the AIs of pesticide molecules (hydrophobic molecules) are randomly distributed with a core-shell structure in a polymer matrix in nanocapsules (known as polymer micelles), and this acts as a reservoir for encapsulation. Polymer nanocapsules provide advantages over larger capsules due to their improved uptake, the stability of the spraying solution, the spraying surface, and the homogenous distribution, which provides better knockdown of pests.

As essential oils and secondary metabolites (phytochemicals) display stability problems (rapid evaporation and degradation of AI in presence of high temperature, moisture, and air) during the application, polymer nanocarriers can play a protective function and can increase their cost-effectiveness by maintaining the minimum effective dosage. For designing polymer formulations, different types of polymers, viz., polysaccharides (chitosan, starch, and alginates), polyesters [PEC (poly-e-caprolactone) and PEG (polyethylene glycol)], and biologically originated biodegradable materials (beeswax, cashew gum, lecithin, and corn oil) have been evaluated.

Among all of these polymers, PEG-based amphiphilic copolymers show better biodegradability, which is also easy for processing. PEG polymer nanoformulation-based pesticides showed significantly slower release of the AIs in comparison to chemical pesticides like thiamethoxam, imidacloprid, beta-cyfluthrin, and carbofuran and better efficacy than the chemical pesticides for control of nematodes.

INORGANIC NANOPARTICLES AS NANOCARRIERS

During the last two decades, for the formulation of pharmaceuticals, solid inorganic NPs are being used due to their various advantages viz., nanoemulsions, polymer NPs, better stability, higher loading, more controlled release, and low-cost production. Similarly, the solid inorganic NPs are used nowadays in agrochemical industries for the development of advanced pesticide delivery systems.

Among all the inorganic NPs, silica NPs are used mostly as nanocarriers for efficient delivery of biopesticides, insecticides, pheromones, and fungicides. Silicon enhances the plant tolerance to different biotic and abiotic stress, and silica NPs are thus a potential candidate for use in pesticides for controlling agricultural pests, including tea pests. Pesticidal formulations based on silica NPs showed slow release of chlorfenapyr and promising results as an insecticide.

Under field conditions, the insecticidal activity of chlorfenapyr associated with silica NPs was found to be two times higher than the chlorfenapyr without NPs. To control insect pests during grain storage and field conditions, the prospective use of nano-silica has been reported against many pests.

Using NPs Alone as Pesticides

NPs with insecticidal properties can be used not only as nanocarriers but also as biopesticides or active agents for pesticides. The amorphous nano silica acquired from the shell wall of phytoplankton, vegetables, rice hulls, and volcanic soil is the most promising example of NPs with insecticidal properties. Similar to the mode of action of diatom particles (used for stored grain protection), the silica NPs can disrupt the protective barrier of insects and finally lead to the death of the same. Earlier studies reported that the application of silica NPs in adult rice weevils, Sitophilus oryzae, caused 100% mortality.

Similarly, a wide range of insect pests in agricultural and animal ectoparasites was successfully controlled using surface-charged modified hydrophobic silica NPs. The application of silica NPs as a thin film on seeds showed a reduction of fungal growth; on the other hand, it boosted germination. World Health Organization (WHO) considered safe for humans in the exploitation of silica NPs as a nano-bio-pesticide. Hence, silica NPs could be used as a pesticide for the control of agricultural and household pests, parasites, and fungi.

Silver and Other Mineral-based Nanoparticles

Similar tothe other nanomaterials, the metal-based NPs can be combined with pesticide molecules, which increases the effectiveness of the pesticide formulation and reduces the application dose for management of tea pests. These NPs can be synthesized solely using chemical compounds or involving living organisms. NPs based on silver, nanostructured alumina, aluminum oxide, zinc oxide, and titanium oxide have shown insecticidal properties. The larvae of case-bearing cloth moths, Tinea pellionella, showed 100% mortality while treated with ethanol-based nanosilver colloid (<20 ppm).

The wettable dust formulation of nanostructured alumina showed more than 95% mortality of S. oryzae and R. dominica adults infesting wheat within 3 days of exposure. Moreover, by combustion of aluminum nitrate and glycine, the prepared nanostructured alumina at 60 ppm dose showed more than 94% mortality for S. oryzae adults when applied as dust on wheat.

Researchers obtained similar results for control of S. oryzae and R. dominica in wheat when novel nanostructured alumina was prepared based on chemical solution method and applied as dust. Researchers while studying the mode of action of the nanostructured alumina dust, reported that, through capillarity mechanism, the epicuticular lipids absorb the dust, which leads to dehydration of the insect and finally causes death.

However, the effectiveness of the metal-based NPs depends on their physical properties, (viz., morphology, size, and surface area of the particle), biotic factors (viz., target species, exposure interval, and dose), and abiotic factors (relative humidity).

ADVANTAGES OF THE USE OF NANO-PESTICIDES OVER CONVENTIONAL PESTICIDES

To develop novel eco-friendly formulations for pest control, nanotechnology offers a tool for target-specific nanopesticides. Usually, target-specific nanopesticides could improve the effectiveness of pesticides and reduce pollution and undesirable residues in tea. They have slow release and protection performance as nanopesticides are prepared using thermo-sensitive, light-sensitive, humidity-sensitive, soil pH-sensitive, and enzyme-sensitive high polymer materials to deliver the pesticides.

On plant surfaces, these formulations improve the adhesion of droplets, conferring improvement of dispersion and bioactivity of AI of target pesticide molecules. For these reasons, compared to conventional pesticides, nanopesticides show better efficacy for the management of tea pests. Nanopestides offer competent and eco-friendly advantages because of their small sizes, their efficacy when spraying in the field improvement of droplet adhesion on the plant surface, their wettability, as well as their rapid absorption by the target. It reduces the quantity of chemical usage, enhances plant protection, and reduces the toxic residues; hence nanopesticides enable the sustainable production of tea.

TYPES OF NANOPESTICIDES

Nanopesticides play a major role in minimizing the footprints left by synthetic pesticides in agroecosystems. Encapsulated nanopesticides (Table 1) and several other nano-forms of silica, silver, iron, copper, and carbon (Table 2) are already in the practice of insect pest management.

Potential of Nano-Pesticides for Agricultural Pest Control

Metabolites (terpenoids, flavonoids, and alkaloids) are plant-derived products (PDP) that consist of a complex mixture of compounds that act in diverse ways on insects, such as repellents, antifeedants, insect growth regulators, oviposition deterrents, and in terms of toxicity, and their mechanisms of actions are diverse (156). Most of the findings on the botanicals or PDP are restricted to the laboratory and have not been validated at field conditions.

Indeed, marginal and small farmers in developing countries have introduced and regularly using the locally available plant products (water extract, crude oils, or seed cakes, etc) and commercial formulations [emulsifiable concentrations, wettable granules, solvent-based products, active ingredient (AI)]. These PDPs are established as effective commercial formulations. The presence of AI and allelochemicals in essential oils (EO) shows better efficacy than the PDP extracts.

As a result of the quick "knockdown effect" on insect life stages and the accessibility of ready-made formulations, farmers use chemical pesticides in agriculture, but in most cases, these pesticides do not reach the target insect and are lost in the air, leached out in soil and water.

Besides, overuse or misuse of doses without following proper recommended mixtures and repeated applications of chemicals pesticides lead to several issues viz., development of insect resistance and undesirable residues in commodities and can harm humans and non-target organisms as well as the environment. Exposure to sunlight would lead to quick degradation of PDP and leave less persistence in the environment, have less of an impact on non-target organisms, and shows less residue.

In comparison to the conventional chemical pesticides, the better properties in PDP nanotechnology are endorsed to controlled systems discharging small-sized molecules at the place of action, decreasing the residual problem, maximizing the action of the AI, increasing the target specificity, and increasing both the physicochemical stability and effectiveness of AI. Nanopesticides are recognized to have less of an impact on human health and the environment and display less toxicity to non-target organisms as well as animals.

TYPES OF STIMULI-RESPONSIVE SYSTEMS

Systems responsive to pH and temperature

Nanocarrier release systems that respond to pH changes are based on the presence of ionizable functional groups in the carrier structure, such as pyrimidines, carboxylates, amines, phosphates, and sulfonates. At certain pH values, depending on the pKa of the carrier or the chemical molecule, the functional groups present a charge, which may be positive or negative, resulting in electrostatic interactions. The sizes of carriers sensitive to pH can be modified by the swelling or shrinking of the macromolecules in response to changes in the pH of the surrounding medium (Fig. 6). There are essentially two types of pH-sensitive materials: those having acidic groups in the structure (–COOH and $-SO_3H$) and those having basic groups ($-NH_2$), which swell at basic and acidic pH, respectively. Hence, the response can be the same, but the applied stimulus differs.

Biopolymers are known for their properties such as biocompatibility, biodegradability, and nontoxicity, which are desirable for the development of nanocarrier systems used in eco-friendly formulations. Many of these biopolymers are ideal for the formation of nanocarriers that respond to pH stimulus.

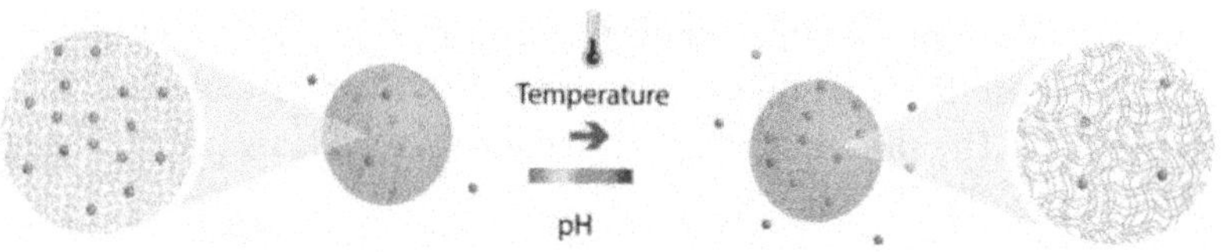

Fig. 6. The release of substances promoted by modification on the carrier structure in response to pH and temperature changes

The literature reports the development of several pH-responsive nanocarriers. For this type of system, a specific chemical molecule, such as a pesticide, can be encapsulated due to chemical (ionic or covalent) interactions, or be trapped in a polymeric mesh within the system during nanoparticle formation. A classic example of a pH-responsive nanocarrier system is based on alginate and chitosan. The alginate polymer chain contains carboxylic groups (pKa 4.8), so at basic pH the polymer is deprotonated and possesses a negative charge.

Chitosan is rich in amine groups (pKa 6.3), so at acidic pH it is protonated, with positively charged functional groups. The ionic gelation method for the formation of nanocarriers composed of alginate and chitosan exploits the interaction between these two biopolymers. Solutions of the biopolymers are generally prepared at around pH 4.7, so that the carboxyl groups are attracted to the amine groups, forming cross-links between the polymers. The formation of this system is highly dependent on pH, which makes it pH-responsive.

For example, pH changes, and consequently the protonation/deprotonation of the polymers, weaken or strengthen the electrostatic interactions, resulting in increase or decrease of the release of a chemical compound. When a system is formed only by a single biopolymer, variation of the pH can result in repulsion between the polymer chains, due to the similar charges, leading to greater or lesser opening of the polymer mesh (swelling), hence affecting the release of the active agent. These mechanisms summarize the way that most of the pH-responsive release systems function.

Other non-polymeric nanocarrier systems may be coated with pH-responsive biopolymers in order to prevent a release burst. Another strategy is functionalization of the surface of the nanoparticles by other compounds, linked by means of electrostatic interactions, in order to generate a response mechanism (gatekeeper) modulated by pH changes. For example, silica nanoparticles present very rapid initial release of the incorporated ingredients, which could limit their applications.

However, their surfaces are rich in hydroxyl groups, which permits the functionalization or coating of these carriers. Generally, pH-responsive nanocarriers are also thermosensitive. Temperature-responsive nanocarrier systems have great potential for applications in agriculture, since the temperature constantly changes in the environment, directly affecting the emergence and development of agricultural pests such as insects, weeds, and fungi.

The release of active agents from temperature-responsive nanocarrier systems occurs due to temperature-dependent changes in the polymer physicochemical properties, allowing the release of the active ingredient and determination of the release kinetics. These thermosensitive polymers are water-soluble at low temperatures but separate from the solution when the temperature exceeds the phase transition temperature, also known as the lower critical solution temperature (LCST).

The main focus of studies of thermosensitive materials as drug delivery agents in controlled release systems is the phase transition temperature. High-viscosity water-soluble carriers are widely employed to achieve the controlled release of active ingredients. These include PEO-PPO-PEO block copolymers that are marketed as Poloxamers® and Tetronics®.

Poloxamers® consist of a central PPO (polyoxypropylene) molecule, surrounded by two hydrophilic polyoxyethylene (PEO) chains. The structure of poloxamines (Tetronics®) is slightly different, with tetrafunctional block copolymers composed of four PEO-PPO blocks linked by a central ethylenediamine bridge. These polymers exhibit sol–gel phase transition below or near physiological body temperature. When dissolved in a liquid at low concentrations, they form monomolecular micelles, with increases in concentration leading to the formation of multimolecular aggregates.

REDOX-RESPONSIVE SYSTEMS

In redox-responsive nanoparticle systems, the stimuli for release of the active compounds are endogenous factors present in the organism, such as antioxidant molecules, including glutathione. Redox-responsive systems first emerged in the medical area, where they are used for the release of chemotherapeutic agents. Tumor cells have a higher concentration of reducing agents, compared to normal cells, and these molecules provide the trigger for the release.

The release mechanism is based on the cleavage of a disulfide bond in the presence of the reducing agent. Therefore, it is necessary to functionalize the nanocarrier with thiol groups that are linked to the gatekeeper by means of disulfide bonds (Fig. 7). This is extremely important, since it is these links that will be cleaved

COMMERCIAL LIMITATIONS AND CHALLENGES

Many publications have shown that nano-based pesticides have great potential in the field of pest management, offering advantages for the environment and human health, compared to the use of conventional pesticides.

Although nanotechnology can bring many advantages to the agricultural sector, there are few nanotechnology-based products available on the market. One of the factors influencing this low level of commercialization is that the great majority of studies are undertaken in universities and research institutes, or by small companies (spin offs and start-ups) set up for specific agricultural purposes. At the same time, large companies possess many patents, with the number increasing annually.

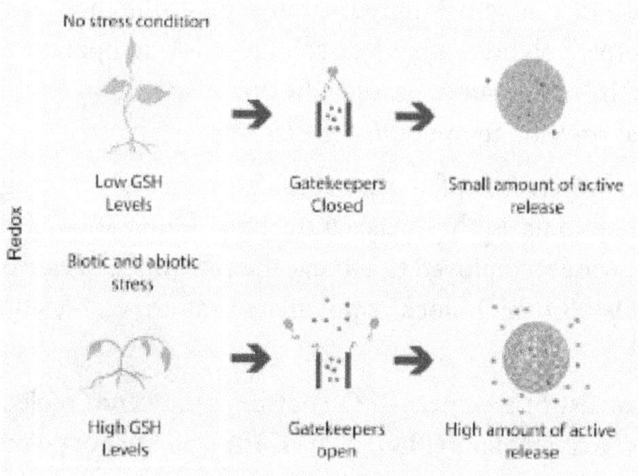

Fig. 7 Release of active substances in response to redox stimuli. In normal environmental conditions the gatekeepers are linked, and the release is low, while under stress condition, the level of antioxidant increases and promote the cleavage of disulfide bonds, which enhance the release of loaded substances

New nano-based products for the agricultural sector do not reach the market, because the large companies are accumulating patents and waiting for opportunities for future exploitation, following the development of promising commercial products. Another issue concern understanding the risks related to the use of nano-based pesticides, which are complex and poorly understood. It is already known that the transport, bioaccumulation, and rate of degradation of nano-based pesticides are different, compared to conventional pesticides, but further work is needed to elucidate these issues. There are limited data available concerning the effects of nano-based pesticides on crop health, soil biodiversity, nontarget organisms, and human health.

Furthermore, the development of nano-based pesticides suffers from a lack of effective strategies to understand their mechanisms of action. It is very important to evaluate the fate of nano-based pesticides in the environment, in order to elucidate the interactions of these materials with terrestrial and aquatic organisms.

In addition to the problems mentioned above, it should be highlighted that economic issues remain a major hindrance for the commercialization of nanopesticides. The initial costs involved in developing nanopesticides are high, with positive financial returns only being possible if large quantities of these products were to be used in crop protection, which is not the currently the case. In addition, the lack of regulation is a major impediment to the expansion of nanotechnology in agriculture.

Another factor that limits the commercialization of nano-based pesticides is the high cost associated with registering a new active compound. The most significant challenges that need to be overcome in order to increase the commercialization of nano-based pesticides can be listed as follows:

a. development of standardized methods for reliable assessment of risk–benefit;
b. advances in knowledge of the interactions between nanomaterials and plants or pathogens;
c. development of strategies to track nanopesticides in the environment and evaluate their impacts on food security and human health;
d. lack of a clear and standardized definition of nanomaterials;
e. development and implementation of legislation at an international level to ensure the safe development and application of stimuli-responsive nanopesticides; and
f. lack of a global network for effective communication among the public and private organizations engaged in the development of nano-based products.

It is clear that further in-depth research will be required in order to address these issues, prior to full commercialization of nano-based products. It is especially important that there should be interaction among all the potential stakeholders, including universities, research institutes, companies, governments, NGOs, and consumers.

BIO-EFFICACY

With the limited bioavailability of nanocarriers, both nanocarriers and AI need to be endowed with a requisite amount for attaining the preferred level of tea pest control. Hence, the durability and bioavailability of these components in PDP-based nanopesticides should be established. Commercial products, specifically the powder formulations confirmed better UV stability than the encapsulated nanoformulations. In practice, liquid formulations are preferred by the farmers.

Hence, escalating persistence and stability in solid formulations is a challenge to the manufacturers. For this, researchers suggested encapsulation of hydrophilic molecules in polymer-based nanoparticles and assessment of risks to non-target organisms with increasing uptake of AI to enhance the bio-efficacy of the same.

Toxicity to Plants

Azadirachtins in nanoformulation cause genotoxic and cytotoxic effects in plants, but during plant development, the toxicity is reduced to an undetectable level under sunlight. In another case, researchers discussed different types of nanomaterials (nanotubes, Al, ZnO, Zn, and multi-walled carbon) that did not show any adverse effect on five different vegetables and ryegrass during seed germination and termination of root elongation. For corn plants treated with oleic acid and nanocapsules separately, researchers observed a negative effect on the physiological parameters against oleic acid, while there was no phytotoxic effect for nanocapsules.

Toxicity to Non-target Organisms in Agro-Ecosystem

Neem oil nanoformulation showed less toxicity to parasitoids of whitefly Bemisia tabaci and a parasitic wasp, Encarsia formosa when neem oil was added on polymeric nanocarriers (beta-cyclodextrin and poly-å-caprolactone) (80). The toxicity of plant-derived metallic nanoparticles on microorganisms and earthworms depends on the concentration, particle size, number, and distribution of nanopesticides bound to the active ingredient.

It is still debatable whether the release of metallic nanoparticles by microorganisms into the terrestrial environment is due to various effects of plant-derived metallic nanoparticles on the metalcore and capped phytochemicals. While researchers confirmed that nanocapsules containing poly-ε–caprolactone treated with neem oil did not affect the soil microbes up to 300 days after exposure. Moreover, soils treated with nanoformulations enhanced the biomass of earthworms even up to 3 months of exposure to the formulation.

Engineered nanomaterials can encourage surface coating and surface charge alongside other characteristics of physicochemical properties in the aquatic environment. As the nanopesticides may not be eco-friendly, it is necessary to understand the environmental fate and the role of surface capped plant-derived molecules in nano-specific processes.

Toxicity to Animals

Nanomaterials are safe at optimum size, but as they approach nanosizes, they often become toxic, especially the non-biodegradable materials (mostly metals). By accumulating and stimulating the immune system, they can cause toxicity in mammals, plants, and soils. Moreover, they can alter the activity of detoxifying enzymes. Aluminosilicate used in nanotubes can stick to leaf surfaces and can increase the stability of the product, even as the ingredients are also able to stick to the body surface of the insect pest, especially in the hairs, and can in due course enter the animal's body and affects physical functions. Apart from direct intake, dermal contact or inhalation can also lead to nanoparticles entering the human body. In pest control operators, chronic exposure and sub-acute causes inhalation toxicity, which is very common, and sometimes this leads to major health risks.

FUTURE PERSPECTIVES

In a future perspective, the development of nanoparticles that can act as fertilizers and pesticides is necessary. Nanotechnology research in agriculture is still at its initial stages, but is developing quickly. The application of nanostructures as agrochemicals for plant growth and protection has to be investigated due to their novel physicochemical characteristics. With clear nano-bio interactions, transport, and fate in the plant and environment, these nanoparticles have sustained release and resistance for plant protection management.

The majority of recently supported projects and future study calls seem to be concentrating on producing safer, environmentally friendly nanomaterials for efficient responses. It is expected that the controlled release of pesticides by nanotechnologies will become essential in the near future to promote cropland fertility and protection with the least effect on the environment and human health. Future studies should be conducted using natural sources rather than standard media and assess the interaction of nano-agrochemicals with other contaminants.

Soon, using nanostructured catalysts will be possible, which will enhance the effectiveness of pesticides and herbicides, allowing the use of more moderate doses. Nanotechnology will also guard the environment obliquely by promoting the use of filters or catalysts, which can decrease the contamination in soil and water. Future opportunities for target delivery of crucial nutrients, pesticides, and genetic materials utilizing nanomaterials will provide new prospects for the agricultural revolution.

However, several features of nanoparticles are supposedly dangerous to human health. Therefore, these substances must be more cautiously studied. For example, for dry nano-products, the relative ease with which nanopowders are suspendable in the air might lead to potential human inhalation and subsequent health hazards. On the other hand, suspensions of nanoparticles in water used as a soil drench or foliar sprays pose risks such as the transformation of the particles into ions or aggregation into submicrometer- or micrometer-sized particles. Experts should evaluate the influences of nanoparticles on human health and the environment, and improve the approaches used to assess and control any hazards they may pose.

Researchers also need to find renewable methods to fabricate nanoscale materials to be employed in agriculture. Moreover, general approaches should be developed in the pioneer institutes or nations to test laboratory-scale nano-products in actual farm settings for more comprehensive technology validation. Finally, educative and developmental initiatives should be established via study plans to bring the users and experts together in addressing possible customer interests. Substantial crop loss occurs annually due to several notorious pests, including insects, mites, and nematodes.

9

BIOPESTICIDES FOR MANAGEMENT OF ARTHROPOD PESTS

Managing arthropod pests with pesticides is a common practice in crop production. Managing arthropod resistance to pesticides is an important part of integrated pest management (IPM.) Under natural circumstances, plants, insects, mites, natural enemies and beneficial or harmful microorganisms continuously co-evolve and adapt to the changes in their environment. When there is a higher selection pressure, such as the indiscriminate use of pesticides, mutations in arthropods enable them to resist the pesticide and increase their chances of survival.

Since there are fewer pesticides for organic crop production, there is a higher chance of their repeated use and increased risk of resistance development. Arthropods, such as insects and mites, develop resistance to chemical pesticides through genetic, metabolic, or behavioral changes resulting in reduced penetration of toxin, increased sequestration or excretion, reduced binding to the target site, altered target site that prevents binding of the toxin, or reduced exposure to the toxin through modified behavior. Through one or more of these mechanisms, arthropods can also develop resistance to biopesticides. When the active ingredient of a biopesticide is a toxic molecule and acts more like a chemical compound, arthropods are more likely to develop resistance even though it is of biological origin.

When the mode of action is due to infection by a microorganism, rather than a toxin, arthropods are less likely to develop resistance. Pest management in organic greenhouses is conceptually the same as pest management elsewhere; however, due to the limitations and restrictions in organic agriculture, the techniques and strategies used tend to focus on augmenting natural processes and hence focus on tactics that produce longer term effects. For example, the land for organic agriculture either must have never been used for crop production or there must have been a minimum of 3 (in some places 5) years without the use of any synthetic soil amendments, fertilizers, pesticides, genetically modified crops, or sewage sludge. The selection of available materials from the preceding list is far fewer than those for nonorganic Integrated Pest Management (IPM) programs, so decisions must be made more carefully.

Finally, meticulous records and periodic on-site inspections ensure that regulations are adhered to and organic certification maintained. In preparation for cultivation, cultural control methods such as soil solarization can be used in warm climates. Crop-free periods, while not an exclusive organic control method, can be used in extremely hot or cold climates to reduce pest levels. In addition to organic pesticides, semiochemicals (behavior-modifying compounds) and biopesticides (microbial control agents including entomopathogenic fungi, bacteria and viruses, and botanical biopesticides) are increasing in importance in organic production strategies. There are a limited number of organic pesticides available and most have short persistence on plants and in the environment.

A primary management tactic is the manipulation of natural enemies: releases of commercially reared predators and parasitoids and occasional augmentation by providing alternative host plants and push-pull methods of directing pests away from economic crops and into areas where they are more vulnerable to management. Greenhouse structures and production practices vary in different regions of the world. In northern Europe and North America, closed ventilated, heated structures are used because of the long periods of cold weather; crops are either planted directly in the ground or in growing media or in soil-free culture systems; and biological control is highly developed.

In Israel, commercial greenhouses are covered with insect netting or plastic sheeting; are not heated because the winters are generally mild by comparison to northern climes and hot in the summers; crops are grown either in the ground or in soil-free culture systems on the ground or suspended, and biological control is used almost exclusively on some crops.

In South America, commercial greenhouses are simple structures covered with plastic materials without nets or heating/cooling systems and biological control is generally underdeveloped but is growing in many countries and currently implemented in countries such as Brazil and Colombia.

Vegetable crops are grown mainly in ground, but soil-free production methods are increasing for conventional crops with the use of hydroponic systems or Nutrient Film Technique in arid or semiarid regions due to the limited amount of water available or to poor local soils. In South America, hydroponics are mainly used in Brazil.

GREENHOUSE STRUCTURE

Greenhouses are simply covered structures in which plants are grown. Initially and for hundreds of years, greenhouses were created to produce flowers and vegetables out of season. In cooler temperate climes, greenhouses are still primarily used for that purpose but also to produce higher quality produce. In southern temperate, subtropical and tropical climes, the primary purpose of greenhouses is for plant protection and to prevent immigration of pests. These structures range from having very simple passive heating to computer-controlled wall and roof openings, illumination, irrigation, and heating.

Commercial greenhouses are constructed from a variety of building and cladding materials and in sizes from walk-in tunnels enclosing relatively few square meters to large structures of a hectare or more. Solid walls, glass or plastic, are effective for retaining heat and are best used in cooler temperate climates.

An unfortunate development in northern climes with short day lengths was the development and utilization of high-pressure sodium lamps. While these lights extended the growing season, they also allowed greenhouses to become warm incubators for a variety of arthropod pests all year round.

At an International Organization for Biological Control meeting in Finland in 2005, there was a special submeeting for the purpose of discussing how to manage Bemisia tabaci (Gennadius), a warm-weather, Mediterranean pest.

Structures covered with netting or screening allow better ventilation and cooling than those covered with glass or plastic. Since these screens are used in warm temperate through tropical climates, mesh hole size and shape are important factors to consider to maintain ventilation. Table 1 provides a range of the screen hole sizes and the insects excluded from an academic and commercial perspective.

Insect exclusion screens are not all interchangeable. It has been demonstrated that screens with higher ultraviolet-absorbing properties are more effective barriers for a variety of pest insects (aphids, leafhoppers, thrips, whiteflies) and can reduce virus spread within a greenhouse. Since a significant proportion of insect vision is in the UV range, research has proceeded apace on the effects of UV deprivation on the ability of predators and parasitoids to function.

Table 1. Insect exclusion screening hole-size recommendations

Insect pest	mm
Aphids	0.34-0.341
	0.266 × 0.818
	0.266 × 0.818
Whiteflies	0.46-0.462
	0.266 × 0.818
	0.230 × 0.900
Dipteran leafminers	0.61-0.64
	0.266 × 0.818
	0.530 × 0.530
Thrips	0.19-0.192
	0.150 × 0.150
	0.135 × 0.135

PHYTOSANITATION

For many plant and animal pest species, greenhouses provide a plethora of food coupled with environmental conditions that favor their population development over time. In organic production, the easiest, least expensive, and most effective practice is to prevent pests from entering the greenhouse is phytosanitation. Internationally, regulations provide most national plant protection organizations with guidance and the measures available to restrict pest movement across borders and in trade. At the greenhouse level, phytosanitation practices include measures aimed at reducing the chance of pest establishment. These guidelines include:

1. Double access doors (interlocking) to limit the entrance of arthropods (and improve thermal insulation).
2. Footbaths filled with disinfectants at points of access to the greenhouse, to reduce pathogen movement.
3. Washing stations with soap and water for sanitization of hands and agricultural implements before movement from one area to another and thus reducing secondary spread of pathogens.
4. Removal of weeds within and immediately surrounding the greenhouse that can be a source of both arthropod pests and plant pathogens.
5. Removal and disposal of infested and infected plant material that may harbor pests and pathogens.

Disposal methods include burial, burning (although this is not permitted in some countries), or compositing at sufficient temperatures to kill pests and pathogens, and 6) avoiding conditions that will cause water condensation on the crop (e.g., overhead irrigation); since the environmental conditions in the greenhouse (high temperatures and relative humidity) favor the development of many diseases. These factors should be taken into account when choosing the irrigation systems and systems to control the greenhouse environment to prevent the dew point being reached.

CULTURAL CONTROL METHODS

Soil or Growing Medium

The advantage of soil-free growing medium is that the incidence of soil-borne pests and pathogens is virtually zero since the growing medium is discarded after the growing season. There are several types of growing media, and they may be inert such as sand, perlite, and rock wool or natural products such as pine bark, coconut coir, or peat, all of which would require organic certification. Soil-free agriculture, as opposed to traditional cultivation, requires a constant supply of water and nutrients and hence a more sophisticated irrigation and feeding system. It should be noted that in the United Kingdom, all organic agriculture must be in soil systems.

Amendments for sanitizing soils in organic greenhouses are few in number and often focus on nematode management or plant pathogens (Bailey and Lazarovits 2003).

Vermicompost, produced through the breakdown of organic matter by earthworms and microorganisms can be used as an amendment to strengthen the plant and reduce pest levels. Vermicompost is high in microbial diversity, contains considerable amount of humic substances, and increases host plant resistance to pests, including aphids, mealybugs, and spider mites. Organic crops that are grown in fertile soils with high organic matter and active soil biology generally exhibit a lower abundance of several pests due to a lower nitrogen content.

Solarization

Solarization, is accomplished by covering soils with clear plastic to allow solar radiation to heat the soil, and can be an efficient and effective nonchemical treatment for pests and pathogen control in warm climates. Moist heat is more effective than with dry soils, and the length of effective solarization varies according to location, soil type, and organic material. Hagimori et al. demonstrated that soil solarization killed eggs, larvae, and pupae of a variety of pests of organic cruciferous vegetables. Athalia rosae ruficornis Jakovlev (Hymenoptera: Tenthredinidae), the striped flea beetle (Phyllotreta striolata (Fabricius)) (Coleoptera: Chrysomelidae) and various Lepidoptera all have soil-borne stages, and these were eradicated by covering wet soil with transparent plastic film for 1 mo during the summer months. They recorded temperatures of 70, 50, and 45°C at soil depths of 0, 10 and 20 cm, respectively.

Researchers investigated the effect of solarization on plant parasitic nematodes, Meloidogyne spp. in tomato, Solanum lycopersicum L. greenhouses. Solarization of moist soil with incorporated organic supplement (e.g., compost or mulch) covered with clear plastic for 52 or 50 d reduced the number of infested plants to 6 and 5%, respectively, compared with 25 and 20% in untreated control. These researchers noted that the solarized greenhouses produced fruit with higher yield and quality compared with untreated and conventional soil treatments, when all other factors remained constant.

Mulching

Mulching is usually used for open fields where insects are not physically restricted and plants are subjected to unfiltered sunlight, but it also has utility in organic greenhouse production. Hagimori et al. (2012) demonstrated that 1.5-m wide plastic sheets circumscribing a greenhouse was effective in preventing weed infestation, and therefore, flea beetles associated with those weeds were prevented from entering the greenhouse.

BIOLOGICAL CONTROL METHODS

One characteristic shared by all major pests of greenhouse production is increasing pesticide resistance due to repeated application of the few pesticide groups available for use in organic greenhouses. This problem is exacerbated for organic growers because there are even fewer registered chemicals. The result of the declining availability of effective pesticides and the expanding market for organic produce is driving the search for and development of effective biological control agents (BCAs). Today, commercial BCAs are commercially available around the world for the control for the major pest species of these groups. In the simplest definition, biological control uses one or more organism against another. For the purposes of this article, the host organism is an arthropod and the BCA another arthropod or pathogen (bacteria, fungus, nematode, or protozoan).

Parasitoids, predators, and entomopathogens are widely used in biological control programs around the world. On occasion there may be negative interactions between BCAs in the form of intraguild predation (IGP), thus reducing the efficacy of pest control. When IGP occurs, one or more predator species may feed on another predator in addition to the host, or a hyperparasitoid (secondary parasitoid) may reduce primary parasitoid populations, e.g., with aphid parasitoids. In addition to IGP, side effects of insecticides and fungicides on BCAs and natural enemies should be taken in account when planning and managing IPM programs, and these interactions must be carefully monitored.

SECONDARY PLANTS

Noncrop plants can serve functions in a greenhouse in the form of insectary plants and banker plants. Insectary plants serve as a source of food, mating sites, and shelter for BCAs. These plants are frequently rich in extra floral nectaries and/or pollen and can serve to sustain natural enemies in periods when host populations are low. Banker plants are open rearing units for producing large numbers of parasitoids and were first developed in the United Kingdom for control of the melon/cotton aphid, Aphis gossypii Glover in cucumber. Banker plants consisted of barley (Hordeum vulgare L.) infested with bird cherry-oat aphid, Rhopalosiphum padi (L.) to which Aphidius colemani Viereck were released (Fig. 1). Large numbers of parasitoids were produced on the banker plants, which then migrated to the crop and led to improved biological control of the pest aphid. Banker plants using barley infested with R. padi or wheat (Triticum sp.) infested with the English grain aphid, Sitobion avenae (F.) have also been used in sweet pepper, Capsicum annuum L., for control of green peach aphid, Myzus persicae (Sulzer) and the foxglove aphid, Aulacorthum solani (Kaltenbach), by A. colemani and Aphidius ervi Haliday or Aphelinus abdominalis (Dalman), respectively.

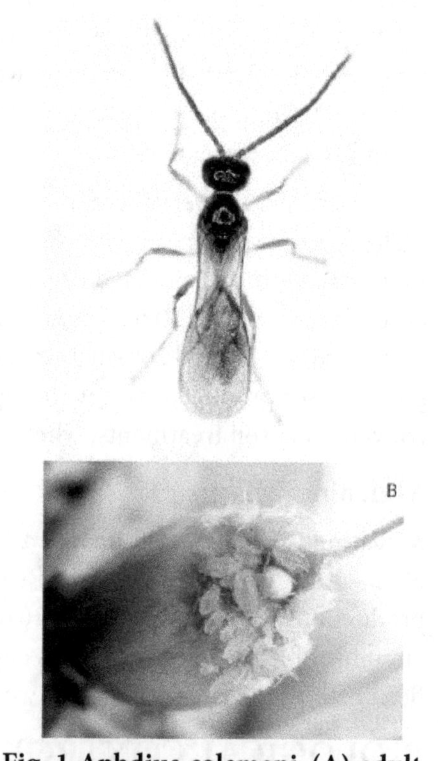

Fig. 1 Aphdius colemani. (A) adult, (B) Mummy in Myzus persicae

In recent years, biological control of aphids on sweet pepper has been threatened by the increasing incidence of hyperparasitoids, including Asaphes spp. High levels of hyperparasitism of the primary parasitoid 'mummies' has occurred in crops both with and without banker plants, but because the banker plants can be a source of large numbers of hyperparasitoids early in the season, the banker plant system is not widely used in the United Kingdom.

Hyperparasitism of primary aphid parasitoids has also been recorded in other protected crops where banker plants are not used, including strawberry (Fragaria x ananassa Duchesne) and ornamentals and on outdoor organic lettuce, and research is needed to overcome this problem for sustainable biological control of aphids within IPM programs. A number of different plants have been used to establish generalist predators and parasitoids. Sweet alyssum (Lobularia maritime L.) Desvaux and coriander (Coriandrum sativum L.) have been used in greenhouse pepper for the generalist predatory hoverfly, Sphaerophoria rueppellii (Wiedemann).

Sweet alyssum has a high sucrose: hextose ratio which is very attractive to hoverflies; however, more eggs were laid when hoverflies had access to coriander but larvae took longer to complete their development on coriander than on sweet alyssum. For the anthocorid pirate bug, Orius sauteri (Poppius), a generalist predator of aphids, thrips, and whiteflies, the presence of black-eyed Susan Rudbeckia hirta L. in tomato greenhouses resulted in significantly higher densities of the predator than with other greenhouse insectary plants. When prey populations were low, the pirate bug was sustained on the black-eyed Susan and migrated to the tomato plants.

Flowering alyssum plants have also shown promise for supporting populations of Orius laevigatus (Fieber) for thrips control in strawberry and could have potential for use in other crops (Fig. 2).

Supplementary Food

The presence of noncrop plants can disrupt planting and harvesting in the limited space in greenhouses; therefore, methods for providing supplementary food to BCAs are being developed. Different types of food and methods of delivery have been shown to enhance BCA populations: flour moth eggs or brine shrimp cysts and pollen have received much attention in recent years. Artificial diets that are liquid based are somewhat problematic because of the sticky residue that remains on plants. Powdered factitious diets have been successful with the generalist predatory bug, Geocoris varius (Uhler).

Fig. 2. Orius laevigatus adult.

The provisioning of supplemental food is not without drawbacks. Some pests, such as thrips, also feed on pollen and their populations may be enhanced and IGP may occur. The development of a nutritious, easily applied factitious diets would be a significant development for the enhancement of BCAs in organic greenhouses.

Semiochemicals

Any chemical substance or mixture produced by an organism that conveys information to another is broadly termed semiochemical; thus, attractants such as allomones, kairomones, pheromone, and repellents are all forms of semiochemicals. Pheromone can be used for monitoring, mass trapping, or mating disruption. Additionally, secondary plant volatiles can be used as: plant traps, in push-pull strategies, or synthesized and used as attractants or lures. Today, there are a number of commercial lures that are used in association with various types of traps. Since the lures are naturally produced, they can be very sensitive and species specific and are compatible with many management practices.

PESTICIDES AND BIOPESTICIDES

The list of amendments, fertilizers, and pesticides qualified for use in organic agriculture varies among countries or regions. In North America, the Organic Materials Review Institute has developed a list for Canada and the United States. In Europe, there are efforts underway to produce harmonized organic agriculture regulations among all of the countries by the end of 2017.

In South America, there is no single accreditation; however, growers exporting to the European Union must be certified by Naturland, BCS Oeko-Garantie, or the Institute fur Marktoekologie (IMO), and to North American it must be certified by The Organic Crop Improvement Association (OCIA) or Farm Verified Organic (FVO). In general, to qualify for use in organic crops, the active ingredient of a pesticide or biopesticides must be a derivative of a living organism (microbe, fermentation product thereof, or plant-based) and the formulation must have no added synthetic ingredients.

Additionally, nonsynthesized minerals (e.g., copper oxide, calcium hydroxide) are generally acceptable. Application does not require the use of special equipment; however, it is necessary to consider that these materials are often less persistent and degrade more quickly within the greenhouse environment than synthetic chemical insecticides. Therefore, these materials are commonly applied during environmental conditions favorable for their efficacy.

For example, entomopathogenic fungi are best sprayed during periods of high relative humidity rather than periods of high UV radiation or at very high temperatures. According to organic farming principles, biopesticides may be used as a curative tool for pest control.

One group of active substances are secondary metabolites produced by plants, commonly known as botanical insecticides, such as antifeedants. For example, neem oil extracted from the seeds of Azadirachta indica Adrien-Henri de Jussieu is currently the most widely used botanical compound to control insects in organic agriculture (including aphids on vegetable crops) in those countries where neem products are approved. The insecticide spinosad, based on secondary metabolites synthetized by soil actinomycetes, is approved for use in organic agriculture to control thrips, as well as Lepidoptera and Diptera. Spinosad is a mixture of two macrolide compounds from the soil bacterium, Saccharopolyspora spinosa Mertz & Yao, through fermentation.

In European countries, approved spinosad products are obtained directly from microbial production, whereas in the United States and Switzerland, purified toxin is used. Other types of biopesticides are based on potassium salts of fatty acids (insecticidal soaps) which can provide some control of aphids and other soft-bodied arthropods by contact action, although thorough plant coverage is needed for optimal effect.

MANAGEMENT SCENARIOS FOR KEY ARTHROPOD PESTS

Most crops can be infested by a wide range of pests that can cause severe economic damage. Mites have historically been a difficult pest group to control, and resistance to conventional acaricides has been reported beginning in the late 1950s. In the second half of the 20th century, outbreaks of aphids, whiteflies, mealybugs, and thrips have been the cause of major yield losses in greenhouse crops as a result of their direct feeding damage and/or the indirect effects of the viruses they can transmit to crops.

Acari

Tetranychidae

One of the most important mite pests of greenhouse crops is the two-spotted spider mite, Tetranychus urticae Koch because it infests many plant families and is cosmopolitan in distribution. This mite reproduces quickly and produces profuse webbing on the leaves and stems that can protect all life stages from pesticides and most predators. Resistance to conventional acaricides was reported more than 60 years ago, and there are no effective organic pesticides. The primary means of spider mite management is through the use of specialist predators such as Phytoseiulus persimilis Athias-Henriot, a fast-moving, voracious long-legged predator that can defeat the webbing. However, P. persimilis is efficacious only at high tetranychid densities and leaves when populations are low.

In northern Europe, P. persimilis can coexist with other predatory mites such as Amblyseius andersoni (Chant), and attempts have been made to release multiple acarine predators including, Neoseiulus californicus (McGregor) and Amblyseius swirskii Athias-Henriot.

In contrast, in Chile, P. persimilis did not improve spider mite control in strawberries (Albendin et al. 2015). Recently there has been some success in managing spider mites with vermicompost in bush beans, Phaseolus vulgaris L., and eggplant, Solanum melongena L.

Tarsonemidae

Two important tarsonemid pests are the broad mite, Polyphagotarsonemus latus (Banks) (Fig. 3) and the cyclamen mite Phytonemus pallidus (Banks). The former tends to occur more on vegetable crops and the latter on ornamental crops although the cyclamen mite is also known as the strawberry mite due to the extensive damage it can cause on strawberry.

Both of these mites only feed on tender young leaves and meristematic tissue thereby causing plant deformity and stunting. In addition, both of these mites can be quickly and efficiently transported throughout a greenhouse by phoresy (by attaching themselves to another organism), especially on whiteflies. Fortunately, a number of phytoseiid predators are also of small size and can therefore physically reach and prey on tarsonemid mites to provide effective control.

Predators include Neoseiulus cucumeris (Oudemans) (Fig. 4), N. californicus, N. fallacis (Garman), and A. swirskii. A. swirskii is a mite that originated from the east Mediterranean region and is considered a generalist predator feeding on many different small insects such as thrips larvae and mites as well as on pollen by piercing and draining their contents using its mouthparts. N. cucumeris is an important predator that feeds on small insects such as thrips larvae, mites, and pollen in northern Europe.

Eriophyidae

An additional group of important plant parasitic mites is the russet or gall mites. Although this group contains a large number of species, the only important greenhouse eriophyid is the tomato russet mite, Aculops lycopersici (Massee) which is host limited. This very small mite completes a generation in 1 wk and large populations rapidly develop. The pest can be treated with predatory mites including, N. cucumeris, A. fallacis, and A. swirskii when 'hot spots' appear.

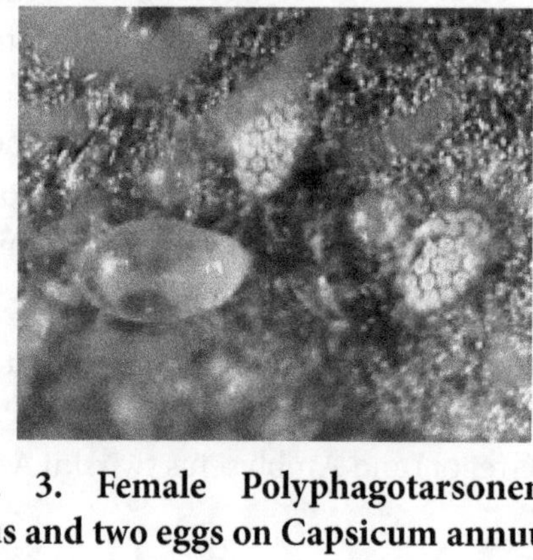

Fig. 3. Female Polyphagotarsonemus latus and two eggs on Capsicum annuum.

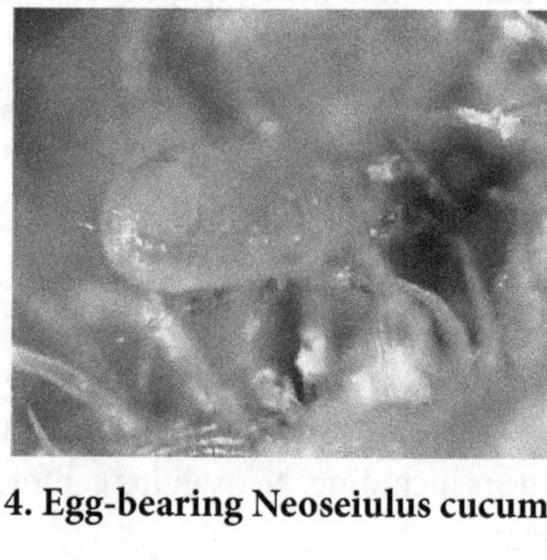

Fig. 4. Egg-bearing Neoseiulus cucumeris

However, the sustained biological control on tomato has not been achieved due to the toxic nature and variety of glandular trichomes. Recent research has shown that Amblydromalus limonicus has good potential for control of tomato russet mite.

Thysanoptera

Thripidae

Thrips species represent a diverse collection of lifestyles including fungivorous, phytophagous, and predatory. The plant-feeding species are believed to have evolved from fungus-feeding thrips that to this day represent ca. 50% of the known species. Greenhouse pest species are in the family Thripidae. The life cycle includes eggs that are laid in the plant tissue that hatch to produce two larval instars, a prepupal stage, and one pupal stage that develops into the adult. In the main pestiferous species, development and reproductive rate are strongly influenced by temperature with a generation time span from egg to adult of about 9–17 d under warm conditions (25 to 30°C) depending on host crop.

The main pest species are polyvoltine, moving between available host plants as long as there are suitable climate conditions. Only a few species of thrips are responsible for significant losses in yield and quality on protected crops. All of the following thrips species discussed here are major pests causing direct damage as well as being competent vectors of tospoviruses (e.g., tomato spotted wilt virus and impatiens necrotic spot virus) except for Scirtothrips dorsalis Hood.

Frankliniella occidentalis (Pergande), known as the western flower thrips, originated from the western United States and has spread worldwide. F. occidentalis has a very broad host-plant range, is strongly thigmotactic, and is often found deep in flowers and under the calyx of solanaceous plants. Thrips tabaci Lindeman, onion thrips, is of Mediterranean origin and is now cosmopolitan in distribution. Although onion and leek are preferred host plants, onion thrips can infest a wide range of vegetable, ornamental, and weed species and their host range includes plants from about 25 families.

In addition to F. occidentalis, T. tabaci is considered a major pest of greenhouse crops such as cucumber and sweet pepper. T. tabaci causes a typical silvering feeding pattern on foliage and flowers. Thrips palmi Karny, melon thrips, originated in southeast Asia and has spread throughout the southern hemisphere. Melon thrips is a polyphagous species and is considered an important pest on ca. 20 plant families but is best known for its preference to Cucurbitaceae and Solanaceae species such as sweet pepper, cucumber, and eggplant in greenhouses. Damage usually consists of a silvery-bronzed appearance on leaves and fruit. T. palmi are also known to feed on stems and flowers (among the petals and developing ovary). Shoots of damaged plants become stunted and fruit scarred and deformed.

S. dorsalis, chilli thrips, also originated in southern Asia and has spread into the northern hemisphere as far north as Israel and the southern United States. Predatory phytoseiid adults and nymphs can feed only on the first or second stage thrips immatures but cannot attack adults due to their small size (see Mites above). After release, the predatory mites aggregate on high-density patches of F. occidentalis immatures where they lay eggs.

Predatory mites have the ability to suppress thrips populations, especially if applied early in the season as a preventative measure. However, once F. occidentalis have established and developed large populations curative releases are not successful. The minute pirate bugs, Orius spp., are generalist anthocorid predators that consume small arthropods as well as pollen.

As with other Hemiptera, their mouthparts are of the piercing sucking type, with the rostrum inserted into the prey's body to remove fluids. Species of Orius are commonly associated with flowers of herbaceous plants, where they feed extensively on thrips, and therefore, a number of Orius species are considered important BCAs in greenhouses. Both adults and nymphs feed on F. occidentalis adults and larvae, with adults consuming five to 20 thrips a day. Due to their variable diet, Orius spp. are used preventively in the greenhouse on flowering crops (e.g., on sweet pepper) where they feed on pollen before the thrips begin to infest and then persist in the greenhouse long after thrips populations have been reduced.

Orius spp. are known to feed on other predators, including phytoseiid mites, when prey populations are low. There are five commercially available Orius species the most widely used being O. laevigatus (used in Europe, North Africa, Asia, and Israel) (Fig. 2), O. insidiosus (Say) (used in Europe exclusive of the United Kingdom, and mainly North America and Latin America), and O. majusculus (Reuter) (used mainly in Europe). The combined use of two predators, O. laevigatus and A. swirskii, against thrips in recent years has increased steadily in Israel and this trend; is representative of the situation worldwide

Hemiptera

Aphididae

Aphids are small, soft-bodied insects that constitute one of the major pests in protected crops. In greenhouse environments, aphids are parthenogenetic and viviparous, building up large populations in a short period. These hemimetabolous insects develop multigenerational colonies on plants that confer an advantage to natural enemies such as some parasitoids and predators because searching time is reduced. There are two types of adult morphs: wingless (apterous) and winged (alate), depending on several factors mainly temperature, crowding, and nutritional conditions of the host plant. Only a few polyphagous and cosmopolitan aphids are the primary species responsible for losses in yield and quality on protected vegetable crops.

M. persicae is common on greenhouse pepper, eggplant, and lettuce, and A. gossypii frequently infests cucurbits, pepper, and eggplant. Macrosiphum euphorbiae Thomas, the potato aphid, is a common pest in protected vegetable crops such as tomato, pepper, and other solanaceous plants and on lettuce. These three aphid species can act as vectors for persistent and nonpersistent viruses in addition to causing direct plant damage. A. solani causes direct damage by feeding on pepper, due to their highly toxic saliva that may produce deformation and yellow discoloration of the leaves resulting in complete plant defoliation and deformed fruits. In recent decades, Nasonovia ribisnigri (Mosley) has become a major pest of protected lettuce, Lactuca sativa L. crops. N. ribisnigri deposits their offspring near the young terminal growth of lettuce and colonizes the inner leaves inside the developing heads, causing leaf distortion and vigor reduction in seedlings.

At harvest, N. ribisnigri presents a significant cosmetic problem because the presence of living aphids on the leaves reduces the market value or makes lettuce unmarketable (Palumbo 2000). Pepper and cabbage, Brassica oleracea L., plants grown in soil amended with 20% vermicompost have shown lower numbers of M. persicae adults and nymphs and lower number of alates alighting on them.

Similar results were observed for M. persicae infesting plants fertilized with animal manure; however, the opposite effects were observed for the specialist cabbage aphid, Brevicoryne brassicae (L.). Physical barriers such as nets can be used around vents in solid structure greenhouses and act as exclusion barriers between the plant and the pest and in consequence are an effective tool to limit the entrance and dispersal of aphids on protected vegetable crops. In addition, the use of UV-blocking plastic materials, reflective and colored mulches applied on the soil can reduce numbers of aphids landing on and colonizing or spreading viruses within greenhouses. Approved practices for biological control of aphids in organic greenhouses are based on routine preventive releases of parasitoids, followed by predators or entomopathogens. Aphid parasitoids usually consume all or most of the aphid body, killing its host at the final nymphal stage and then pupae.

At this stage, the dead aphid takes on a beige, brown, or black color and swollen appearance, commonly known as a mummy, which is easily recognizable within an aphid colony (Fig. 1B). Free-living adult wasps emerge from the pupae and begin a new generation of parasitoids. The adult wasps require nectar, pollen, or honeydew as food. If banker plants are used to augment numbers of parasitoids, and to act as a preventative measure, timing of their introduction into the greenhouse is crucial to achieve successful control of aphid populations. In a recent study (in Argentina), the introduction of a banker plant system was compared with inoculative releases of A. colemani in Eruca sativa Miller (arugula) and C. annum greenhouses. The banker plant strategy was the most efficient method to control M. persicae on arugula; however, no significant differences between these two strategies were found on pepper crops. Researchers introduced aphid-infested barley as banker plants in pepper greenhouses and concluded that they were also effective in attracting natural populations of aphidophagous hoverflies to control aphids.

The main parasitoids of aphids are A. colemani (attacks A. gossypii and M. persicae), A. ervi, and A. abdominalis (attack M. euphorbiae and A. solani). Other common wasp species parasitizing aphids in protected crops are Lysiphlebus testaceipes (Cresson) and Praon volucre (Haliday). Mummies of P. volucre are easily recognizable from other parasitoids species because the dead aphid remains attached on a white silk cocoon, while the parasitoid is developing within it.

Parasitoid mixes containing up to six species of parasitoids (A. colemani, Aphidius matricariae Haliday, A. ervi, A. abdominalis, P. vulucre, and Ephedrus cerasicola Stary) are now commercially available, and these mixes can be very useful in a wide range of crops where several aphid species can occur and thus need a range of parasitoid species for control. It should be noted that the efficacy of any parasitoid species could vary depending on a number of factors.

For example, and focusing only on A. colemani, researchers showed that stressors on different trophic levels could affect fitness and efficacy of the parasitoid. Biotic factors such as crop morphology, crop quality, production of volatile organic compounds, pest density, presence of other natural enemies, including hyperparasitoids and abiotic factors such as temperature, humidity, and lighting in the greenhouse must be considered. Recently, even the type of grain used in banker plants for A. colemani has been shown to affect the parasitoid efficacy. Ladybirds, lacewings, hoverflies, and the gall midge, Aphidoletes aphidimyza (Rondani), represent the main aphid predators.

Predatory insects differ from parasitoids in that they catch and feed on their prey, and their juvenile stages require several to many prey individuals to attain maturity. Several species of ladybirds are commonly found naturally in protected crops, such as Adalia bipunctata (L.), Coccinella septempunctata (L.), and Hippodamia spp. Lacewings are generalist predators feeding on various pests affecting greenhouse vegetables crops; however, they have a preference for aphids as was observed for Crysoperla carnea (Stephens) larvae when aphids and thrips were provided simultaneously (Shrestha and Enkegaard 2014). Hoverfly larvae also feed on aphids by puncturing the cuticle and imbibing their content, whereas adults are diurnal flower visitors feeding on nectar and pollen. Entomopathogenic fungi attack the integument of a variety of sucking insects, including aphids.

Species of fungi-infecting aphids belong mainly to the Entomophthorales and Hypocreales. Entomophthorales are obligate pathogens and have a narrow host range of arthropods causing death of aphids by the invasion of different tissues and organs. Hypocreales are generalist pathogens that also cause fungal diseases on aphids, but they combine a parasitic phase with a saprophytic phase colonizing the body after the death, which is mediated by toxin production. Within the Entomophthorales, Pandora neoaphidis (Remaudière & Hennebert) Humber is the most abundant naturally occurring aphid-pathogenic fungus and frequently cause epizootics in several protected crops in Argentina.

In addition, other species of Entomophthorales such as Zoophthora radicans (Brefeld) Batko, Conidiobolus obscurus (Hall & Dunn) Remaudière & Keller, and Entomophtora planchoniana Cornu were found killing M. persicae and N. ribisnigri in lettuce greenhouses in Argentina. Several species of Hypocreales such as Beauveria bassiana (Balsamo-Crivelli) Vuillemin, Paecylomices sp., Metarhizium anisopliae (Metchnikoff) Sorokin and Lecanicillium lecanii R. Zare & W. Gams were observed infecting aphids.

As was mentioned previously, aphids produce large quantities of sugar-rich honeydew, which is very attractive for ants. Ants actively milk/stroke, herd/move, and protect honeydew-producing aphids in a symbiotic relationship. These behaviors can be highly disruptive to biological control because ants fend off predators and parasitoid as well as translocate the aphids to other parts of the plant.

However, ants can be managed in organic crops with baits, pheromone disruptors, or insecticides. Boron compounds are often used as an abrasive but can be incorporated into baits. Borate (disodium octaborate tetrahydrate), boric acid, or sodium borate in sugar water can disrupt activity for tens of meters from the trap through trophallaxis. Additionally, ants can be managed by disruption of trail pheromone. Finally, pyrethrum, diatomaceous earth, and silicon dioxide could be applied for management.

Aleyrodidae

The lifecycle of a whitefly is comprised of an egg, four nymphal stages, and winged adults. The first instar, termed 'crawler', has well-developed legs and walks short distance in search of suitable settling sites. The crawler is the only active nymphal stage and once settled then molts. The next three sessile nymphal instars feed on the phloem tissue available within reach of their stylets throughout their development. Unlike the gradual metamorphosis of other Hemiptera, whitefly metamorphosis is intermediate. During the latter part of the fourth instar, colloquially referred to as the 'red-eyed nymph' and technically a puparium, the four wings and antennae of the adults develop inside the pupal case. Two whiteflies specie have been the focus of many studies due to the extensive damage they cause to protected crops. These are the sweetpotato whitefly, B. tabaci and the greenhouse whitefly, Trialeurodes vaporariorum Westwood. These two species can be easily distinguished as the nymphs of T. vaporariorum have long hair-like projections while B. tabaci does not, and B. tabaci adults hold their wings in a roof-like manner over their body whereas T. vaporariorum holds its wings more horizontally at rest.

A. swirskii is a predatory phytoseiid mite. The first published study on its biology showed that A. swirskii can be maintained on a diet of B. tabaci eggs and young larvae; however, it was not proven successful as a BCA for B. tabaci in the greenhouses of Spain until early 2000s. A. swirskii are able to establish on many vegetable crops before pests are present because they feed on pollen. Since A. swirskii is adapted to warmer and humid subtropical climates, it may be less effective in cooler dry conditions. Currently, A. swirskii is marketed in more than 20 countries worldwide including Europe, Africa, North

America, Latin America, and Asia. The predatory mirid bug, Macrolophus pygmaeus, originated in the Mediterranean region.

It is a highly polyphagous predator feeding on several soft-bodied pests such as whiteflies, aphids, mites, thrips, and moth eggs with a preference for whitefly eggs and nymphs. The predator has long legs which allow movement even on hairy leaves. The life cycle includes eggs that are oviposited deep in the plant tissue, five nymphal stages, and the adult. To ensure the predator's population establishment in the greenhouse, it is important to make releases early in the season. However, populations of the predator need to be carefully managed since once prey populations are low they can feed on the crop to be protected causing premature fruit and flower drop and direct damage to fruit.

The two primary aphelinid parasitoid wasps used for whitefly management in greenhouses are Encarsia formosa Gahan, a solitary endoparasitoid (which lays its eggs and develops inside the body of the host) and Eretmocerus mundus (Mercet) (Fig. 5) (which lays eggs under the nymph with subsequent larvae penetrating and developing in the whitefly nymph). E. mundus is indigenous to the Mediterranean Basin and is one of the most effective natural enemies of B. tabaci, although it is also known to parasitize and control T. vaporariorum in greenhouses. Both species can be released as parasitized whiteflies affixed to cards, and E. mundus can also be spread as loose parasitized pupae mixed in sawdust. Eretmocerus eremicus Rose & Zolnerowich is used instead of E. mundus because in some European countries, including the United Kingdom, E. mundus is not approved for release.

Pseudococcidae

There are a few polyphagous mealybug species with worldwide distribution that cause damage to greenhouse crops, yet these species are not generally considered key pests in organic greenhouses.

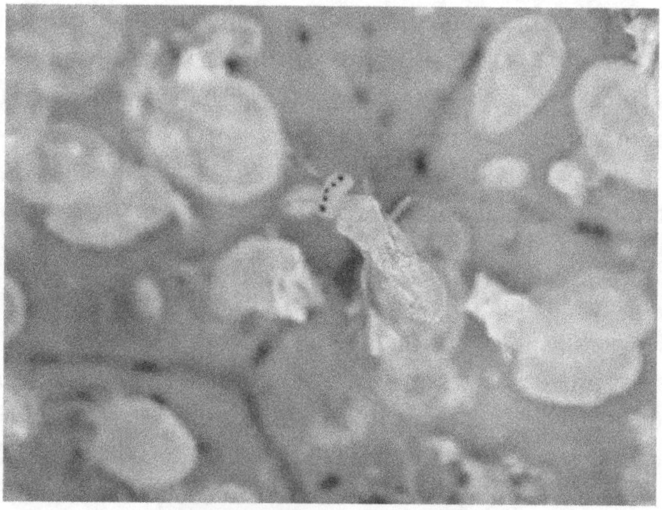

Fig. 5. Eretmocerus mundus

In South America, Phencoccus solenopsis Tinsley was first recorded in northern Argentina on sweet pepper and 2 yr later in tomato in northern Brazil but does not cause economic damage. In northern Europe, mealybugs are not generally pests in organic greenhouses although they occasionally occur in tomatoes. The species discussed below do cause economic damage in organic herbs and peppers in parts of Israel.

The citrus mealybug, Planococcus citri (Risso), is highly polyphagous and considered the most cosmopolitan and destructive species of the family.

In recent years, the citrus mealybug has infested organic herbs: including: tarragon, Artemisia dracunculus L., basil, Ocimum basilicum L., mint, Mentha sp. and arugula, Eruca sativa Miller in the Israeli regions of Judea and Samaria/West bank. Phenacoccus solenopsis Tinsley, is also a polyphagous species from all zoogeographic regions of the world. In Israel P. solenopsis infests organic sweet pepper in the Shfela region of south-central Israel.

Phenacoccus solani Ferris, is commonly found in the New World, expanding from the Mediterranean into Israel in 2003. No males have been found, so it has been assumed to reproduce parthenogenetically. In recent years, P. solani has increased its presence and while it causes damage in organically grown tarragon, mint, and sweet peppers, it is easily controlled through the management of ants (see Aphididae above).

Biological control of mealybugs is difficult due to their cryptic behavior, clumped distribution pattern, protective wax secretion, and protection by ants. As with other pest species, the international plant trade has been a major contributor to the invasive status of many multivoltine mealybug species, which are able to develop on a wide range of host plants.

In Israel, additional BCAs are used against mealybugs such as the encyrtid parasitoids Leptomastix algirica Trjapitzin for the management of Ph. solenopsis and Ph. solani mealybugs and Anagyrus sp. near pseudococci against Pl. citri.

Lepidoptera

Gelechiidae

Tuta absoluta (Meyrick), the tomato leaf miner, is native to South America. The moth invaded Europe through Spain in 2006 and since then has spread to the Mediterranean Basin, the Middle East, and Europe. T. absoluta is a major pest of tomato, although there is evidence that it can complete its life cycle on other Solanaceae plants. The pest status of T. absoluta changed from being a local South American pest to being a key pest that threatens the world tomato industry. T. absoluta established in greenhouses in western Europe regardless of the cold winters typical of this region. The supercooling point of the adult pest is e17.8°C and e18.2°C for the larvae and at 0°C, 10% of the larvae may survive for more than 3 wk and adults may survive more than one month. There is no evidence of an obligatory diapause, and if food is available, the pest may still be active in winter. Maintaining greenhouses free of tomato long enough to eradicate the pest is not commercially feasible in northwestern Europe.

Since T. absoluta was detected in the Mediterranean Basin, a few indigenous predators and parasitoids have been reported. Predators include the zoophytophagous predators Nesidiocoris tenuis (Reuter) and M. pygmaeus (Rambur) whose nymphs prey on T. absoluta eggs and adults consume eggs and larvae. Egg parasitoids as Trichogramma spp., are also known especially T. pretiosum Riley and T. achaeae\. Several parasitoids species of various families were recorded from T. absoluta larvae in the Mediterranean basin.

Microbial agents, such as the entomophathogenic bacteria Bacillus thuringiensis (Bt), are used in greenhouses against T. absoluta first instar larvae. In some cases, weekly repeated applications have resulted in good control of the pest. The combination of Bt with B. bassiana has also demonstrated better results than when either was applied alone. T. absoluta larvae and pupae were also found to be susceptible to three species of entomopathogenic nematodes. These microbial agents may be used in combination with other control agents in IPM programs against the pest.

The female sex pheromone is often used for mass trapping of males although when used alone the method was not effective enough in reducing moth population and damage to the crop. The pheromone has also been used for mating disruption, but studies in protected crops revealed mixed results. The failure to achieve effective control of the pest using the pheromone can be attributed to the efficacy of the formulated pheromone, the initial pest population level, isolation of plots from invading males and gravid females and the moth mating behavior.

In T. absoluta, the males and females mate more than once. Male sensitivity to the pheromone assures efficacy in mate finding and remating and female remating tendency increases the chances of encountering high-quality males.

Trials using the above mentioned methods, natural enemies, and semiochemicals are currently being studied to manage T. absoluta. Reduced crop damage was observed a few years after the first invasion of the pest to the new region but is likely attributed to increased grower awareness of the pest biology and behavior. Additionally, damage reduction has been the result of augmentation of opportunistic native natural enemies and applications of microbial agents such as foliar-applied Bt.

Noctuidae

Lacanobia oleracea (L), the bright-line brown-eye moth, can be a major pest of organic sweet pepper, due to larval feeding on leaves and internally within fruit. The pest can be difficult to control with foliar Bt products because only early instars are sensitive and Bt must be consumed by larvae prior to entering the fruit.

If a greenhouse is properly closed, larger lepidopterans pests should not gain access. However, there are several polyphagous predator species that contribute to the control of lepidopteran eggs and larvae, such as, O. laevigatus, N. tenuis, C. carnea, and Nabis spp. Microhymenopteran parasitoids such as Cotesia spp., Hyposoter didymator (Thungberg), and Chelonus oculator (Linnaeus) are also effective. Microbial insecticides based on B. thuringiensis var. aizawai, on nucleopolyhedrosus virus and certain formulations of spinosad have been approved for use in organic certified crops.

BOTANICAL PESTICIDE RESISTANCE

Botanical insecticide pyrethrum, extracted from the flowers of Chrysanthemum cinerariaefolium and C. cineum, contains insecticidal esters known as the pyrethrins. Pyrethrins are nerve poisons disrupting the sodium channels in neurotransmission and are commonly used for controlling pests in agricultural, structural and public and animal health sectors.

Arthropod resistance to pyrethrins and their synthetic analogs, pyrethroids, is very common and has been reported for several decades. Insect mutations that reduce the penetration of the toxin through the cuticle, reduce its binding to the target site, or alter the properties of the sodium channels, and other mechanisms impart pyrethrin resistance.

Another botanical insecticidal compound, azadirachtin, is a tetranortriterpenoid limonoid from neem (Azadirachta indica) seeds, which acts as an insecticide, antifeedant, repellent and insect growth regulator.

While neem oil, which has a lower concentration of azadirachtin, has been used in the United States as a fungicide, acaricide and insecticide for a long time, several azadirachtin formulations in powder and liquid forms have become popular in recent years. While there was a report of artificially induced resistance to azadirachtin in the green peach aphid, several studies demonstrated the efficacy of azadirachtin against various pests and recommended it as an option in IPM and for managing resistance to other pesticides. Azadirachtin is also thought to reduce the production of detoxification enzymes and known to improve the efficacy of other biopesticides. While arthropod resistance to botanical pesticides other than pyrethrins is not commonly reported, it is known that plant allelochemicals (e.g. alkaloids, phenolics, terpenoids, etc.) can contribute to cross-resistance to certain chemical pesticides. It is important to monitor the potential risk of resistance development and cross-resistance when botanical pesticides are used.

BACTERIAL BIOPESTICIDE RESISTANCE

Bacillus thuringiensis (Bt) is a gram-positive soil bacterium, which contains crystalline toxic protein that is activated upon ingestion by an insect host, binds to the receptor sites in the midgut, and eventually causes insect death. Bt pesticides are used against a variety of lepidopteran (Bt subsp. aizawai and Bt subsp. kurstaki), dipteran (Bt subsp. israelensis and Bt subsp. sphaericus) and coleopteran (Bt subsp. tenebrionis) pests and are very popular in organic farming. Since the mode of action involves toxins rather than the bacterial infection itself, resistance to Bt pesticides or transgenic crops that contain

Bt toxins is also very common. Some examples of resistant insects to Bt include the beet armyworm, the cabbage looper, the corn earworm, the Colorado potato beetle and the diamondback moth. Mutations in insects that regulate the immune system or interfere with the activation of Bt toxins and their binding to the target site led to insect resistance.

Spinosad is a mixture of macrocyclic lactones, spinosyns A and spinosyns D, derived from Saccharopolyspora spinosa, an actinomycete gram-positive bacterium, and is used against dipteran, hymenopteran, lepidopteran, thysanopteran, and other pests. Although naturally derived, some spinosad products are not registered as biopesticides. Insect resistance to spinosad later led to the development of spinetoram, which is a mixture of chemically modified spinosyns J and L. Both spinosad and spinetoram are contact and stomach poisons and act on insect nervous system by continuous activation of nicotinic acetylcholine receptors. The American serpentine leafminer, the beet armyworm, the diamondback moth, the tomato borer, the onion thrips, and the western flower thrips are some insects with spinosad resistance. Production of detoxifying enzymes (metabolic resistance) and changes in the target site are the most common mechanisms of spinosad resistance in insects. Cross-resistance between spinosad and some chemical insecticides has also occurred in some insects.

VIRAL BIOPESTICIDE RESISTANCE

Baculovirus infections in Lepidoptera have been known for centuries, especially in silkworms. Currently, there are several commercial formulations of nucleo-polyhedroviruses (NPV) and granuloviruses (GV). When virus particles are ingested by the insect host, usually a lepidopteran insect, they invade the nuclei of midgut, fatbody, or other tissue cells and kill the host. Baculoviruses are generally very specific to their host insect species and can be very effective in bringing down the pest populations. However, variations in the susceptibility of certain insect populations and development of resistance to viruses has occurred in several host species. Mutations in one or more genes with complex mechanisms are involved in insect resistance to viruses.

FUNGAL BIOPESTICIDE RESISTANCE

There are several fungi that infect insects and mites. The fungal infection starts when fungal spores come in contact with an arthropod host. First, they germinate and gain entry into the body by breaching through the cuticle. Fungus later multiplies, invades the host tissues, kills the host and emerges from the cadaver to produce more spores. Entomophthoralean fungi such as Entomophthora spp., Pandora spp., and Neozygites spp. can be very effective in pest management through natural epizootics, but cannot be cultured in vitro for commercial-scale production.

Hypocrealean fungi such as Beauveria bassiana, Isarea fumosorosea, Metarhizium brunneum and Verticillium lecanii, on the other hand, can be mass-produced in vitro and are commercially available.

These fungi are comparable to broad-spectrum insecticides and are pathogenic to a variety of soil, foliar and fruit pests of several major orders. Since botanical, bacterial and viral biopesticides have insecticidal metabolites, proteins, or viral particles that have specific target sites and modes of action, insects have a higher chance of developing resistance through one or more mechanisms.

Although fungi also have insecticidal proteins such as beauvericin in B. bassiana and I. fumosorosea and dextruxin in M. anisopliae and M. brunneum, their mode of action is more through fungal infection and multiplication, and arthropods are less prone to developing resistance to entomopathogenic fungi. However, insects can develop resistance to entomopathogenic fungi through increased melanism, phenoloxidase activity, protease inhibitor production and antimicrobial and antifungal peptide production.

It also appears that the production of detoxification enzymes in insects against fungal infections can also impart resistance to chemical pesticides.

These examples show that insects can develop resistance to biopesticides in a manner somewhat similar to chemical pesticides, but due to the typically more complex and multiple modes of action, at a significantly lesser rate depending on the kind of botanical compound or microorganism involved. Resistance to entomopathogenic fungi is less common than with other entomopathogens. The risk of biopesticide resistance is much higher in organic farming because of the widespread use and limited biopesticide choices.

Avoiding the repeated use of biopesticides reduces the risk of resistance development and can reverse the resistance mechanism in some cases. IPM recommends the use of host plant resistance, biological control, cultural control and other management options before using pesticide applications. When pesticides (both biological and synthetic) are to be considered, using them only when needed, applying at recommended rates and alternating/rotating with other modes of action can reduce the selection pressure on arthropod populations and resulting mutations.

10

BIOHERBICIDES FOR MANAGEMENT OF WEEDS

INTRODUCTION

Weeds are plants that create serious constraints in agricultural production. They compete with crops for water, gases, nutrients, space, light, and other growth resources, and can become hosts to pests and diseases. The growth factors of crops are challenged by weeds and cause on average 15 to 66% yield losses in direct-seeded rice, 18 to 65% in maize, 50 to 76% in soybean, and 45 to 71% in groundnut. Crop yield losses due to weeds vary considerably depending on the crop, weed management strategies, weed composition, infestation period, and abiotic factors (e.g., climate and soil edaphic factors). In agricultural farming, weed management is a key agronomic practice.

Owing to the labor shortage in the agriculture sector, the practice of using herbicides to control weed densities is increasing around the world. Inevitably, the constant use of herbicides on the same field to control weeds over a prolonged period has been shown to cause herbicide resistance, residue in crops, ecological imbalance between harmful and beneficial organisms, and environmental pollution. However, time constraint, advances in pest control technology, as well as a continuous 'enticement' from the current agricultural system have encouraged farmers to keep using conventional herbicides which have been found to be effective and time- and cost-efficient. The application of synthetic herbicides for effective weed control has thus become indispensable despite the unwelcome side effects.

Recently, there has been a growing interest in organic fruits, vegetables, dairy products, and beverages all over the world, particularly in developed countries. Organic products make up a small percentage of the food industry, but their rapid growth has created considerable interest among consumers and businessmen, as well as researchers. In 2013, there were almost two million produces, and 36% of global organic farmers are in Asia, followed by Africa (29%) and Europe (17%).

Organic product sales have consistently increased over the last decades. To cope with the rising demand from consumers, farmers are shifting from harmful chemical-dependent conventional agriculture to more sustainable and greener farm practices. Such development has led to the advent of more sustainable and environmentally friendly weed control alternatives.

The fundamental philosophy of sustainable weed management is based on the idea of preventing the spread of weeds rather than controlling them until they have developed and started to cause harm. Sustainable weed management comprises a suite of weed management options such as crop rotation, inter-cropping, crop competitiveness tillage, mulching, biological control agents, and green/bioherbicides which preclude the use of chemical herbicides. Biological weed control is a mechanism to suppress the germination and growth of weed populations to an economic threshold level by utilizing natural enemies, natural substances, or biotic agents.

Bioherbicide and conventional herbicide application methods are similar, although for mycoherbicides the pathogenic fungi are 'inoculated' by spraying the pathogens onto target weeds. Recently, bioherbicides have been regarded as a crucial weed control element, albeit not as a total replacement but rather as an alternative to chemical herbicides. Sustainable weed management does not rely on any single weed management technique, unlike synthetic herbicides in conventional agriculture; therefore, bioherbicides should be used concurrently with other weed management techniques to control weeds.

Bioherbicides that are thought to be safer and 'greener' have drawn attention, as scientific reports provide increasing evidence of their efficacy. However, their commercial presence in comparison to conventional herbicides is relatively new. Therefore, rigorous testing and validation is necessary to evaluate their efficacy and reliability for weed control. This review explores the impact of bioherbicides, particularly plant-based, on weed physiology and the factors that influence their efficacy as well as the limitations of their use.

Biological control of weeds is the intentional use of living organisms (biotic agents) to reduce the vigor, reproductive capacity, density, or impact of weeds. The strategies of biological control can be classified in two broad categories: (i) classical or inoculative, and (ii) inundative or mass exposure. The classical strategy is based on introduction of host-specific organisms (insects, pathogens, nematodes) from the weed's native range into regions where the weed has established and become a widespread problem.

The biotic agents, after quarantine to assure host specificity, are released into weed-infested sites and are allowed to adapt and flourish in their new habitat over time to eventually establish a self-perpetuating regulation of the weed infestation at acceptable levels. Thus, classical biological control requires a time period of one to several years to achieve adequate control while the agent population builds up to levels to impact the weed population.

The inundative strategy attempts to overwhelm a weed infestation with massive numbers of a biotic agent in order to attain weed control in the year of release. In contrast to classical biological control, inundation involves timing of agent release to coincide with weed susceptibility to the agent and formulation of the agent to provide rapid attack of the weed host.

A development of the inundative strategy is the bioherbicide approach, which involves application of weed pathogens in a manner similar to herbicide applications. Since most bioherbicides have been developed using selected plant pathogenic fungi that cause such diseases on weeds as anthracnose and rust, the term mycoherbicide is often used in reference to these fungal preparations.

BIOHERBICIDES

Bioherbicides consist of microorganisms such as pathogens and other microbes or phytotoxins derived from microbes, insects, or plant extracts that act as a natural means of weed control. According to Bailey, bioherbicides are naturally originated products which can be used to control weeds. But one must remember that, although bioherbicides comprise nature-derived compounds, this is not to say they are completely harmless.

Plants produce natural toxins that could affect the health of non-flora organisms in the environment or certain bacteria, viruses, and fungi that could cause health problems to animals and humans. Therefore, these natural toxins must be carefully managed to avoid any unintended impact on crops or beneficial fauna and flora.

The first evidence of bioherbicide development was documented in the mid-1970s with the discovery of mycoherbicides. Since then, numerous bioherbicides have been registered and become available in the global market. The earliest bioherbicide project involved simply the application of Fusarium oxysporum Schlecht, a fungus, against Opuntia ficus-indica (L.) Mill.

In the 1950s, the parasitic weed Cuscata spp. was controlled with Alternaria cuscutacidae Rudakov. In the meantime, the use of registered and unregistered bioherbicides has also increased significantly. In the late 1960s, an ambitious program was conducted to identify pathogens from Rumex spp. in the United States and Rubus spp. in Chile for weed control.

Table 1 collates the development of several bioherbicides from different sources with their target weed species.

Advantages and Disadvantages of Bioherbicides

The growing public demand for safe 'green' products has resulted in many new environmentally friendly products becoming available for controlling pests, including weeds. Bioherbicides developed from plant extracts, phytopathogenic microorganisms, or microbial phytotoxins (i.e., mycoherbicides) are a useful approach to weed control.

They usually do not possess persistent characteristics—in other words, bioherbicides do not remain active in the environment for long periods, are less likely to cause soil and water contamination, and do not cause any adverse effects on non-target organisms. Bioherbicides prepared from allelochemicals are thus negligibly harmful to the bio-ecosystems and human health. Some allelochemicals are soluble in water, making them easier to apply without adding surfactants.

Table 1. Development of bioherbicides from different sources with their target weeds.

Source	Phytotoxic Effects	Target Weeds
Plants		
Achillea santolina L.	Inhibit growth and changes in the metabolic process	Medicago polymorpha L.
Brassica napus L.	Suppress germination and root length	Phalaris minor Retz., Convolvulus arvensis L., Sorghum halepense (L.) Pers.
Carum carvi L.	Leaf lesions and biochemical changes in plant tissues	Echinochloa crus-galli (L.) P.Beauv.
Chrysanthemum coronarium L.	Inhibit germination and Growth	Sinapis arvensis L., Phalaris canariensis L.
Cynara cardunculus L.	Suppress germination and growth and cause necrosis or chlorosis chlorophyll and protein content	Trifolium incarnatum L., Silybum marianum (L.) Gaertn., P. minor
Cymbopogon nardus (L.) Rendle.	Inhibit germination and plant development and reduce	Digitaria horizontalis Willd., Cenchrus echinatus L.
Mimosa pigra L.	Root growth retardation	Lactuca sativa L., Ruellia tuberosa L.
Parthenium hysterophorus L.	Seed germination, growth, and development	Oryza sativa f. Spontanea Roshev., Echinochloa colona (L.) Link., Euphorbia hirta L., Ageratum conyzoides L.
Pinus densiflora Siebold & Zucc.	Suppressed shoot and root growth	Lolium multiflorum Lam., Digitaria sanguinalis (L.) Scop.
Pinus nigra J.F.Arnold	Inhibited germination and growth	P. canariensis, Trifolium campestre Schreb., S. arvensis
Sinapis alba L.	Reduced dry biomass	Amaranthus powellii S.Watson, Setaria viridis (L.) P.Beauv.
Fungi		
Ascochyta agropyrina	Reduced root growth	Chenopodium album L., Cirsium arvense (L.) Scop., Mercurialis annua L., Sonchus oleraceus L., Setariavirdis (L.) P.Beauv.
Diaporthe gulyae	Necrosis	Papaver rhoes L., Ecballium elaterium (L.) A.Rich., Urtica dioica L., Hedysarum coronarium L.
Fusarium fujikuroi	Chlorosis, necrosis, and decrease in plant height and root length	Cucumis sativus L., Sorghum bicolor (L.) Moench.

Phoma herbarum	Maximum toxicity	P. hysterophorus, Lantana camara L.,
		Hyptis suaveolens (L.) Piot., Sida acuta Burm.f.
Lasiodiplodiapseudothe-obromae	Inhibited germ activity	Solanum lycopersicum L., Amaranthus hybridus L., E. crus-galli
Myrothecium roridum	Necrosis	Eichhornia crassipes (Mart.) Solms
Sclerotium rolfsii	Decrease in population density	Solidago canadensis L.
Bacteria		
Pseudomonas aeruginosa	Inhibited seed germination, growth, and germ activity	A. hybridus, S. lycopersicum, E. crus-galli,
		Pennisetum purpureum Schumach.
Pseudomonas fluorescens	Suppress germination and root growth	Bromus tectorum L., Aegilops cylindrical Host, Taeniatherum caput-medusae (L.) Nevski

The chemical structures of allelochemicals are more environmentally friendly compared to those of synthetic herbicides. Allelochemical bioherbicides typically have short-lived environmental persistence and low toxicity, and they often employ multiple modes of action, which reduces the risk of herbicide resistance. As a result, allelochemicals serve as good candidates for the development of bioherbicides, antimicrobial agents, and growth regulators.

Despite all the benefits associated with bioherbicides in controlling weeds in a more sustainable manner, there are some drawbacks that make bioherbicide application less appropriate than the current synthetic herbicides, particularly at the field scale. Bioherbicides have a relatively short environmental half-life—while this is an ideal attribute for reducing environmental toxicology, an effective herbicide must persist long enough to exhibit the desired effect on weed species. Plants from the same area or from the same taxonomic group do not produce the same quantity or content of secondary metabolites, and therefore may not exude the same amount or quality of allelochemicals. Furthermore, many allelochemicals are too expensive to be seriously considered for use as agrochemicals because of their structural complexity.

For example, the cyclic tetrapeptide toxin is an excellent herbicide, but it is very expensive. Some natural products that are extremely phytotoxic are often very poisonous to mammals such as the AAL-toxins, that are fairly toxic to mammalian cells. These aspects of some natural phytotoxins have reduced the interest to develop them into herbicides for weed management.

Types of Bioherbicides and Currently Marketed Products

Bioherbicides were first introduced into the market in 1980, and, since then, several biopesticides comprising bioinsecticides, biobactericides, biofungicides, and bionematicides have been introduced globally, but the market share of bioherbicides still makes up less than 10% of all biopesticides. Of the registered bioherbicides on the global market, many are derived from microorganisms.

CURRENT STATUS OF BIOHERBICIDE DEVELOPMENT AND USE

Several selected microorganisms have been extensively evaluated and developed or are under development for commercial application. The discovery, development, practical application, and commercialization of early mycoherbicides (i.e., 'Collego', 'Devine', 'Biomal') have been thoroughly described (Charudattan, 1991; TeBeest, 1996). The early mycoherbicides consisted of highly virulent fungal plant pathogens that infected the aerial portions of weed hosts resulting in visible disease symptoms.

These fungi could also be mass cultured in artificial media to produce large quantities of inocula needed for field application. Microbial agents comprising more recent bioherbicides can include obligate fungal parasites, soil-borne fungal pathogens, nonphytopathogenic fungi, pathogenic and nonpathogenic bacteria, and nematodes. Many of these organisms have different cultural and application requirements compared to the early mycoherbicides.

This presents a curious dilemma in that even though the number of potential bioherbicides and target weeds in diverse habitats has expanded over the past 25 years, the production and formulation requirements and application methods have become more complex.

This is a disadvantage for developing production facilities devoted to bioherbicides because standard culturing and processing techniques cannot be used in the preparation of the various biological control agents necessary for attacking several target weeds that typically infest crop production fields.

Unfortunately, production systems for bioherbicides, especially mycoherbicides, will continue to follow empirical processes illustrated by evaluation of candidate pathogens on a case-by-case basis.

The bioherbicides listed in Table 1 include organisms that have been extensively evaluated for commercial development including several that have undergone field testing and evaluation as required by regulatory agencies. The original bioherbicide or mycoherbicide concept was based on mass artificial culture of organisms to obtain large quantities of inoculum for inundative application to the weed host to achieve rapid epidemic buildup and high levels of disease.

Because many of the recent bioherbicide candidates differ from the original definition in requirements for mass production and application, the bioherbicide concept has been redefined as living products that control specific weeds in agricultural systems within a specific, short timeframe.

Most mycoherbicides consist of fungal pathogens such as the Colletotrichum species able to exist saprophytically in the absence of the plant host, which is important in developing artificial culture strategies for production of inocula. However, some fungal pathogens are obligate parasites that can only proliferate directly on host plants.

Such is the case of the rust fungus Puccinia canaliculata, a foliar pathogen of yellow nutsedge (Cyperus esculentus), that is cultured en masse on the weed host planted in small field plots or greenhouses from which uredospores are vacuum-harvested and stored in bulk prior to preparation of the mycoherbicide 'Dr. Biosedge'.

The harvested uredospores can be applied in the field through center-pivot irrigation systems. The mycoherbicide 'Eco-Clear', containing the fungal pathogen Chondrostereum purpureum, must be applied to wounded branches or stumps of weedy tree species to inhibit re-sprouting by enhancing decay of the woody tissues.

Soilborne fungi have become important bioherbicide candidates because these fungi are applied directly to soils to reduce weed populations through decay of seeds prior to emergence or kill seedlings shortly after emergence. Trichoderma virens inoculated in composted chicken manure significantly reduced broadleaf and grass weeds in horticultural crops.

Plant pathogenic bacteria including Xanthomonas campestris pv poannua (Xcp) and Pseudomonas syringae pv tagetis (Pst), were developed as bioherbicides for control of annual bluegrass (Poa annua) and Asteraceae (composite) weeds, respectively.

The Xcp bioherbicide ('Camperico') must be applied by spraying the bacterial suspension while mowing to allow bacterial cells to invade wounded tissue of the grass. The Pst bioherbicide is prepared with organosilicone surfactant to enhance bacterial infection of leaf and stem tissue and onset of disease.

Deleterious rhizobacteria (DRB) are bacteria that differ from bacterial pathogens based on their nonparasitic nature and ability to aggressively colonize plant roots and suppress plant growth without invading the root tissues. The DRB are under intensive investigation for bioherbicidal potential.

Pseudomonas fluorescens D7 is a soil-applied DRB bioherbicide formulated as a liquid suspension or encapsulated in clay that effectively suppresses downy brome (Bromus tectorum) in cereal grain crops. Selected DRB formulated as semolina flour granules suppressed velvetleaf (Abutilon theophrasti) seedling growth and root growth of leafy spurge (Euphorbia esula) established in field infestations.

Mechanically-transmitted viruses (i.e., Tobacco Mild Green Mosaic Virus), produced on surrogate host plants, harvested, and freeze-dried, are under investigation as bioherbicides of invasive weeds including tropical soda apple (Solanum viarum). Plant-specific nematodes have been reared and formulated for preliminary evaluation for wide-scale management of the rangeland weed, Russian knapweed (Acroptilon repens).

Limiting factors of bioherbicide adoptions and use

Factors of narrow host range, specific requirements for culturing and formulation to assure biotic agent efficacy, and possible by-products of potent mammalian and avian toxins by some fungal agents have limited commercial development of bioherbicides because of their likely low market potential.

ENHANCING HOST RANGE OF BIOHERBICIDES

Successful incorporation of bioherbicides into conventional agriculture will only be achieved if they are able to effectively and consistently suppress multiple weeds of economic importance on a very large scale. Currently, the use of a bioherbicide to control one species in a mixture of weeds in the field is questionable if a producer has the option of a broad-spectrum herbicide to control multiple weed species.

Fungal pathogens of the same species containing strains or subspecies with activity against several weeds might be developed through selective screening or through genetic recombination or hybridization into broad host-range pathotypes for use against multiple weed targets. These are longterm tactics, however, as intense evaluation will be required to meet stringent regulations before approval is granted for release of genetically-altered organisms into the environment.

Alternatives to genetic manipulation for improving the spectrum of weeds controlled include using a "multiple-pathogen strategy" and manipulating the formulation to aid infectiveness of the pathogen. For example, a mixture of three pathogens, Drechslera gigantia, Exserohilum longirostratum, and Exserohilum rostratum, which were isolated from three different host weeds, successfully suppressed growth of seven weeds of citrus groves in Florida.

The multiplepathogen strategy also shows promise for controlling weeds in row crops illustrated by the successful control of eight weedy Amaranthus species in soybean by a mixture of Phomopsis amananthicola and Microsphaeropsis amaranthi conidial suspensions.

Screening trials of the bacterial pathogen Pst prepared as a post-emergence bioherbicide revealed efficacy on several economically important weeds in the Asteraceae family including wild sunflower (Helianthus spp.), common cocklebur (Xanthium strumarium), common ragweed (Ambrosia artimisifolia), and Canada thistle (Cirsium arvense) in soybean field trials. Amendment of Pst aqueous suspensions with surfactants to aid the bacteria to efficiently invade plant leaves broadens the host range when it was found that weeds outside the Asteraceae family, green foxtail (Setaria viridis), and velvetleaf (Abutilon theophrasti), were also severely injured.

FORMULATION OF BIOHERBICIDES

Many of the foliar and stem fungal pathogens require specific levels of humidity (dew periods) and temperature for full effectiveness, which necessitates development of special formulations to assure effectiveness after delivery of the agent in the field. Researchers outlined the factors that must be addressed in bioherbicide formulation technology in order to maintain or enhance efficacy of the biocontrol agent as well as to be compatible with conventional field application systems. The main types of formulations currently in use include various emulsions, organosilicone surfactants, hydrophilic polymers, and alginate-, starch-, or cellulose encapsulated granules, all of which have advantages and disadvantages in promoting virulence and efficacy of the biotic agents and ease of application. The primary functions of the formulation should include predisposal of the weed to infection by the pathogen and protection of the infecting pathogen against environmental constraints while promoting disease development.

POTENTIAL HEALTH RISKS OF BIOHERBICIDES

Some mycoherbicides exhibit strong herbicidal activity against a range of economically important weeds, however, the fungal pathogens may also produce undesirable mammalian and avian toxins. Myrothecium verrucaria is highly effective in controlling a number of weeds through production of herbicidal metabolites during infection of the host plant; however, mammalian-toxic macrocyclic tricothecenes are simultaneously produced, thereby presenting a human health hazard. Similarly, tricothecenes are produced by the fungal pathogen Fusarium tumidum, under study as a potential bioherbicide for gorse (Ulex europaeus) and broom (Cytisus scoparius) in New Zealand. Researchers suggests that fungal pathogens that produce both herbicidal metabolites and mammalian toxins must be fully investigated to assess their role and risks as potential bioherbicides.

BIOHERBICIDES IN CONVENTIONAL CROPPING SYSTEMS

Conventional cropping systems are characterized as large-scale production enterprises that utilize highyielding crop varieties generally in monoculture or short term rotations planted on the most fertile, productive soils available with large, costly inputs of chemical fertilizers and pesticides. The introduction and rapid adoption of geneticallymodified (GM) varieties of the major crops (soybean, maize, cotton, canola) that are resistant to herbicides allow control of a broad spectrum of weeds with a single herbicide.

Conventional cropping systems using either GM or non-GM crop varieties are based on the seasonal use of herbicides to reduce weed infestations to acceptable levels so that crops can be grown profitably with little consideration of a more long-term approach for weed management. Despite the advancements in herbicide technology and development of GM herbicide-resistant crops, surveys continue to indicate that farmers perceive annual and perennial weed infestations as the most serious crop pest problem that affect their enterprises.

These findings suggest that many farmers might consider incorporating alternative, effective, nonchemical weed control methods into their overall weed management plans.

Despite limitations of current bioherbicides, chemical herbicide use will level off and possibly decrease in importance due to social and environmental concerns, development of herbicideresistant weed biotypes, and reduced availability of new, environmentally compatible herbicides while emphasis on "biologically-based pest management" will increase. Bioherbicides may be effective in conventional cropping systems under several circumstances including: (i) weeds evolved resistance to a broad-spectrum herbicide; (ii) excessive rates of herbicides are required to control one weed species (bioherbicides would allow less herbicide use); (iii) parasitic weeds are not controlled due to lack of selective herbicides; (iv) herbicides cannot be used due to cost or environmental limitations; and (v) invasive weed species become established and few herbicide options are available.

BIOHERBICIDES AND HERBICIDE-RESISTANT WEEDS

Although management of the multiple-weed complex in row crops may be addressed in the short term through use of herbicide-resistant transgenic crops, herbicide-resistant weed biotypes will eventually develop after repeated applications of the same herbicide in a given field. For example, glyphosate-resistant rigid ryegrass (Lolium rigidum) developed after repeated applications of glyphosate in an orchard to control grass weeds.

Even more serious is the development of weed biotypes with resistance to multiple herbicides that are widely used in row crop production. Researchers reported a biotype of common waterhemp (Amaranthus rudis), a major weed in maize and soybean in the Midwestern U.S., with resistance to two widely-used herbicides. Also, a downy brome biotype was characterized with resistance to eight chemically dissimilar herbicides.

As herbicide resistance becomes more problematic with many common weeds, strategies using bioherbicides will become more important in maintaining adequate weed control in conventional systems. The potential for successful use of bioherbicides in managing herbicide-resistant biotypes was demonstrated where growth of an imazaquin-resistant common cocklebur biotype originating in soybean fields was suppressed with the mycoherbicide, Alternaria helianthi.

BIOHERBICIDES AND PARASITIC WEEDS

Many crop production regions throughout the world are infested with parasitic weeds that attack specific crop plants causing drastic yield reduction. There are no selective herbicides for satisfactorily controlling parasitic weeds; therefore, bioherbicides could be potentially highly effective on these weeds. Indeed, some of the most recent successful bioherbicides have been for control of dodders (Cuscuta spp.) in soybean and cranberry.

Also, a soil fungus under evaluation suppresses germination and attachment of witchweed (Striga sp.) seedlings to grain sorghum roots, contributing to grain yield increases. Development of this pathogen as a biotic agent could have significant impacts on food production in regions where Striga spp. are dominant weed problems. Recently, a strategy for selecting rhizosphere bacteria or fungi that inhibit broomrape (Orobanche spp.) germination and infection of horticultural crops by overproducing amino acids has been proposed as an effective bioherbicidal approach.

Bioherbicides and Developing Weed Problems

Bioherbicides may also have a place in weed species that have not yet reached a competitive threshold level. An example might be their use against perennial weeds that are increasing under reduced-tillage farming systems. Shifts in weed composition in response to changes in cropping practices, such as tillage, result in development of sub-competitive populations during the early years of the shift. The general tendency is for perennial species to increase as the amount of tillage decreases.

For example, common milkweed (Asclepias syriaca) infestations increased in the Midwestern United States in response to continued use of reduced tillage. Although common milkweed typically infested wheat, corn, and soybeans under reduced or no-tillage, it probably only slightly reduced crop yields.

Bioherbicides might provide a way of keeping this weed from becoming an economically important weed in the future. The discovery of a bacterial disease affecting common milkweed may lead to the development of the causal agent, Xanthomonas campestris pv. asclepiadis, as a biotic agent for maintenance of common milkweed infestations below economic threshold levels. The disease is a systemic blight and reduces overall plant vigor and stand density.

The report of milkweed bacterial blight is significant because a practical "preventive biological control approach" can be developed and because discovery of similar agents on other perennial weeds for which no options for chemical control is possible.

Bioherbicides may have an important role in managing invasive weeds, defined as those alien plants spreading naturally in natural ecosystems and producing significant changes in terms of composition, structure, or ecosystem processes. Invasive weeds are an emerging management challenge because from an economic or agricultural productivity perspective most are not undesirable but are considered very disruptive from an ecological and conservation standpoint. Although classical biological control strategies may be most appropriate for these situations, some bioherbicides developed for weeds in these habitats have shown promise in effective control. The phytopathogenic bacterium Pseudomonas syringae pv. phaseolicola was shown to suppress growth of the invasive weed kudzu (Pueraria lobata).

Researchers have demonstrated the potential of Ralstonia solanacearum as biological control agent of alien kahili ginger (Hedychium gardnerianum), and invasive weed in tropical forest ecosystems. Bioherbicides will be of significant value in managing weeds in areas where herbicides are not effective due to regulations that severely restrict or prohibit herbicide use and where a primary management goal is preservation of the environment. These special situations include restoration of native ecosystems, wetlands, national parks, wildlife refuges, and areas bordering waterways (riparian borders).

For example, red alder (Alnus rubra) is a forest weed that interferes with timber production in the Pacific West forest region of Canada. Red alder infestations can be suppressed using biocontrol fungi that are inoculated into woody stems with injecting devices.

This mycoherbicide is useful for control of red alder along streams, where herbicide application is prohibited, by causing a "slow kill" and allowing simultaneous release of nutrients from dying vegetation for use by desirable tree species as well as gradual re-population of the site by the timber trees.

Plant-Based Bioherbicides

Bioherbicides from Plant Extracts

Plant extracts, which are traditionally used for medical or nutritional purposes, may serve as an alternative for developing bioherbicides for sustainable agricultural practices in weed management. Bioherbicides produced from the extracts of natural sources have shown promising potential against weeds. Several plant extract compounds possess a specific inhibiting activity against weed growth but cause no detrimental injury to crops.

This may be explained by the difference in sensitivity in the target enzymes or the existence of specific receptors in weeds that recognize and react with the compounds. Certain plant species have the capacity to secrete different metabolites known as allelochemicals, such as alcohols, fatty acids, phenolics, flavonoids, terpenoids, and steroids, that reduce the reproduction, growth, and development of adjacent vegetation, including weed species.

The phenolic extract of C. cardunculus possesses phytotoxic properties due to the presence of an aromatic ring in its composition, which consists of multiple hydroxyl groups. These phytotoxic properties destroy the plasma membrane, contributing to the initial and fundamental impacts of oxidative stress. Phytotoxic water extracts from S. bicolor are a known example of bioherbicides with the ability to control weeds without yield losses.

The application of sorghum water extract caused a 40% reduction in the biomass of E. crus-galli, resulting in an 18% rice yield increase. In another scenario, the extracts from the leaf, stem, flower, and root of Brassica nigra (L.) K. Koch. were found to strongly inhibit the germination, growth, and radicle length of Avena fatua L..

The inhibitory effect is believed to be mainly due to the high concentrations of glucosinolates, the bitter sulfur-containing compounds which are largely present in Brassica sp. These compounds can be enzymatically hydrolyzed to isothiocyanates, thiocyanates, and nitriles, yielding new biologically active particles with the potential to reduce various weeds.

The isothiocyanates, when interacting with the cell-damaging sulfhydryl-containing enzymes, besides acting as an inhibitor to several fungi and pathogenic and food spoilage bacteria, also showed an inhibitory effect on the germination of Matricaria inodora L., E. crus-galli, Sonchus asper L., A. hybridus, and Alopecurus myosuroides Huds.. Other plants with extracts exhibiting strong phytotoxic inhibition include Pisum sativum L. against the germination, growth, and development of Polygonum persicaria L., A. hybridus, Galinsoga parviflora Cav., C. album, and also Medicago sativa L. against the germination of Artemisia vulgaris L., with up to 83% efficacy in petri dish assays and up to 89% under field conditions.

Essential oils are natural volatile compounds derived from different plant parts, such as leaves, bark, flowers, fruits, seeds, roots, and also from the whole plant. Terpenoids (mainly mono and sesquiterpenes) are the main compounds of essential oils' activity which could be potential candidates for the development of new bioherbicides because they have a strong phytotoxic activity toward different weed species.

The phytotoxic potential of essential oils involved chlorosis, the burning of leaves, and plant growth reduction, as well as mitosis inhibition, membrane depolarization, decrease of chlorophyll content, cellular respiration, and oxidative damage. According to Verdeguer et al., the essential oils extracted from Cistus ladanifer L. were found to inhibit the germination and growth of A. hybridus, Portulaca oleracea L., C. album, Conyza Canadensis, Parietaria judaica L.

Bioherbicides from Allelochemicals

Allelochemicals are nature's own herbicides, which offer several benefits over synthetic compounds. There is a great diversity in the chemical structures of allelochemicals, and some of them can serve as lead molecules for the development of herbicides. Some natural compounds are water-soluble and non-halogenated molecules. They have short half-lives relative to synthetic herbicides and are thus considered safe to the environment.

Researchers discussed the approaches to select sources of natural products as potential herbicides: (i) acquiring pure compounds from other laboratories, (ii) using the previously unexploited biological material, and (iii) using the ethnobotanical and/or chemical ecology data to select the herbicidal material. In the past, many herbicides such as Cinmethylin, AAL toxins, Mesotrine, Artemisinin, Biolaphes, Glufosinate, and Dicamba have been developed from plant allelochemicals. After much evaluation, it became clear that plant phytotoxic extracts could be a key tool in integrated weed management.

Bioherbicides from Natural Byproducts

Byproducts coming from natural sources have also been observed to suppress weed growth. Boydston et al. reported that a byproduct of ethanol production called distillers' dried grains with solubles (DDGS) successfully controlled Stellaria media L. and P. annua and inhibited the germination of Oxalis corniculata L.

A byproduct named corn gluten meal (CGM), obtained from the wet-milling corn process, has long been acknowledged to have natural herbicidal activity. Applying CGM on the soil surface in a greenhouse caused physiological changes in C. album, S. nigrum, Agrostis palustris Huds., P. oleracea, and R. crispus. Another byproduct possessing herbicidal activity is mustard seed meal (MSM), which is produced from the mustard oil pressing method.

Glucosinolates are the main compounds of MSM, which can deliver a promising plant inhibitory effect. In one study, Brassicaceae seed meals (BSMs) showed a phytotoxic inhibitory effect on weeds and at the same time increased carrot yield by enhancing the inorganic nitrogen content in the soil. According to Intanon et al. the Limnanthes alba Hartw. seed meal not only enhanced the yield and nitrogen (N) content in leaves but also controlled weeds.

EFFECTS OF PLANT-BASED BIOHERBICIDES ON WEED GROWTH

Seed Germination

Seed germination is considered as an important factor in plant development and productivity. During germination, the physiological, biochemical, and morphological changes are strongly linked with the survival rate of seedlings. Plant-based bioherbicides can inhibit seed germination by blocking the hydrolysis of nutrient reserves and cell division. The inhibition of the germination process by plant extracts involves osmotic effects on imbibition rates, which eventually inhibit germination and, in particular, cell elongation. The phytotoxicity of plant extracts, residue, or mulch may affect weed germination and growth.

Plant residues release biochemical substances on the soil that can hinder seed germination and the seedling growth of weeds. Aslani et al. reported that a leaf extract of Tinospora tuberculata Beumee inhibited the germination and growth of E. crus-galli, although it also showed phytotoxicity in the crops (O. sativa, Dacus carrota L., L. sativa, Cucumis sativa L. and S. lycopersicum). Likewise, the extracts of S. nigrum, C. album, and Matricaria chamomilla L. exhibited suppressive action on the germination and seedling growth of H. vulgare, Phaseolus vulgaris L., Cicer arietinum L., Zea mays L., Allium cepa L., Capsicum annuum L., S. lycopersicum, and Triticum aestivum L.. The seed germination of C. arietinum, T. aestivum, B. nigra, and Lens culinaris L. was significantly reduced by the application of Butea monosperm L..

Shoot and Root

Shoot length is one of the significant growth parameters considered for the growth and development of plants. Generally, shoot growth is less sensitive to phytotoxic plant extracts compared to radicle growth. The greater sensitivity of radicle growth to plant extracts is due to the radicle being the first organ to be exposed to the phytotoxic substances and having a more permeable tissue than other organs and/or a low mitotic division in the root apical meristem.

Moreover, phytotoxic substances can affect genes responsible for the cellular characterization of radicle tissues and the endoderm, inhibiting their development. The leaf extract of D. stramonium at 25 to 100% concentration significantly reduced the shoot and root growth of Cenchurus ciliaris L. and Neonotonia weghtii Wight & Arn.. Motmainna et al. reported that the methanol extracts of P. hysterophorus, Cleome rutidosperma DC., and Borreria alata (Aubl.) DC. showed a promising inhibitory effect in the shoot and root growth of Oryza sativa f. spontanea Roshev. (weedy rice).

Leaf Area

The leaf area of a plant is an important parameter for assessing growth. It is a variable that relates the atmospheric condition of the plant through the process, such as transpiration, respiration, and photosynthesis. It is a fundamentally essential tool used in scientific disciplines such as agronomy, plant physiology, entomology, ecology, plant pathology, and many others. The cuticle is present on the upper and lower surfaces of the leaf that line the sub-stomatal cavities. The epidermal surfaces of the plant are covered by a cuticle to protect against water loss and desiccation.

Herbicide movement or absorption into leaves depends on the spray retention of a herbicide on the leaf surface and the diffusion through the cuticle. In wheat, the leaf area index was significantly affected by sorghum extract at different application times. The leaf area index decreased with each increase in concentration, and a lower leaf area index (3.00) was recorded with 1:5 concentrations. Sorghum water extract applied at tillering gave a higher leaf area index, whereas sorghum water extract applied 50% at emergence + 50% at tillering resulted in a reduced leaf area index.

Stomatal Conductance

Stomata are small pores that absorb CO_2 and remove water vapor from the top and/or bottom of a leaf. When plant stomata open, water escapes through transpiration, and carbon dioxide is taken up via photosynthesis.

The connection between the plant hydraulic system and photosynthesis is governed by a gaseous exchange and certain biochemical processes. The measurement of the gaseous exchange is important, as it is linked to photosynthesis and transpiration, which may be manipulated by environmental elements such as carbon dioxide, light, humidity, temperature, and wind speed.

The rhizome yield of Zingiber officinale Roscoe was significantly low when treated with Tamarindus indica L. (512 g/plant) leaves and mulched Mangifera indica L. (521 g/plant), attributed to drastic decreases in stomatal conductance, leaf production, rhizome thickness, rhizome spread, root length and spread, tiller production, and plant growth.

Chlorophyll Pigment

Chlorophyll molecules play an important role in photosynthesis and serve as a green pigment embedded in photosynthetic membranes, which is why a chlorophyll reduction normally results in a decrease in photosynthesis. Many plant extracts have been reported to directly or indirectly affect the chlorophyll content. Phenolic compounds such as p-coumarin, ferulic, and o-hydroxyphenyl acetic acids have been shown to inhibit chlorophyll biosynthesis and stimulate its degradation within 1 h after treatment. A crude powder of M. polymorpha significantly decreased the chlorophyll content and total photosynthetic pigments of the recipient species. Plant extracts inhibit certain enzymes associated with chlorophyll and ultimately affect the integrity of chloroplasts and thylakoid membranes.

Photosynthesis

Plant extracts affect protein metabolism in plants, with the result that protein binding in chlorophyll a/b is reduced by two-fold. They also affect the photosynthesis by suppressing chlorophyll synthesis. Oxygen evolving enhancer protein 1 (OEE1) plays an important role in increasing the thioredoxin movement, defends the tetra-manganese cluster, and helps to release oxygen by splitting water. Plant-based bioherbicides have the ability to affect the gas and nutrient exchange of weeds by reducing OEE1 biosynthesis. Magnesium (Mg) is an essential part of chlorophyll, involved in carbon fixation and a variety of biochemical processes. Bioherbicides from plant extracts have the ability to reduce the Mg concentration in weed species, which has a major effect on chlorophyll synthesis. Sousa et al. reported that photosynthesis-inhibiting herbicides like bentazon can limit electron transfer and carbohydrate synthesis.

Plant Hormones

Plant hormones are essential to plant growth through different metabolic activities. Gibberellin (GA) is a plant hormone that assists in the development of plant shoot growth. Lee et al. reported that the allelochemicals isolated from the seed of Sicyos angulatus L., a broadleaf weed, contains a high amount of 2-linoleoyl glycerol, which can inhibit the gibberellin pathway. On the other hand, 2-linoleoyl glycerol can stimulate the accumulation of abscisic acid (ABA), jasmonic acid (JA), and salicylic acid (SA). Increased ABA production causes stomatal closing, catalyzes plant senescence, and lowers the rate of photosynthesis, further inhibiting plant growth and development. Moreover, higher accumulations of JA could greatly reduce photosynthesis. Janda et al. also reported that higher SA accumulations inhibit photosynthesis.

EFFECTS OF PLANT-BASED BIOHERBICIDES ON WEED BIOCHEMISTRY

One of the important characteristics of plant tissues is their ability to respond to variable changes in the environment, which allows them to effectively manage and adjust to the altered ecosystem. Abiotic stress conditions, which include drought, heat, cold, salinity, nutrient deficiency, and oxidative stress, are among the environmental changes that greatly influence plant productivity, leading to morphological, physiological, and biochemical responses in plants.

Oxidative stress is one of the main consequences of biotic and abiotic stresses which affect physiological and biochemical metabolism in plants; hence, a balanced amount of reactive oxygen species (ROS) scavenging through proteins and antioxidant enzymes is important. Many reports have shown that ROS are the key cause of cell damage in biotic and abiotic stresses. These oxidative molecules are typically formed when oxygen is reduced by 1, 2, or 3 electron transfer to form hydrogen peroxide, superoxide, and hydroxyl radicles, which are highly reactive species that are cytotoxic to biomolecules such as nucleic acids, proteins, and lipids, thereby causing protein denaturation and lipid peroxidation.

The plasma membrane is the primary site of cellular and organelle injury by ROS, since they can react with unsaturated fatty acids to cause the peroxidation of the lipid bilayer in both cellular and intercellular structures. Cellular damage consequently leads to the leakage of cellular contents, rapid desiccation, and, inevitably, cell death, while intercellular damage affects the respiratory activity in mitochondria and causes pigment breakdown in chloroplasts. ROS are equally produced by normal cell metabolism in organelles such as chloroplasts (photosynthesis), mitochondria (photorespiration), and peroxisomes (respiration), all powerful generators of ROS.

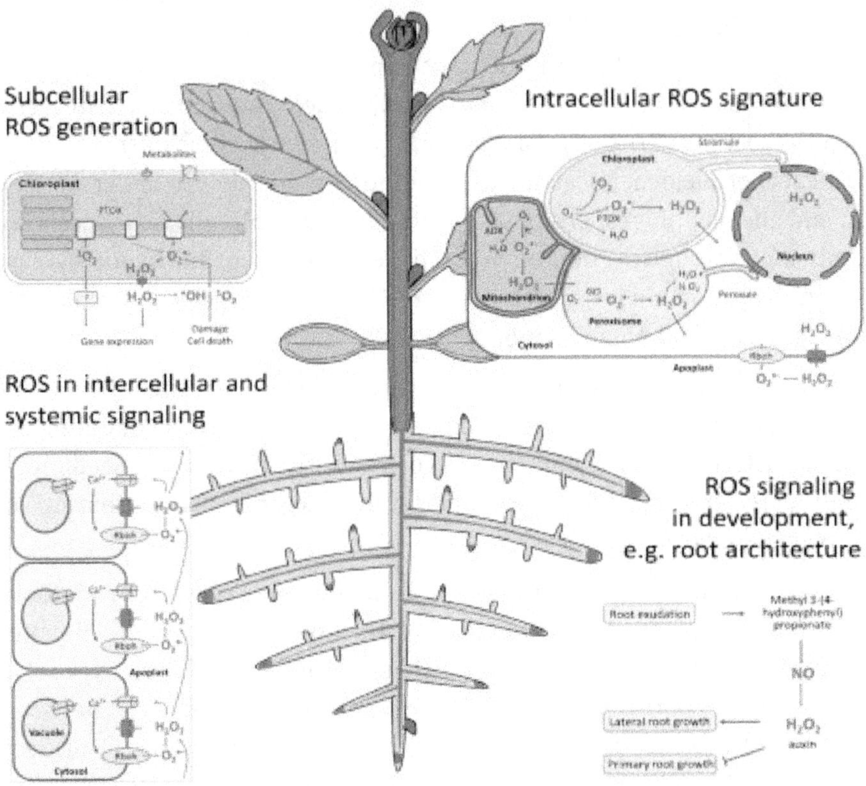

Fig. 1. Roles of reactive oxygen species (ROS) involving multiple regulations in different cellular compartments of a plant

Generally, ROS function as an oxidant of proteins and lipids by changing their functions through the release of single active compounds that regulate photosynthesis, flower senescence, pollen growth, root formation, and root hairs. Therefore, the activity of the enzymes catalase (CAT), monodehydroascorbate reductase (MDA), peroxidase (POD), superoxide dismutase (SOD), guaiacol peroxidase (GPX), and glutathione reductase (GSH) is an important indicator of plants under stress conditions.

SOD is one of the first protection lines for ROS and stomatal reactions involved in the O_2 metabolism in plants. It regulates lipid peroxidation and inhibits ROS damage to the membrane system, while hydrogen peroxide (H_2O_2) could be a strong oxidizing agent, producing highly reactive OH^- that may decompose into H_2O and O_2 by POD and CAT, thereby preventing possible ROS oxidative damage in plants. SOD, POD, and CAT highly suppressed physiological activity in rice that was exposed to volatile allelochemicals (octane and undecane), and this activity was gradually enhanced as their concentrations increased. This shows that the SOD and POD activity of a plant can be decreased by allelochemicals, thereby increasing the number of free radicals in the cells, which in turn leads to lipid peroxidation and the damaging of cellular and intercellular membranes.

To avoid self-destruction or necrosis in the cells, plants generally respond to oxidative stress by maintaining antioxidant defense compounds and continuously engage in removing the ROS at levels that reflect safety. Similarly, the antioxidant assay provides an insight into the metabolic phenomena that reflect physiological and biochemical responses in the plant's ecosystem. The physiological and biochemical mechanisms underlying the inhibition effect of allelochemicals are of paramount importance to understand the mode of action and allelopathic response in the plant. For example, allelochemicals restrained seed germination through the inhibition of radicle development and hypocotyl elongation, hindered seed germination by the synthesis of ROS, and damaged sub-cellular structures, protein metabolism, and phytohormones.

Peroxidases bind to cellular polymers through covalent interactions, and these bindings are thought to be involved in lignin biosynthesis and the development of the cross-linking within the cell membrane. Sometimes, their activity in seedlings initiates mitochondrial respiration, and this accounts for the oxygen molecules that are converted into hydrogen peroxides.

A marked increase in POD production in lettuce seedlings, from 13 to 15%, was evident at both chloroformic and methanolic fraction concentrations. Brassica napus extract increased POD, SOD, and lipid peroxidation in the radicle and hypocotyl of rice, sorghum, and rape compared to control. Similarly, researchers reported a substantial reduction in the POD activity of T. aestivum (monocot) and B. napus (dicot) recorded at 10,000 ppm with the methanol extract of Phytolacca latbenia Moq. H. Walter.

Catalases are important regulators of ROS in plant homeostasis. Although their regulatory activity is not fully understood, their reported functions include relieving cells from oxidative stress and improving organelles' integrity. The enzymes have been documented to play an important role in plant detoxification and anti-oxidative processes that are closely linked to ROS generation throughout photorespiration and photosynthesis. The inhibitory effect of Mentha piperita L. caused an increase in catalase activity from 6 to 10% in the aerial part of radish and tomato, which is an indication of the induction of stress.

Proline Content

Proline is the amino acid that has been linked to different plant stresses. Under abiotic stress, an accumulation of proline in the plant may be an adaptive and metabolic measure of stress, an inhibitor of lipid peroxidation, and a defense against toxicity. Proline is used by numerous organisms against the cellular imbalance caused by environmental stress. It is an inhibitor of lipid peroxidation and is usually synthesized by the plant in a high amount under stress conditions, where it serves as a scavenging molecule that protects the plant against abiotic stresses.

Proline was reported to increase ROS production, propel signaling, promote cellular apoptosis, and balance water stress in plants. The foliar application of M. sativa leaf extracts significantly raised the proline content of three wheat varieties. The proline contents of tomato, wheat, and cucumber treated with the aqueous extract of Calotropis procera (Aiton) Dryand were stimulated with an increase in the concentration of the extract.

MALONDIALDEHYDE CONTENT

Malondialdehyde (MDA) content serves as an important indicator of membrane damage due to oxidative pressure. ROS induce the oxidation of unsaturated fatty acids through the overproduction of MDA in the cells. The excessive secretion of ROS leads to the peroxidation of unsaturated lipid components, resulting in the damage of membrane rigidity and, eventually, cell dehydration. ROS induce a free radical reaction in the cell membrane which triggers the peroxidation of lipids, the escalation of which exacerbates the damage to the membrane, eventually leading to a leakage of electrolytes from the cells. Saidi et al. reported that bean plants exposed to 20 μM cadmium showed increased MDA levels compared to control plants. Kapoor et al. reported that the application of an aqueous extract of Artemisia absinthium L. and Psidium guajava L. resulted in an increased MDA content in P. hysterophorus seedlings which support free radical production and the occurrence of lipid peroxidation.

BIOHERBICIDES IN INTEGRATED WEED MANAGEMENT

An expanded and long-term approach to weed control is integrated weed management in which all available strategies including tillage, cultural practices, herbicides, allelopathy, and biological control are used to reduce the weed seedbank in soil, prevent weed emergence, and minimize competition from weeds growing with desired plants. Like chemical herbicides, bioherbicides may be most effective as a component in an overall management program rather than as a single tactic approach. This may be the most promising situation for bioherbicides as a practical management option in cropping systems. When considered as a three-part system, weed management offers several opportunities for integration of bioherbicides at critical stages during weed development: as seeds in soil, as growing and competitive plants, and during seed production.

INTEGRATING BIOHERBICIDES WITH CHEMICAL HERBICIDES

Several scenarios for integrating bioherbicides into weed management programs can be developed. Because most biological control agents are specific toward one weed and most production fields contain several predominant weed species, the use of bioherbicides for control of a single species in conjunction with herbicides selected for control of other weeds present is a logical approach.

Compatibility of the bioherbicide Fusarium solani f. sp. cucurbitae, which controls Texas gourd (Cucurbita texana), a problem weed in soybean and cotton in the southern United States, demonstrated that it could be integrated into a weed management strategy to broaden the spectrum of weed control within the crop. Integration with reduced rates of herbicides can successfully improve activity of mycoherbicides toward weeds.

For example, Phoma proboscis was more effective in controlling field bindweed (Convolvulus arvensis) when combined with sublethal doses of 2,4-D than when applied alone. Application of either of three herbicides at one-fourth rate with a sub-lethal dose of the pathogen Pyricularia setariae achieved complete control of green foxtail, which demonstrated a pathogen-herbicide synergy.

The fungus Colletotrichum gloeosporioides f. sp. malvae, endemic on round-leaved mallow, adequately controls this weed (about 75% kill) when applied alone as a bioherbicide. Several chemical herbicides are only effective on round-leafed mallow in the early seedling stage. Combinations of the bioherbicide with several herbicides at recommended rates were evaluated for postemergence control at the 4- to 5-leaf stage of growth.

Tank mixtures of the fungus with either metribuzin or imazethapyr greatly enhanced control and reduced biomass production over the fungus or the herbicide alone. These results clearly demonstrate that in some cases no single method is adequate for weed control and that combinations of methods are most effective.

INTEGRATING BIOHERBICIDES WITH CULTURAL PRACTICES

Cultural practices offer convenient application methods for integrating bioherbicides in cropping systems. Crop rotation is a practice that may also be manipulated to encourage development of specific inhibitory bacteria on roots. Tillage can influence the frequency of inhibitory bacteria occurring in soil and their growth-suppressing activity.

Greater proportions of indigenous rhizobacteria inhibitory to downy brome and jointed goatgrass were detected under either conventional or reduced tillage compared to no-till. This finding suggests that application of selected DRB during tillage may be effective in integrated weed management. Vegetative residues at or near the soil surface could serve as substrates for production of weed-suppressive chemicals by DRB applied as bioherbicides directly to the residues.

Previous work reporting a rotation effect in corn was due partly to certain rhizobacteria specifically associated with corn roots illustrates the potential for using DRB to achieve suppression of weeds in crop rotation systems.

Increasing crop interference in the field by manipulating row spacing, seeding rates and other cultural practices to suppress early weed growth has been proposed as a viable component of integrated weed management. Selection of highly competitive and allelopathic soybean varieties and matched with compatible bioherbicides may provide early-season weed suppression and require only minimal subsequent post-emergence weed control.

BIOHERBICIDE AND MANAGEMENT OF SEED BANKS AND SEEDLINGS

Prevention of seed germination and seedling emergence is fundamental to maximal effective, longterm weed management. Thus, bioherbicides can play a significant role in reducing weed infestations by attacking seeds and seedlings before they become competitive with crop plants. Several approaches for managing the seed bank and seedling emergence have been described including direct application of biotic agents to soil or crop residues, or to crop seeds to prevent emergence of weeds in the crop seed-germination zone, and in combination with solarization for enhancing seed deterioration in soils. Also, certain agrochemicals that stimulate seed imbibition or germination can be combined with bioherbicides containing selected seed-attacking microorganisms, incorporated into soil, and kill germinating weed seeds.

SUSTAINABLE AGRICULTURAL SYSTEMS

Crop production enterprises that avoid or restrict the use of chemical herbicides both in production fields and in environmentally-sensitive areas favor the adoption of bioherbicides. For simplicity, sustainable agriculture encompasses similar systems known as alternative, natural, organic, biological, ecological, and biodynamic farming. The sustainable agricultural systems involve a range of technological and management options to reduce costs, protect health and environmental quality, and enhance beneficial biological interactions and natural processes. In nearly all cases, little, if any, synthetic chemicals (including herbicides) are used.

Thus, sustainable agricultural systems provide the greatest opportunities to study, refine, and implement nonchemical weed management that will yield valuable information for developing improved bioherbicides and advancing their use in broader biologically-based weed management systems. Because the demand for bioherbicides for sustainable agriculture and ecosystem management is currently small relative to chemical herbicides, such products may be provided most efficiently through small-scale, specialized industries or even on-farm production facilities focused on "niche markets".

BIOHERBICIDES AND BIOLOGICAL WEED MANAGEMENT

Bioherbicides may be most effective in managing weeds as a component in a biological weed management system that is associated with sustainable agriculture. Biological weed management involves the use of diversity of biological agents such as bioherbicides and other biopesticides, and biological approaches including allelopathy, crop competition, and other cultural practices to obtain similar dramatic reduction in weed densities often associated with herbicide use.

Because some sustainable agricultural systems demand pesticide-free crop production (i.e., organic farming), the approach to weed control resembles a biological weed management system in which bioherbicides are major components. Many of the approaches are similar to those for integrated weed management except that herbicides are not involved.

Therefore, bioherbicides that would not be used in conventional integrated weed management because of unacceptable efficacy or too much time required for realization of effects would be under practical use in biological weed management.

For example, prevention of weed seed production and reduction of the seed bank could be attained using a bioherbicide consisting of seed pathogens applied to weeds infesting the crop.

However, the impact of the bioherbicide would not be evident for one to two years when noticeable decreases in weed seedling densities occur due to reduced weed seed populations in the soil.

Cover crops and mulches as components of sustainable management systems may be used for integrating bioherbicides by delivering the agents on seeds and promoting their establishment in soils for attack on weed seeds and seedlings prior to planting the main crop. Recent research demonstrated that several cover crop species inoculated with a DRB bioherbicide at planting maintained DRB populations on their roots and in the rhizosphere thereby promoting root colonization of giant foxtail (Setaria faberi) seedlings that emerged early in the growing season of the main crop after the cover crops were terminated.

Combined effects of the DRB and allelopathic activity of the cover crop residues suppressed the growth of the weeds. Similarly, a "system management" approach where a specific weed growing with a crop is sprayed with a post-emergence bioherbicide and the crop is underseeded with a living green cover crop has resulted in successful control of the target weed as well as the remaining weed flora by the cover crop.

Also, agents in formulation applied at planting can attack weed seeds and seedlings through delivery of bioherbicides to soil by either direct inoculation of crop seeds or by promoting colonization of crop roots. Crop roots not only might deliver microbial agents to adjacent weed roots but may also maintain or even enhance the agents' numbers for attack of seedlings emerging later in the season. In sustainable agricultural systems the use of synergisms can be explored in which the combined use of two or more methods enable bioherbicides to control weeds more effectively than when used alone. Bioherbicide efficacy on hemp sesbania (Sesbania exaltata) was increased by combining selected bacteria with the fungal pathogen, Colletotrichum truncatum.

Combinations of a Colletotrichum sp. bioherbicide with a naturally-occurring rust fungus allowed the bioherbicide to infect the weed host (Xanthium sp.) through rust lesions resulting in death of the plant. A seed-feeding insect combined with seed-attacking fungi significantly decreased velvetleaf seed viability and seedling emergence and increase seed infection compared to either the insect or fungus alone. Pre-dispersal seed mortality of weeds escaping herbicide control may be effective in manipulating and reducing seed banks in soil. A practical application of soil-applied detrimental bacteria and fungi combined with insects would be in situations where the insect feeds on roots or crowns of target weeds.

The very nature of high inputs of organic amendments and green manure crops in sustainable agricultural systems promotes the ability of crops to compete more vigorously with weeds, which intuitively suggests that efficacy of bioherbicides would also be enhanced when used with these amendments.

Indeed the bioherbicidal fungus, Trichoderma virens, produced the phytotoxin viridiol when grown in organic substrates such as peat and composts making this bioherbicide ideal for use in biological weed management. This bioherbicide combined with herbicidal compounds released by a rye cover crop extended weed suppression throughout the growing season of selected horticultural crops.

Furthermore, Trichoderma virens also produces biocidal compounds effective against fungal plant pathogens, which suggests the possibility of developing biotic agents with efficacy toward multiple pests.

Sustainable agricultural systems are positioned for implementation of novel tactics and approaches for manipulating the field environment to enhance the activity of indigenous pathogens of weeds present as natural bioherbicides. This strategy of weed suppression, also known as "conservation biological control," is an ecological concept that has only recently received attention from an agroecosystems standpoint. Management activities that affect soil organisms can directly affect weeds as demonstrated with reduced tillage, maintenance of high soil organic matter, reduced inputs of agrichemicals, which lead to high populations of deleterious soil microorganisms that contribute to natural weed suppression.

Specific examples are limited, however, researchers reported a natural population of the fungal pathogen Verticillium sp. significantly suppressed velvetleaf growth in soybean under reduced tillage. Additions of composted swine manure to soil inhibited germination and seedling emergence of three weed species possibly by enhancing soilborne weed-suppressive microorganisms. As pointed out by researchers, conservation biological control should not be considered the primary weed control strategy but should be investigated as a component of an integrated weed management program.

FUTURE DIRECTION OF PLANT-BASED BIOHERBICIDES

Despite the recent developments in the research of plant-based bioherbicides, there is still much room for scientists to explore new solutions and improve existing ones. Taking an example in Malaysia, a new bioherbicide product—WeedLock, which was developed from a plant extract and marketed locally by EntoGenex Industries Sdn. Bhd. since 2017—is a contact non-selective herbicide that controls a wide range of weed species.

Moving forward, more emphasis should be given to advancing the formulation of bioherbicide active substances, which is a key factor influencing the efficacy of a bioherbicide product. In particular, techniques employing multiple surfactants or nano-formulation could be considered to enhance the penetration and absorption of active herbicidal compounds in plants. Moreover, we should look into isolating and identifying new allelopathic compounds from invasive weeds and, upon validating their bioactivity under laboratory and field conditions, recommending them for herbicide development for sustainable agriculture.

Due to their easy degradability and high cost, the direct use of allelochemical extracts as bioherbicides in the field is not feasible in many cases. This could be resolved if the chemical industries elaborated appropriate formulations of the bioherbicides to improve their activity and synthesized the relevant allelochemical compounds whenever this is more convenient that obtaining them from nature. Researchers working on biostimulants shall continue to identify new and potential plant species containing a wide range of phytotoxic compounds acting at the molecular or cellular level of target plants and investigate the antioxidant enzymatic activity related to the mechanism of inhibition and tissue injury of these compounds.

11

BIOPESTICIDES FOR MANAGEMENT OF AQUATIC WEEDS

INTRODUCTION

Aquatic weeds such as muskgrass (Chara spp., Charales: Characeae), water hyacinth (Eichhornia crassipes, Commelinales: Pontederiaceae), water lettuce (Pistia stratiotes, Alismatales: Araceae), hydrilla (Hydrilla verticillate, Alismatales: Hydrocharitaceae), filamentous algae (Lyngbya wollei, Cyanobacteria: Oscillatoriales), and duckweed (Lemna minor, Alismatales: Araceae) are widespread in the Everglades Agricultural Area (EAA) of South Florida. This situation is especially prevalent during the late spring to early fall when warm temperatures promote the growth of the aquatic weeds. Within the EAA, the total drainage water conveyance area that is not subjected to weed control can be covered on average by up to 50% of aquatic weeds.

This coverage drops to 20% in the areas where aggressive aquatic weed control is applied. Although floating aquatic weeds are capable of sequestering and storing high concentrations of phosphorus (P) from the drainage water and sediment, they release this nutrient back into the ecosystem once they senesce and die. Large populations of floating aquatic weeds can reduce light penetration to the deep layer of the waterbody, thus resulting in a reduced primary production. Moreover, excessive presence of aquatic weeds in farm canals can block water outlets during irrigation, clog canals, and impede water flow, resulting in low oxygen and poor water quality.

Allelopathy is a biological phenomenon during which an organism releases allelopathic chemicals into the environment. Allelochemical compounds can influence the germination, growth, survival, and reproduction of neighboring organisms, and these effects can either be beneficial or harmful. Biological functions that are affected include cell division, membrane permeability, respiration, and photosynthesis. In the past decades, the potential applications of allelopathy in agriculture have been hypothesized and investigated. Pesticidal effects of several plants were demonstrated. The aqueous extract of neem (Azadirachta indica, Sapindales: Meliaceae) killed a wide range of harmful insects. Water hyacinth reduced the feeding rate of an insect pest, Spodoptera litura (Lepidoptera: Noctuidae), and was considered as a potential bioinsecticide for the management of this insect.

Among 13 different Chara populations screened against microalgal strains including cyanobacteria and eukaryotic microalgae, 10 inhibited growth of cyanobacteria. After one day, water lettuce and duckweed had a significant negative effect on germination and root growth of a weed, common lambsquarters (Chenopodium album, Caryophyllales: Amaranthaceae). These findings revealed that some plants, including aquatic weeds, contain effective allelopathic compounds that can possibly be used as biopesticides.

The overall purpose of this study was to investigate the allelochemical properties of six aquatic weeds from Florida towards 100 bacterial strains, an insect pest, and three terrestrial weeds. The six aquatic weeds were chosen because they are the most prevalent and invasive species in the EAA watershed. More specifically, the research objectives were to determine the effects of extracts of these weeds on (i) the growth of bacteria collected from plants and soil samples in Florida; (ii) the survival, growth and feeding preference of larvae of the fall armyworm (FAW, Spodoptera frugiperda, Lepidoptera: Noctuidae), an insect pest of corn in Florida; and (iii) the germination and growth of nutsedge (Cyperus esculentus, Poales: Cyperaceae), amaranth (Amaranthus tricolor, Caryophyllales: Amaranthaceae), and common ragweed (Ambrosia artemisiifolia, Asterales: Asteraceae), three terrestrial weeds in Florida.

PLANT MATERIALS IN WATER

Entire plants of muskgrass (a genus of alga), water hyacinth, water lettuce, hydrilla, filamentous algae, and duckweed were collected from multiple farm canals within the EAA of South Florida (26°39'58" N; 80°37'45" W) during summer of 2018. These canals were either located on University of Florida's property or on private properties for which authors had permission to access and collect samples.

Preparation of powdered aquatic weeds and aquatic weed extracts

All six aquatic weeds were washed utilizing a 1.0-L beaker and deionized (DI) water. The beaker was filled with aquatic weeds to the 0.5 L mark, then DI water was added until the resulting mixture reached the 1.0 L mark. The mixture was then stirred 10 times; after which the DI water was disposed. This process was completed twice, each time utilizing fresh DI water. The plants were then dried in a hot room at 50°C for 7 days until they were completely dry. The dried materials were finely ground to 1 mm particle-sized powder with a Thomas Model 4 Wiley laboratory mill and then stored in the dark in a plastic container at 25°C until further use.

Aquatic weed extracts were prepared using 5 g of powdered aquatic weed mixed with 50 ml DI water and 10 ml of 97% ethanol in a 200 ml wide-mouth conical flask. The mouth of the conical flask was sealed with aluminum foil, and the mixture was heated at 75°C for 4 hours in a Precision Thelco oven model 16. After heating, each flask was placed in a sonic bath for 1 hour and the aqueous residue was first filtered through cheese cloth followed by a Fisherbrand P5 5–10 ¼m filter paper. Ten to 30 ml of weed extract were obtained per batch. The nutrient composition of aquatic weed extracts and aqueous ethanol solvent was determined by the Soil, Water and Nutrient Management Laboratory of the University of Florida at Belle Glade. Aquatic weed extracts used in bacterial growth assays described below were filtered through a 0.22 ¼m syringe filter (Millipore) before incorporation into the culture media.

Media used for isolation and culturing of bacteria

The basic medium (BM) used to isolate and grow bacteria contained yeast extract (5 g), bactopeptone (5 g), sucrose (5 g), agar (15 g), and 1 L DI water. Medium pH was adjusted to 6.8–7.0 with the Fisher Scientific buffer solution pH 10.00 or pH 4.00 before autoclaving the medium at 120°C for 20 minutes. After autoclaving, when the agar medium had cooled down to around 50–60°C, 0.1 g cycloheximide mixed in 5 ml of 97% ethanol solution and 1 µl Tilt® 250 EC fungicide (propiconazole, Syngenta Crop Protection, Greensboro, NC) were added and mixed evenly. Approximately 20 ml of the BM supplemented with the anti-fungal compounds were poured into sterile Petri dishes of 10 cm diameter, and maintained at room temperature under a laminar hood until solidification of the medium.

Blank plates contained BM supplemented with the aqueous ethanol solvent used to prepare aquatic weed extracts: 10 ml and 1 ml of the 16.17% ethanol solution was added to 1 L of BM after autoclaving to obtain a final solvent dilution of 1:100 and 1:1,000, respectively. Extract plates contained BM supplemented with one of the six microfiltered aquatic weed extracts (muskgrass, water hyacinth, water lettuce, hydrilla, filamentous algae, and duckweed): 10 ml and 1 ml of a plant extract were added to 1 liter of BM after autoclaving to obtain a final weed extract dilution of 1:100 and 1:1,000, respectively.

Isolation and storage of bacteria

During fall 2018, 100 bacterial strains were isolated from 25 plants (sweet corn, sorghum, papaya, etc.) and soil samples collected from the Everglades Research and Education Center (EREC) at Belle Glade and other locations in Florida (Milton, Gainesville, Ft. Pierce, and Homestead). Plants for isolation of bacteria were collected on University of Florida property or on public domain where no sampling permits were required. Under a laminar floor hood, an approximately 2-cm^2 area of each leaf sample was cut in a Petri dish into small pieces in 1 ml of sterile DI water.

Approximately 0.5 g of soil sample was added into a 14-ml polystyrene tube and mixed with 8 ml of sterile DI water. One hour after homogenization of the leaf or soil material in water, a drop of liquid was transferred with a smear loop to a plate containing BM. The liquid sample containing bacteria was spread evenly on the medium with the smear loop before sealing plates with parafilm (M PM996 All-Purpose Laboratory Parafilm Film).

Plates were incubated at 28°C for 2–3 days. For each plant or soil sample, single colonies with different colors or different growth rates were sub-cultured on fresh BM plates. Once bacteria were sub-cultured successfully, each BM plate was inoculated with 16 bacterial isolates.

For storage of the bacterial collection, a bacterial suspension (>10^{10} colony forming units/ml) was prepared for each isolate in a 2 ml tube containing 0.5 ml of 50% glycerol and 1.5 ml sterile DI water. Each bacterial suspension was stored in a freezer at -80°C until further use.

EFFECT OF WEED EXTRACTS ON BACTERIAL GROWTH.

Each bacterial isolate was streaked with an inoculation loop on three media: BM, BM + solvent (1:100 and 1:1,000 dilution) and BM + weed extract (1:100 and 1:1,000 dilution). BM was considered the positive control growth medium, BM + solvent was the blank medium without weed extract, and BM + weed extract was the testing medium. Each plate was inoculated with 16 bacterial isolates. All the plates were incubated at 28°C for three days and growth of bacterial cultures was observed after this incubation time.

When bacteria did not or only partially grew on the testing medium (corresponding to possible growth inhibition), a dilution plating method was performed. This method allowed us to determine the concentration (number of colonies) of an unknown sample by counting the number of colonies cultured from serial dilutions of the sample, and then back track the measured counts to the unknown concentration.

The first step was to take a single bacterial colony of stock and to place it into 0.9 ml DI water, and this was the original stock. The technique used to make a single dilution was repeated sequentially using more and more diluted suspensions, at each step: 0.1 ml of the previous suspension was added to 0.9 ml of DI water, each step resulting in a further 10-fold change in the concentration from the previous concentration. The first tube after the original stock was a 1:10 dilution, the second a 1:100, the third a 1:1,000, etc., to the sixth, a 1:10^6 dilution. The second step was to determine the number of viable cells in each of the dilutions. A pipette was used to transfer 50 ¼l of each dilution, starting from 1:10^6 (the most diluted) to 1:10 and finally from the original stock onto an agar plate of each of the three tested media.

The plates would have different numbers of colonies depending on the dilution of the sample. When too many colonies grow, it can be very difficult to count them. Thus, we counted the number of colonies only for dilutions for which less than 50 colonies grew. The bacterial growth was considered to be partially inhibited when the bacterial colony number on BM + weed extract at 1:10^n dilution was more than 50 percent less than the bacterial colony number on BM at 1:10^{n-1} dilution.

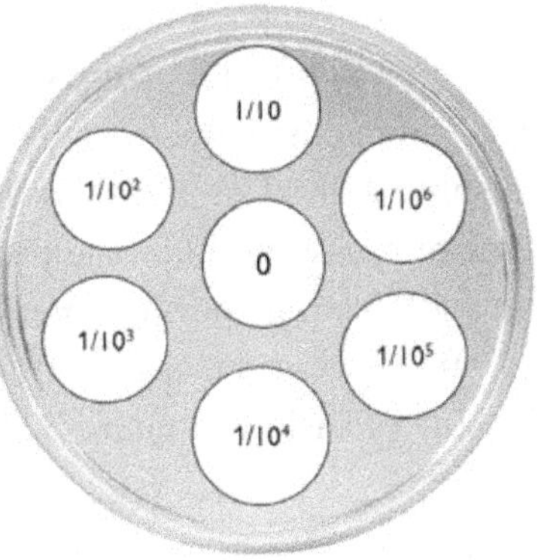

Fig 1. Schematic aspect of a serial dilution plate to determine the concentration of a bacterial suspension.

INSECT MATERIALS

7-day old FAW larvae obtained from a colony maintained in the laboratory at the EREC were used in biossays. The colony was initiated in October 2018 using approximately 150 larvae collected in the field on sweet corn at the EREC. The colony has been maintained in a climate-controlled room at 25.5±1°C, 40–60% RH (relative humidity), and L14:D10 h (light: dark hour) photoperiod. Larvae were individually kept in 37-ml translucent plastic cups (Dart Container Corp., Mason, MI) filled with Stonefly Heliothis Diet (Ward's Science, Rochester, NY). Upon larval development completion, 15–30 pupae were placed in moist vermiculite at the bottom of a 3.7-L carton container lined with cheesecloth and paper towels. Adult eclosion and oviposition was monitored 3 times a week, and cheesecloth and paper towels with FAW eggs were placed in transparent plastic bags. Newly emerged neonate larvae were transferred using a paintbrush from plastic bags to cup with diet until pupation or use in bioassays.

EXPERIMENTAL TREATMENTS.

The organic insecticides used as commercial standards were pyrethrins (PyGanic Crop Protection 5.0 ECII, Valent U.S.A., Walnut Creek, CA) for larval dip bioassays and Bacillus thuringiensis subsp. kurstaki (Bt, Dipel DF, Valent BioSciences, Libertyville, IL) for larval diet incorporation bioassays. PyGanic 5.0 ECII is registered for foliar applications in the United States at 329–1,242 ml/ha and a concentration of 4.2 ml/L was used in this study. Dipel DF is registered for application in the United States at 560–2,242 g/ha and a dosage of 7.5 g/L was used in this study. In addition to commercial standards, bioassays were conducted with six aquatic weed extracts and one control solution at 1:10 dilution.

EFFECT OF WEED EXTRACTS ON FAW SURVIVAL AND GROWTH.

Laboratory experiments were conducted to determine the effect weed extracts on FAW larval survival and growth. Contact activity of the extracts was determined in larval dip bioassays, whereas ingestion activity was determined in larval diet incorporation bioassays. Each type of bioassay was conducted four times using two different FAW generations and two different FAW cohorts two days apart per generation in order to reduce the impact of a specific cohort on the results. A total of 130 FAW larvae (replications) were used for each treatment in each type of bioassay (50 larvae per treatment for each of the two cohorts of the first generation and 15 larvae per treatment for each of the two cohorts of the second generation). FAW larvae were starved overnight to void their gut and then weighed immediately on a TS400D Precision Standard Balance (Ohaus Corp., Parsippany, NJ) before each bioassay.

For larval dip bioassays, each larva was individually dipped for 1 second in the treatment solution. The larva was subsequently placed on filter paper for 10 seconds, and then in a cup with diet. Dipped larvae were reared in the insectary for 4 days. For larval diet incorporation bioassays, each larva was placed in a cup with treated artificial diet prepared with a treatment solution. Transferred larvae were reared in the insectary for 4 days. Larvae were considered alive if they could right themselves after being flipped on their back and prodded for a maximum of 10 seconds with the blunt tip of a metallic pin. The live larvae were subsequently starved overnight and weighed to determine their relative growth rate (RGR). The RGR of a larva is its daily growth rate relative to its average size over the duration of an assay. The RGR was calculated for each live larva using the following formula:

$$RGR = \frac{\text{Final weight - Initial weight}}{(\text{Initial weight } + \text{ Final weight }/2) * \text{Feding duration}}$$

with weight expressed in grams and feeding duration expressed in days.

EFFECT OF WEED EXTRACTS ON FAW FEEDING PREFERENCE

A laboratory experiment was conducted to determine the effect of aquatic weed extracts on FAW larval feeding preference. This experiment consisted of a choice bioassay conducted twice using two different FAW and sweet corn plant cohorts. For this experiment, we used plastic Petri dishes (15 mm x 100 mm diameter, Cole-Parmer Instrument Company, LLC, Vernon Hills, IL). A total of 40 Petri dishes (replications) were prepared (20 per bioassay). For each bioassay, 2-week old sweet corn plants ('Cabo' hybrid, Syngenta, Greensboro, NC) grown in a greenhouse were randomly selected and leaves were cut into leaf discs with a disc cutter (1.6 cm in diameter).

Each leaf disc was dipped for 1 second into the aquatic weed extract or the control solution. The Petri dishes were lined with one layer of wet filter paper to prevent leaf disc and FAW larva desiccation. The treated discs were placed on filter paper to remove the excess liquid, and then in the Petri dishes. Each Petri dish included all experimental treatments along a circular layout. One FAW larva was placed at the center of each Petri dish. All the Petri dishes were kept in the climate-controlled room and subsequently observed 24 and 48 hours after the FAW larva was released (Fig 2).

TERRESTRIAL WEED MATERIALS

During fall 2019, the tubers of nutsedge (Cyperus esculentus) and the seeds of amaranth (Amaranthus tricolor) and common ragweed (Ambrosia artemisiifolia) were collected in the field at the Southwest Florida Research and Education Center (SWFREC) in Immokalee, FL.

Effect of weed extracts on terrestrial weed germination and growth in greenhouse

In the greenhouse, for each species of terrestrial weed, 10 seeds or tubers were planted 1–2 cm deep in 1-L plastic pots. The soil used for planting was Black Gold all-purpose potting mix with a pH of 5.5–6.5 and N-P-K of 0.13–0.04–0.13. Each pot was filled with approximately 340 g of potting mix.

Four rates (0, 0.1, 0.5, and 1.0 g) of ground powder of aquatic weeds were added on the surface of the pot (103 cm^2) and mixed evenly with top soil. Pots with no aquatic weed extract were used as control. Four replications were prepared for each treatment and each application rate. All the pots were kept in the greenhouse at 27°C and 50% RH, and sprinkle-watered every day to the field capacity of each pot. The sprinkler was set to mist level to avoid the leaching of aquatic weed powder by the heavy drops of water. The germination rate of plants in each pot was recorded for five weeks after planting (days 8, 12, 18, 22, and 39). Visible coleoptiles on the soil surface were used as an indicator of plant emergence.

Data were determined as the total number of germinated seeds/tubers divided by the total number of seeds/tubers in each pot. After six weeks, the fresh biomass weight of each pot was assessed by cutting plants at the soil level and weighing them using a laboratory scale (Ohaus; model: PX124).

Effect of weed extracts on terrestrial weed germination in laboratory.

The effects of different dilutions of the aquatic weed extract solutions (dilution at 1:100 and 1:10) were studied using a laboratory bioassay. Five tubers of nutsedge, 10 seeds of amaranth and common ragweed were germinated in a 100 mm diameter polystyrene Petri dish lined with two layers of 90 mm size Whatman 1 filter paper. The filter paper was moistened with 5 ml of each extract solution. The same amount of aqueous ethanol solution was used for the aqueous ethanol control, and the RO water solution was used as water control. Four replications were prepared for each treatment and each dilution. All Petri dishes were sealed with PM996 All-Purpose Laboratory Parafilm to reduce the loss of water and then maintained in a growth chamber with light from 7 am (25°C, 50% RH) to 7 pm. The temperature of the growth chamber at night was 22°C with 50% RH. All Petri dishes were randomly distributed in the growth chamber. Visible coleoptiles on the soil surface were used as an indicator of plant emergence. Germination values were calculated by dividing the total number of germinated seeds or tubers by the total number of seeds or tubers in the Petri dish. Germination values were determined after one week of growth.

STATISTICAL ANALYSIS

For the antimicrobial streaking assay, the evaluation of bacterial growth was only qualitative (yes or no or weak) and not quantitative. These qualitative data can therefore not be statistically analyzed. The second assay performed with three isolates was based on quantitative data (colony counts) and has been statistically analyzed using a one-way ANOVA test.

Statistical analyses were conducted using SAS PROC GLIMMIX. For the larval dip and diet incorporation bioassays, linear mixed models were applied to compare survival rate and RGR of FAW as affected by treatment. Treatment was a fixed effect, whereas generation and cohort(generation) were random effects. For the leaf disc assays, linear mixed models were used to compare leaf area consumed by FAW larvae as affected by treatment and observation time. Treatment, observation time, and their two-way interaction were fixed effects. Assay, dish(assay), and treatment x dish(assay) were random effects. Thus, a variance component covariance structure was used to model the effects of repeated measures of the leaf disc assays. The Kenward-Roger adjustment for denominator degrees of freedom was used to correct for inexact F distributions. Least square means ± SEs from the least square means statement output are reported unless stated otherwise. The Tukey-Kramer adjustment (Alpha = 0.05) was used for pairwise separation of least-square means when a fixed effect was detected ($P<0.05$).

Statistical analyses were conducted using SAS PROC GLM to compare germination rates and biomass affected by each treatment rate of the powdered aquatic weed experiment. For the aquatic weed extract experiment, the PROC GLM procedure was used to compare germination rates affected by each treatment and dilution. Pairwise mean separation was performed when using the Tukey-Kramer adjustment ($P < 0.05$).

ANTIMICROBIAL ACTIVITY OF AQUATIC WEEDS

A total of 100 different bacterial strains were isolated from 18 plant and 7 soil samples. After culturing these bacteria on culture medium with six different weed extracts, the growth of 97 of them was not affected by any of the six aquatic weed extracts. Only three of the 100 strains showed a partial growth on the medium with extracts at 1:100 dilution. These three strains included isolate 11 and isolate 72 on BM + muskgrass extract, isolate 58 on BM + water hyacinth extract, and BM + filamentous algae extract. After further testing of these isolates by the dilution plating method, the bacterial colony counts for isolate 11 on BM + muskgrass were not significantly different ($P = 0.386$) from the colony counts on the control plates. Similarly, colony counts for isolate 72 on BM + muskgrass and for isolate 58 on BM + water hyacinth were not different from those on their respective control plates ($P = 0.540$ and $P = 0.056$).

Only bacterial counts for isolate 58 on BM + filamentous algae were significantly lower (P = 0.013) compared to the control plates (54% growth reduction). This suggested that the partial growth observed in the first assay for these isolates (with the exception of isolate 58 on filamentous algae extract) was due to insufficient inoculum during streaking and not an inhibitory effect of the culture medium.

Consequently, there was overall no visible difference in bacterial growth of the 100 bacterial strains on plates containing weed extracts and on control plates. These results suggested that the six aquatic weed extracts did not contain any antibacterial compounds or only at very low concentrations, or that biologically active molecules were inactivated during the extraction process.

In a similar study, water lettuce was shown to have an antimicrobial activity based on inhibition zones of Escherichia coli, Bacillus subtilis, and Klebsiella pneumonia. However, only fresh samples of water lettuce had a growth inhibitory effect, whereas an aqueous extract did not show this activity. Antibacterial activity of medicinal plants after extraction with different solvents suggested that extractions made with acetone and methanol were more efficient than water to recover molecules with antibacterial activity.

Methanol might be more effective at recovery of higher concentrations of antibacterial compounds. Abraham et al. tested antibacterial activity of different extracts of water lettuce using clinical pathogens. Compared to chloroform and hexane, methanol extracts resulted in higher yields of chemicals such as alkaloids, phytosterols, phenols, flavonoids and tannins that possess bioactive properties.

The studies mentioned above confirmed the hypothesis that aquatic weeds can contain allelochemicals with antimicrobial properties. These studies also showed that the methods of extraction are critical to obtain a positive result. The negative results obtained in our study can be associated with the extraction method or the bacteria used for testing.

Therefore, in the future, instead of using water and ethanol extracts, different organic solvents such as methanol and acetone could be investigated to identify antibacterial compounds in aquatic weeds of the EAA. Furthermore, we tested diverse bacteria isolated from an agricultural environment. Another option could be to test bacteria from a clinical environment.

INSECTICIDAL ACTIVITY OF AQUATIC WEEDS

For the larval dip bioassay, differences in survival rate were not detected among treatments, including the control (P = 0.994) and commercial treatment (P = 0.979;). However, differences in RGR were detected among treatments (P < 0.001). The highest RGR, 0.415 g g^{-1} day^{-1}, was observed with filamentous algae and this RGR value was 1.05-fold higher than the RGR values for the commercial treatment and muskgrass (P < 0.001).

However, none of the treatments were different than the control (muskgrass: P = 0.951; water hyacinth: P = 0.828; water lettuce: P = 1.0; hydrilla: P = 0.987; filamentous algae: P = 0.164; duckweed: P = 0.829). Therefore, the dermal contact of aquatic weed extracts did not have an apparent effect on FAW survival or growth. It was unexpected for the commercial pyrethrins to not affect FAW, but according to Abrahams et al., FAW has become resistant to insecticides in modes of action groups 1A (carbamates),1B (organophosphates), and 3A (pyrethroids and pyrethrins) in several parts of the Americas, thus higher doses or alternative chemicals or methods have to be applied. This may explain why the commercial treatment used in our study was not effective against FAW.

For the larval diet incorporation bioassay, average survival ranged from 5.5 to 93%. Differences in survival were detected among treatments (P < 0.001;). Survival in the commercial treatment was 93 to 94% lower than in any other treatment (P < 0.001), including the control. Survival in water lettuce was 14% lower (P < 0.001) than in the control, but also 12% lower than in hydrilla, and 15% lower than water hyacinth. All other survival rates of aquatic weed extracts were not significantly different from the control (muskgrass: P = 0.928; water hyacinth: P = 1.0; hydrilla: P = 1.0; filamentous algae: P = 0.132; duckweed: P = 0.829). The RGR of FAW larvae in the diet incorporation bioassay differed among treatments (P < 0.001).

The commercial treatment had a negative RGR (-0.0215 g g^{-1} day^{-1}), which was lower than any other RGRs determined in the bioassay. This observation indicated that in general, the larvae that survived this treatment lost weight. However, hydrilla caused 11.3% reduction in RGR compared to the control (P < 0.001). The RGR of larvae feeding on the diet with hydrilla was also significantly lower than for water lettuce (1.1-fold; P < 0.001) and water hyacinth (1.1-fold; P < 0.001). Duckweed reduced RGR of larvae by 9.4% (P < 0.001) relative to the control. The RGR associated with duckweed was also 8.9% lower than water lettuce (P < 0.002), and 11.3% lower than water hyacinth (P < 0.001). The RGR for other aquatic weed extracts were not significantly different from the RGR of the control (muskgrass: P = 0.791; water hycinath: P = 0.994; water lettuce: 1.0; filamentous algae: P = 0.554). B. thuringiensis showed a strong insecticidal effect on FAW larvae with high mortality and low RGR.

The average value of RGR was negative, which indicated that the larvae that survived generally lost weight and lived in a very poor health after consuming B. thuringiensis. This result is consistent with the previously reported effectiveness of insecticidal properties of toxins produced by this bacterial species. Water lettuce showed a negative impact on FAW larvae survival, and hydrilla and duckweed demonstrated reduction in FAW larvae growth. A chemical analysis by Tripathi et al. revealed that the biologically active chemical constituents of water lettuce are alkaloids, glycosides, flavonoids, tannins and steroids. Hydrilla contains phenolic and hydroxy acid.

Loliolide was also detected in ethyl acetate and ethanol extracts. An earlier study identified tannins as inhibitor of the growth of various species of pest insects including Lepidoptera larvae. Those tannins can impede the activity of proteases and thus result in negative effects on larvae health and growth. Loliolide was reported to have diverse biological properties including repellency against insects.

In the leaf disc bioassays, the percentage of leaf area over the two observation times was between 67% and 77% among treatments, but there was no difference among treatments including the control (F = 0.87; df = 6,234; P = 0.517;). There was a significant difference between the two observation times (F = 162.51; df = 1,273; P < 0.001). An interaction between treatment and observation time was not detected (F = 1.06; df = 1,273; P = 0.388). Thus, the feeding activity of FAW larvae was not affected by any of the treatments in our study. Compounds that possess antifeedant activity can be found in all chemical classes in plants. Terpenes and alkaloids are the two classes that are especially effective in inhibiting feeding of a variety of insects. In a similar study, the leaves treated with methanol and n-hexane extracts of water hyacinth suggested high antifeedant efficacy on Spodoptera litura. More than 20 types of chemical compounds (including alkaloids, phenols and flavonoids) with potential insecticidal activity were identified by gas chromatography-mass spectrometry (GC-MS) analysis of methanol extract of water hyacinth. This suggested that some other organic solvents such as methanol and n-hexane might be efficiently used to obtain higher yields of effective compounds from aquatic plants against insect larvae.

Overall, water lettuce, hydrilla and duckweed showed significant effects on FAW larvae survival and growth by ingestion. However, the observed differences were small, and none of the aquatic weeds possessed contact and antifeedant effect towards FAW larvae. In the future, different solvents can be used to isolate higher contents of the allelopathic compounds in the aquatic weed that are likely to have better insecticidal activity.

HERBICIDAL ACTIVITY OF AQUATIC WEEDS

In the greenhouse, application of duckweed powder resulted in a growth reduction of nutsedge. At the rate of 0.1 g, 0.5 g and 1.0 g, powder of this aquatic weed reduced the biomass of nutsedge by 36.6%, 40.5% and 36.3%, respectively (P = 0.042). The other aquatic weed powders did not have any effect on nutsedge germination or growth as compared to the control.

Use of 1.0 g of filamentous algae powder reduced the germination of amaranth by 22% from day 17 to day 22, and by 21% on average (P = 0.037). The other aquatic weeds tested had no effect on germination of terrestrial weeds at any treatment rate and at any time. In terms of biomass, the application of muskgrass powder resulted in a significant reduction on amaranth growth (P = 0.002); 0.1 g muskgrass reduced the biomass by 58%, 0.5 g by 55%, and 1.0 g by 68%. Use of 0.1 g and 1.0 g of water hyacinth powder reduced amaranth biomass by 50% and 44%, respectively (P = 0.011).

However, the result obtained with 0.5 g was not different from the control. The same result was obtained with water lettuce: 0.1 g application resulted in 56% reduction, and 1.0 g in 53% reduction respectively (P = 0.026) while the biomass obtained after application of 0.5 g was not different from the biomass of the control. Both 0.5 g and 1.0 g of filamentous algae powder reduced biomass of amaranth significantly by 48% and 55%, respectively (P = 0.004). Hydrilla and duckweed powder showed no significant effect on amaranth growth at any rate.

The response of seed germination and growth of terrestrial weed species to six powdered aquatic weeds suggested that duckweed had an overall negative effect on nutsedge growth. Filamentous algae and muskgrass inhibited the germination and growth of amaranth, while water hyacinth and water lettuce have an inhibitory effect on the growth of amaranth.

These results are similar to the results by Bhadha et al., that the application of powdered water lettuce and filamentous algae demonstrated overall negative germination and root growth of plant species, including common lambsquarter, sorghum, snap bean and corn. Mode of action of some allelochemicals can be similar to synthetic herbicides, and different allelochemicals may have different modes of action or physiological target sites. Some allelochemicals, including flavonoids and fatty acids present in the aquatic weed powder, can act on different cellular processes, such as disrupting plasma membrane function and interrupting intracellular enzymes.

Besides directly targeting the weeds, the metabolites such as phenolics in the powder may release into the soil and alter its properties such as pH, organic matter, microorganisms, and nutrient status, resulting in a negative effect on plant germination and growth. The results also showed no significant difference between 0.1 g to 1.0 g of duckweed on the growth of nutsedge, and also in terms of muskgrass on the growth of amaranth. This might due to that all the rates are at the same dose-response level of inhibitory effect on a nutsedge. For future work, higher application rates should be used to investigate their potential effect.

In contrast, common ragweed showed a positive response of germination with the application of powdered muskgrass and also of growth with powdered water hyacinth and hydrilla. This might suggest that the allelochemical in muskgrass and hydrilla had a stimulatory effect on the growth of common ragweed. The stimulatory effect of allelopathy was described in; to show that the aqueous extract of mungbean (Vigna radiata) had a significant increase on seed germination and growth of plant species including sweet corn and okra. The result might also suggest that common ragweed had high tolerance to the allelochemicals in these aquatic weeds and used the nutrients of powder for its own growth.

Previous study had shown that aqueous extracts of the whole common ragweed plant inhibited seed germination of many plants including onion, oat, ryegrass, and Palmer amaranth. Some allelochemicals such as phenolic acids and sesquiterpene-lactones were produced by ragweed in order to compete for nutrients and resist other allelochemicals.

Allelochemicals are known to induce hormesis in several plant species. Hormesis refers to biphasic dose-response with a low dose stimulatory or beneficial effect and a high dose inhibitory or toxic effect. The current study showed that the lower rate (0.5 g) of hydrilla and muskgrass had a higher stimulatory effect on common ragweed than their higher rate (1.0 g) and suggested a hormetic response. A similar study also described the hormetic effect of aquatic weeds, that low application rates (0.03 g and 0.06 g) of water lettuce and filamentous algae promoted the root growth of corn, sorghum, and snap bean while high application rates (1.0 g) had an inhibitory effect on their root growth. To further confirm the existence of hormesis, more application rates should be used in order to generate a more reliable dose-response model.

In the Petri dish experiment with aquatic weed extracts, aqueous ethanol control at 1/10 dilution showed no germination for amaranth, and nearly 70% less than the water control for ragweed. This suggested that the ethanol at 1/10 dilution had a potentially negative effect on seed germination of amaranth. Previous studies demonstrated that ethanol had different effects on the germination of different types of plant seeds.

For instance, germination percentage of ryegrass (Lolium) seed decreased with ethanol concentration over 0.8% (v/v), while for bermudagrass (Cynodon), germination rate was increased with concentrations of ethanol from 0.1% to 3% (v/v). This also demonstrated that different concentrations of ethanol result in different effects on seed germination. However, filamentous algae at 1:100 dilution showed an inhibitory effect on seed germination of amaranth compared to aqueous ethanol control, which indicated the allelochemicals in the extract had an herbicidal effect on amaranth.

In this study, the presence of ethanol caused a significant reduction in seed germination of terrestrial weeds tested at 1/10 dilution, which made it hard to tell the activity of aquatic weed extracts at the same time. For future work, aqueous extraction could be used to avoid the impact of ethanol. In addition, other different types of solvents might be able to isolate the effective compounds. Previous studies described that aqueous methanol extract of duckweed was tested to have a strong inhibitory effect on root and shoot growth of terrestrial weeds including cress (Lepidium sativum L.), alfalfa (Medicago sativa L.), and timothy (Phleum pratense L.), and water lettuce aqueous methanol extract showed an inhibitory effect on shoot growth of cress and root growth of ryegrass (Lolium multiforum L.) and timothy. Nutsedge tuber failed to germinate in the Petri dish, and this may be due to the lack of sufficient moisture for the tuber germination in the Petri dishes. Vermiculite could be a potential alternative medium for testing tuber germination instead of filter paper as it can hold relatively more moisture and is sterilizable.

CONCLUSIONS

The objective of this study was to evaluate six aquatic weeds, including muskgrass (Chara spp.), water hyacinth (Eichhornia crassipes), water lettuce (Pistia stratiotes), hydrilla (Hydrilla verticillate), filamentous algae (Lyngbya wollei), and duckweed (Lemna minor), as potential biopesticides. The aquatic weed extracts were tested at dilutions 1:100 and 1:1,000 against 100 bacterial strains to evaluate their antimicrobial activity. Except for one bacterial isolate and one plant extract at the lowest dilution, the growth of these bacteria isolated from 25 plant and soil samples was not affected by the aquatic weed extracts in our testing conditions. The same aquatic weed extracts were used at 1:10 dilution to determine their insecticidal and antifeedant effects on the FAW.

No significant treatment effect on FAW survival or growth was observed with the dip bioassay. However, using a diet incorporation bioassay allowed us to show that water lettuce extracts reduced the survival rate of FAW by 14%, and hydrilla and duckweed caused 11% and 9% average reduction of FAW growth, respectively. The aquatic weeds had no antifeedant effects on FAW. Four rates of powdered aquatic weeds and two dilutions of aquatic weed extracts (1:10 and 1:100) were used to determine their herbicidal effect towards seed germination and growth of nutsedge, amaranth, and common ragweed. In a greenhouse study, duckweed at all rates showed an inhibitory effect on the growth of nutsedge, ranging from 36% to 41%.

Filamentous algae at 1.0 g reduced the germination rate of amaranth by 20% and biomass by 56%. Growth reductions of 68%, 49%, 53%, and 56% were observed in amaranth after application of muskgrass, water hyacinth, water lettuce, and filamentous algae, respectively. However, powdered muskgrass and hydrilla increased the growth of common ragweed by 200% and 145%, respectively. This suggested that common ragweed has a high tolerance to the allelochemicals present in the tested aquatic weeds.

The powder used as a source of nutrient or the allelochemicals released from aquatic weed had a stimulatory effect on common ragweed. In the laboratory study, the highest reduction of seed germination (27–28%) was obtained for 1:100 dilution of filamentous algae extracts applied to amaranth. All the treatments at 1:10 dilution (including aqueous ethanol control) resulted in a significant reduction of seed germination of both amaranth and common ragweed. This indicated that the presence of ethanol inhibited seed germination, which interfered with allelochemicals present in the aquatic weed extracts.

This research contributed to pushing the boundary for future studies focused on reducing the dependence on synthetic pesticides and finding alternative strategies within the framework of promoting sustainable agriculture. Future studies could include identification of the chemicals contained in aquatic weed extracts in order to characterize the possible effective allelochemical compounds and their mode of action.

Gas chromatography-mass spectrometry (GC-MS) analysis is a major analytical method to identify organic compounds in many studies. To further identify the molecules, the GC-MS spectrum of aquatic weed extracts could be compared with the known components stored in the National Institute Standard and Technology (NIST) library. Since no universal extraction method exists for all allelochemicals, different solvents such as water, methanol, and acetone should also be evaluated in the future to identify the best extraction methods of allelochemicals from aquatic weeds.

12

BIOPESTICIDES FOR IPM AND IRM

Biopesticides contain active ingredients of natural or biological origin that include plant extracts, microorganisms, microbial metabolites, organic molecules, minerals, or other such natural materials that have pesticidal properties. Pests such as herbivorous arthropods, pathogens, parasitic nematodes, mollusks, rodents, and weeds cause significant crop damage when they are not managed. Pest suppression is a critical part of crop production to maintain plant health, prevent yield losses, and optimize returns. As agriculture advanced from subsistence farming to a global enterprise, crop protection also evolved over millennia.

When farming was less organized, nature maintained a balance and provided solutions initially. Then natural solutions were actively implemented until industrialization led to the use of synthetic inputs in the 20th century. While synthetic fertilizers and pesticides contributed to a tremendous improvement in the yield potential, the indiscriminate use of some of them and the resulting damage to the environment and human health steered food production in the recent past towards organic farming with the use of nature-based solutions.

Although biopesticides have been around for a few decades, the growth of organic farming gave an impetus to the biopesticide industry during the past few years resulting in the development of new active ingredients and improved formulations. Now, biopesticides are considered an important part of integrated pest management (IPM) strategies in both organic and conventional systems. With a considerable industry investment in research and development, the quality and efficacy of biopesticides have also significantly improved. This has also contributed to optimizing the cost of some formulations.

However, there is still a need to fill the knowledge gaps in biopesticides and their use. Depending on the active ingredient, the mode of action for biopesticides, their target pests, their storage and handling, and the use strategies are quite diverse, and a thorough understanding of these aspects is critical for their successful use. As emphasized in the new

IPM model, while biopesticide use is an integral part of crop protection, understanding the pest biology, using biopesticides appropriate for the target life stage of the pest, applying them at the right time and rate using the right technology, avoiding incompatibility issues, building and sharing effective use strategies, and continuously investing in research and outreach are essential elements of biopesticide use. Biopesticides also play an important role in insecticide resistance management (IRM) to address resistance issues associated with synthetic pesticides. This article provides an overview of various biopesticide categories and general strategies for their successful use for IPM and IRM.

Biopesticides can be used for managing arthropod pests, bacterial or fungal pathogens, plant-parasitic nematodes, weeds, and snails and slugs. Some formulations or active ingredients have multiple roles and can be effective against more than one category of pests. While some active ingredients are very specific to a particular pest or related species, others have a broad-spectrum activity. Based on the source, biopesticides can be placed in four broad categories: i) botanicals, ii) microbials, iii) toxins, and iv) minerals and other natural materials.

BOTANICAL EXTRACTS

Plants are a rich source of numerous phytochemicals or secondary metabolites that have a wide range of properties including pesticidal activity. Acids, alkaloids, flavonoids, glycosides, saponins, and terpenoids in plant extracts or oils obtained from seeds and other plant parts are some of the compounds present in various biopesticides. Azadirachtin, BLAD (polypeptide from sweet lupine seeds), citric acid, essential oils, pyrethrins, soybean oil, and extract of the giant knotweed are used for their acaricidal, insecticidal, fungicidal, nematicidal, or herbicidal properties.

MICROBIALS

Some of the microbial pesticides have live microorganisms (such as entomopathogens, Bacillus spp., Streptomyces spp., and Trichoderma spp.) while others (such as Burkholderia rinojensis and Chromobacterium subtsugae) have heat-killed microorganisms and fermentation solids as the active ingredients. Ento-mopathogenic microorganisms [Bacillus thuringiensis (bacterium), Beauveria bassiana and Cordyceps fumosorosea (fungi), Heterorhabditis spp. and Steinernema spp. (nematodes), and granuloviruses and nucleopolyhedroviruses] primarily kill their hosts through infection; microbe-based fungicides antagonize plant pathogens through competitive displacement and production of toxic metabolites; nematophagous fungi parasitize plant-parasitic nematodes; and plant pathogenic bacteria, fungi, and viruses infect and suppress weeds.

Bacteriophages, which are viruses that parasitize bacteria, are used against the plant pathogenic species of Clavibacter, Erwinia, Pseudomonas, Xanthomonas, Xylella, and other genera.

TOXINS AND OTHER ORGANIC MOLECULES

There are multiple examples of toxic organic molecules derived from various organisms. Avermectins from the bacterium Streptomyces avermitilis and spinosad from the bacterium Saccharopolyspora spinosa, strobilurin from the mushroom Strobuluris tenacellus, and cerevisane from the yeast Saccharomyces cerevisae are some of the microbial toxins that are effective against insects, plant-parasitic nematodes, or snails and slugs. A venom peptide from the Blue Mountains funnel-web spider, Hadronyche versuta, from Australia is a recently developed insecticide active ingredient with its unique mode of action class. Chitosan, a polysaccharide from the exoskeleton of shellfish, is a fungicide.

Minerals and other natural materials

Diatomaceous earth, mineral oil, and minerals such as sulfur are used for controlling multiple categories of pests. Potassium salts of fatty acids of plant or animal origin, known as insecticidal soap, have insecticidal and fungicidal properties. Organic acids such as acetic acid and citric acid are derived from plants and have fungicidal and herbicidal properties. Since these are different from other botanical extracts, they are placed in this category.

Except for the microbial pesticides that have live microorganisms, most biopesticides have chemical molecules of microbial, fungal, botanical, or mineral origin and work through various modes of action similar to synthetic pesticides. Several synthetic pesticides are developed from natural molecules. Abamectin, pyrethroids, neonicotinoids, spinetoram, and storbulurins are synthetic analogs based on avermectins, pyrethrins, nicotine, spinosad, and strobulurin, respectively, and were developed for improved stability, safety, or ease of commercial-scale production.

BIOPESTICIDES AND INTEGRATED PEST MANAGEMENT

Biopesticide refers to a pesticide which originates from animals, microorganisms, or plants. In addition to preventing yield losses through pest and disease control, biopesticides improve environmental and human health by contributing to the reduction of chemical pesticides as well as by improving the quality of produce. Additionally, these products have the potential to improve harvest and shipping flexibility, assist with environmental stewardship, and assist growers to achieve sustainability goals. Biopesticides are also important tools in integrated pest management (IPM) programs and reducing the risk of resistance to chemical pesticides, improving worker safety through short restricted entry intervals, conserving natural enemies, and maintaining environmental health.

Biopesticides are inherently less toxic than conventional pesticides. Most affect only the target pest and closely related organisms, in contrast to broad spectrum conventional pesticides that may affect nontarget organisms such as beneficial insects, birds, wildlife, aquatic animals, and mammals. The majority of biopesticides often rapidly decompose, resulting in decreased exposure as well as preventing many pollution problems commonly associated with conventional pesticides. Although relatively safer than chemical pesticides, users or applicators should follow safety guidelines and wear personal protective equipment according to the label directions. It is also important to follow guidelines for spray volume, application rates, droplet size, water pH, compatibility with tank-mix partners, time and frequency of application, and other details to ensure efficacy of the biopesticides.

Biopesticides use has been increasing in the recent years. They can be used as standalone treatments or combined or rotated with other pesticides in both organic and conventional production systems. The fact that there are no residues is a huge benefit for exported commodities, as maximum residue limit issues continue to be a challenge in this arena.

In expanding upon the role of biopesticides in biocontrol, the topic of resistance management is a key consideration. Pest resistance to conventional chemical pesticides is a significant concern. Scientific research has repeatedly demonstrated that continuous use of the same class of pesticides, especially those reliant on a single mode of action, will result in the emergence of a pest population resistant to those products. Populations of insect pests, plant pathogens, nematodes, and weeds all have the ability to develop resistance quickly, even to different types of functionally similar chemistries. This phenomenon is called cross-resistance and is caused by multi-chemistry detoxification mechanisms present in many pest populations.

Because of the increasing number of novel, low-impact chemistries available, educators and growers have additional tools to manage resistance within IPM programs. Biopesticides have long been used in combination with synthetic chemistries to provide the basis for excellent control programs that effectively manage resistance. Additionally, they typically have modes of action that are different from synthetic pesticides and do not rely on a single target site for efficacy. Properly used, these products have the potential to extend the effective field life of all products by curtailing the development of resistant pest populations.

According to the United States Environmental Protection Agency (EPA), "IPM is an effective and environmentally sensitive approach to pest management that relies on a combination of common-sense practices". The University of California Statewide Integrated Pest Management Program defines the IPM approach as combining prevention, cultural, physical, biological and chemical means to control pests, all the while minimizing economic, public health, beneficial as well as non-target organism, and environmental risks. Biopesticides are noted among the low-risk and most highly effective tools for achieving crop protection in IPM systems. The challenges of farming require that IPM systems actively integrate multiple management approaches to balance optima productivity with sustainability.

Biopesticides should be considered as a component of a holistic total program and used at an appropriate time and pest density. Today, many forward looking IPM professionals are incorporating biopesticides into traditionally conventional pest management strategies.

However, education and training are needed to address biopesticide best use practices, the methods of integrating them into IPM programs; as well as instruction to promote an understanding of their unique modes of action. Part of the educational process involves research through fair and realistic field trials that evaluate biopesticides both as standalone treatments as well as in combination and rotation with other options with an objective of improving IPM practices. All of these learning experiences are useful in demonstrating the science of biopesticide use and establishing best use practices.

A better understanding of biopesticide potential and the mode of action of different active ingredients, increased grant support to promote biopesticide research, and productive grower-industry-researcher collaborations to generate applied research data and design IPM strategies are necessary to make the best use of biopesticides and for environmental sustainability.

INTEGRATED PEST MANAGEMENT AND RESISTANCE MANAGEMENT

Biopesticides are very diverse in their origin and mode of action and have been successfully used in several cropping systems for managing a variety of pests. They have complex interactions with plants, soil microbiota, pests, and environmental conditions. It is critical to have a good understanding of the source of biopesticides and how they act on their target pests. Certain biopesticides may have special storage and handling requirements or tank-mixing restrictions. It is essential to refer to the manufacturer's guidelines or label instructions to avoid incompatible tank-mix combinations, understand proper application sequences, and to store, transport, and apply under unfavorable conditions. While it is very important to use biopesticides as a part of the IPM program and tools for IRM, caution is warranted to avoid repeated use of the same or a similar type of biopesticide. Pests can develop resistance to biopesticides just as they do to synthetic pesticides.

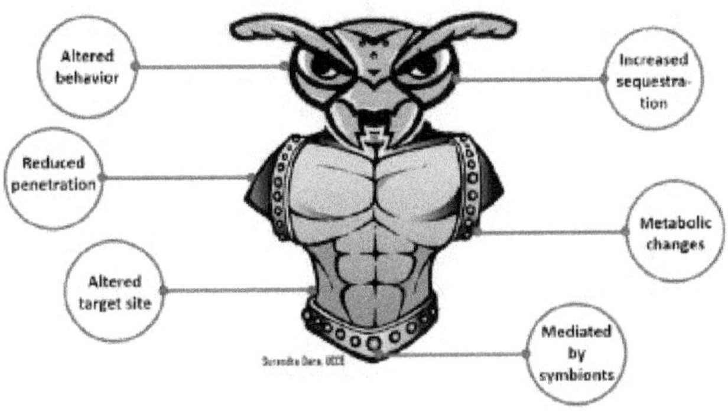

Fig. 1. Mechanisms of resistance to pesticides

STRATEGIES FOR USING BIOPESTICIDES

From the seed or transplant treatment to soil or foliar application, biopesticides can be used throughout crop production. Certain combinations can have an additive or a synergistic effect on pest suppression. At the same time, certain inputs or practices can negatively impact biopesticide efficacy. For example, alkaline tank-mix components breakdown the protein coat of entomopathogenic viruses and Bacillus thuringiensis. Botanical oils can be incompatible with cold water. Some fungicides such as captan and thiram are incompatible with entomopathogenic fungi like Beauveria bassiana while several others are compatible.

Investing in biopesticides

Environmental safety and resistance development are two major concerns for excessive use of synthetic pesticides and incorporating biopesticides into IPM will help address both issues. Substituting biopesticides for synthetic pesticides will reduce the total amount of the latter during a production season and their potential negative impact on the environment and human health. Several biopesticides are not harmful to pollinators and in some production systems, pollinators are used to deliver biopesticides to the crops they pollinate. Adding biopesticides to the standard crop protection program will also increase pest control efficacy. Additionally, by not continuously using synthetic pesticides, the risk of resistance will be reduced and thus their efficacy will continue to be maintained. Although some biopesticides can be more expensive than synthetic pesticides, investing in them will be a good strategy for both the short-term benefit of effective pest suppression and the long-term benefit of a healthy and resilient ecosystem. Since pests do not have boundaries, area-wide implementation of good agricultural practices with a balanced use of synthetic and natural inputs is necessary for maintaining the productivity of the cropping systems.

Productive collaborations among the pesticide industry, researchers, extension educators, and the grower community are critical for successfully using biopesticides for sustainable food production. While research helps to develop effective formulations and their use strategies, outreach helps with the implementation of those strategies.

13

RESISTANCE TO BIOPESTICIDE

INTRODUCTION

The awareness of secondary effects of pesticide use on the environment and human health is currently causing a green revolution. Modern agriculture is changing; it tends to spread less pesticide and generally use more-selective and less-polluting products. Bio-pesticides fit this definition, and in recent years, pesticide firms have invested significantly in companies producing bio-pesticides.

Bio-pesticides have a reputation of being very expensive and often not reliable. To date, the price of bio-pesticides is still high because of their low market share compared to the chemical compounds. The market size of bio-pesticides increased 9.9% between 2005 and 2010. In the same period, the market size of synthetic pesticides decreased by 1.5%. For all crop types, bacterial bio-pesticides represent 74% of the market, fungal bio-pesticides represent approximately 10%, viral bio-pesticides 5%, predator bio-pesticides 8% and other bio-pesticides 3%.

In the 1990s, products based on Bacillus thuringiensis (Bt) accounted for 95% of the global microbial insecticide market. Currently, the apparition of new biological control agents, such as beneficial insects, viruses and entomopathogenic fungi, has lowered the prevalence of Bt, which still represents more than 60% of the bio-pesticide market. The panel of biological agents that are available on the market is gradually increasing and should continue to increase over the next few years due to their variety (not only microorganisms but also predators and parasites).

One of the major issues in bio-pesticide development is their lack of robustness and the high-tech level that is necessary for their utilization. Indeed, these products are generally more sensitive to variations in temperature, humidity or light radiation than are their synthetic counterparts. Formulation technologies have been used to improve the number of commercial biological control products, the delivery, the shelf life and the field efficacy of bio-pesticides by adding various compounds to the active agents. Although there is limited information on the research in this area, due to concurrent issues, recent studies have permitted the emergence of (i) new formulations, (ii) formulations with precise time or location delivery, and (iii) formulations helping with long activity persistence.

However, the shorter persistence of bio pesticides can be considered a quality as it reduces their environmental risk, but this often results in the more frequent pulverization of those products to maintain an acceptable level of control. The quantity of products that are released into the environment is therefore more important, and the costs of these multiple interventions make biological control less competitive than chemical control.

Moreover, the maintenance of selection pressures and the relatively low efficiency levels (70%) of biocontrol products favor resistance development. In recent years, many studies have been conducted to improve the efficiency, ease of use and persistence of these products.

The possible side effects of the use of bio-pesticides have often been neglected or minimized compared to those of chemical pesticides, and cases of resistance have been described. Thus, the question of the sustainability of these strategies over time arises. At least 27 pest species have developed resistance to the most used bio-pesticide in the world: Bacillus thuringiensis.

We propose in this review a detailed inventory of cases of resistance to bio-pesticides to understand this phenomenon and anticipate it better. We rely on experience gained from numerous studies on resistance to chemical insecticides and the best ways to manage it to propose new solutions thanks to the intrinsic properties of bio-insecticides.

To investigate this problem, we decided to present modes of action and resistance mechanisms for the main bio-insecticides. Then, we evaluated the durability of biocontrol agent efficacy to determine the risks factors of resistance against future biocontrol agents.

Bio-pesticide is a term that was first applied with a narrow focus on preparations containing living microorganisms and then expanded to a broader definition that encompasses botanical compounds and semiochemicals (e.g., pheromones). In this review, bio-pesticide will include botanical insecticides (plant extracts), microbial pest control agents (bacteria, viruses, fungi and protozoa) and/or bioactive compounds (such as metabolites) that are produced directly from these microbes and used to suppress populations of insect pests.

In addition, because the aim of this review is to provide a critical point of view on the sustainability and effectiveness of bio-pesticides, we present some examples of commercialized and currently used products (Genetically modified (GMO) crops and human disease vectors excluded).

RESISTANCE DEVELOPMENT: A BRIEF REVIEW

Weeds, insect pests and disease-causing fungi and bacteria are living organisms capable of evolving in response to selection. And one source of selection can be the pesticides used to control them. Most pest populations include a few individuals with heritable traits that make them more able to survive exposure to certain types of pesticide active ingredients. For example, a genetic mutation resulting in a slight change to an enzyme that is the target of a particular active ingredient (AI) might make individuals expressing the altered enzyme no longer susceptible to that AI.

Application of pesticides containing that AI acts as a selective agent favoring those individuals (and their offspring) over individuals lacking the mutant genes for resistance to that mode of action. A resistance trait that was initially very rare in the pest population can become much more prevalent after repeated exposure of the pest population to the same mode of action over multiple generations, leading to control failures in the field. Seemingly different active ingredients may share similar target sites or be sufficiently similar in chemical structure that pests resistant to one AI are already preconditioned for resistance to the other, even if they have not yet been exposed to it - a condition known as cross resistance.

THE TRUTH ABOUT BIOPESTICIDES

Generally, it takes relatively longer (although this depends on the frequency of application) for an insect or mite pest population to develop resistance to a selective pesticide or biopesticide. This is because most selective pesticides and biopesticides have broad or non-specific modes of action, which means these compounds are active on either multiple target sites in or on the insect or mite pest body. Or, the active ingredient attacks a variety of enzymatic or metabolic systems.

As already mentioned, a common claim or misperception published in both trade magazines and scientific-based literature is that biopesticides and selective pesticides are "less susceptible" to resistance. This is not entirely untrue, either. The consistent use of any pesticides or pesticides with similar modes of action will likely lead the development of resistant individuals in pest populations. So, regardless of whether a pesticide is considered a conventional or a biopesticide, insect and mite pest populations may develop resistance due to selection pressure. This is associated with the frequency of applying pesticides and dosage used (rate response).

Why Pests Build Resistance

There are a variety of reasons why insect (and mite) pests may develop resistance to selective (alternative) pesticides or biopesticides. For example, buprofezin (e.g., Talus) has demonstrated to be less effective against the greenhouse whitefly, which may be associated with the "high" volatility of the active ingredient resulting in inadvertent selection pressure. Thus, there is an increased probability of resistance. Furthermore, greenhouses, and even nurseries, tend to retain or restrict insect and mite pest populations with minimal migration of susceptible individuals. This results in the population interbreeding and "passing on" traits to its next generation, which will enrich the gene pool with resistant genes. This increases the proportion of resistant individuals in the population.

Still, it is important to note that this varies depending on the time of year and openness of the greenhouses and nursery. In addition, some insect and mite pests may not disperse very far, which means that concentrated populations of resistance individuals are continually present and may be exposed to pesticide applications. And if pesticides with similar modes of action are used and applications are made on a frequent basis, this will, in all likelihood, allow for more resistant individuals in the population - regardless of whether a conventional or biopesticide is involved.

MANAGING PESTICIDE RESISTANCE

Managing pesticide resistance starts with prevention through adoption of basic IPM tactics. These include field monitoring and use of economic injury levels to guide deployment of multiple control measures, such as chemical and biological pesticides, beneficial insects, cultural practices, host plant resistance and crop rotation.

But another important component of resistance management is the use of alternatives, rotations or sequences of different modes of action (MoA) to avoid selecting for resistance or cross-resistance by repeated use within the crop cycle—or year after year—of products in the same MoA class.

The Resistance Action Committees (RAC) of CropLife International classify pesticide active ingredients based on MoA as a tool for developing resistance management programs. Many pesticide labels indicate the FRAC (for fungicides) or IRAC (for insecticides) MoA group of the active ingredient. By following RAC guidelines, PCAs can avoid consecutive applications of the products in the same MoA group and rotate between these groups reducing the risk of product failure due to resistance or cross-resistance.

Constant reliance on pesticides having the same modes of action creates more risk of resistance. Using different modes of action means you don't expose the target population to the same "bullet" again and again.

RESISTANCE MANAGEMENT WITH BIOPESTICIDES

The use of biopesticides has become a common practice in many horticultural crop protection programs. Biopesticides are effective tools in integrated pest management (IPM) programs for helping to manage resistance to synthetic chemical pesticides and reduce worker and environmental exposure to synthetic pesticides. Multiple studies have documented the development of resistance by pathogen and pest populations to chemical pesticides. Conversely, the risks of developing resistance to biopesticides are extremely low — even as biopesticide use continues to increase.

The efficacy of each synthetic pesticide developed over the past 60 years is generally based on one mode of action (MOA) that disrupts or inhibits a specific biochemical pathway or biological function of the target pest. Continuous, extensive and intensive use of synthetic pesticides can inflict very strong and widespread selective pressure on pest and pathogen populations, resulting in rapid emergence of resistant or tolerant pests and pathogens. This phenomenon is also commonly experienced with human and veterinary pathogens where antibiotics have been used extensively, and, often, unnecessarily. In fact, pesticides are grouped into classes based on their MOAs, and end-users are specifically instructed not to over-use certain classes of chemical pesticides.

Nearly all biopesticides are most effectively used as preventive treatments. Since they deploy multiple MOAs to suppress pests and pathogens, development of resistance to these multiple factors by target organisms is extremely unlikely. Combinations of two or more of the characteristics described below prevent the development of resistance to biopesticides.

- **Competitive exclusion:** Many microbial biopesticides not only grow very effectively in the environments from which they were originally isolated, they can also physically occupy these niches to prevent establishment of pests and pathogens in these spaces. If these microorganisms grow well on plant leaves or in close association with plant roots, they can also provide host plants with a protective barrier against pathogens and certain pests.
- **Production of secondary metabolites:** Since bacterial and fungal biopesticide strains are isolated from very competitive environments, each produces numerous secondary metabolites for protection, survival and competition for nutrients.
- Predation and parasitism: Some biopesticide agents physically attack and completely consume specific pathogens and pests. Other biopesticide agents actually act as parasites that feed on detrimental organisms, leading to weakening and eventual death.
- **Induced host resistance and enhancement of plant vigor:** Certain microorganisms used as biopesticides can stimulate subtle biochemical responses in host plants that enhance their abilities to resist or better tolerate pest attack and diseases. In addition, certain microbial biopesticides can also promote plant growth or enhance plant vigor by production of particular secondary metabolites and increasing availability of plant nutrients.
- **Alteration of the soil or plant host environment:** Biopesticides made from soft chemicals or biological extracts chemically or physically change conditions in the soil or on plant surfaces to be unfavorable to the establishment and growth of pests and pathogens.
- **Disruption of fundamental biological functions, development, and structures of target organisms :** Biorational chemicals and biological extracts can directly inhibit biochemical processes, interfere with developmental pathways of pathogens, and compromise the physical integrity of pests and pathogens.

Because of the nature and diversity of the MOAs presented above, plant pathogens and pests would need to undergo immense physical, biochemical, physiological, and genetic changes to develop resistance to biopesticides. Therefore, the loss of effective biopesticides due to development of resistant pests and plant pathogens is highly unlikely.

RESISTANCE TO BOTANICAL PESTICIDES

Nicotine, an alkaloid from Nicotiana spp., is one of the earlier botanical pesticides known. Although nicotine is not currently used as an insecticide, its synthetic alternatives - neonecotinoids - are commonly used against several pests. Botanical insecticide pyrethrum, extracted from the flowers of Chrysanthemum cinerariaefolium, contains insecticidal pyrethrins (synthetic pyrethrins are referred to as pyrethroids).

Although insect resistance to pyrethrum or pyrethroid compounds has been known, they have been effectively used against a number of pests through careful placement in IPM, organic, or conventional management strategies. Additionally, pyrethrin products have been effectively used along with piperonyl butoxide, which acts as a synergist and resistance breaker.

Another botanical insecticidal compound, azadirachtin, is a tetranortri-terpenoid limonoid from neem (Azadirachta indica) seeds, which acts as an insecticide, antifeedant, repellent and insect growth regulator. While neem oil, which has a lower concentration of azadirachtin, has been used in the United States as a fungicide, acaricide, and insecticide for a long time, several azadirachtin formulations in powder and liquid forms have become popular in recent years and were found effective in managing important pests.

Researchers reported that the green peach aphid, Myzus persicae developed resistance to pure azadirachtin under artificially induced selection pressure after 40 generations, but did not develop resistance to a refined neem seed extract.

They suggested that natural blend of azadirachtin compounds in a biopesticide would not exert selection pressure that could lead to resistance. Additionally, researchers discussed that azadirachtin can play a role in insecticide resistance management because it reduces the detoxification enzyme production as a protein synthesis inhibitor. Azadirachtin also improved the efficacy of other biopesticides in multiple studies.

Insects feeding on plant allelochemicals can develop cross-resistance to insecticides. For example, overproduction of detoxification enzymes such as glutathione S-transferases and monooxygenases in the fall armyworm, Spodoptera frugiperda,when it fed on corn and cowpea, respectively, imparted cross-resistance to various chemical pesticides. It is important to keep this in mind when botanical pesticides are used to detect potential resistance issues.

RESISTANCE TO BACTERIAL BIOPESTICIDES

Bacillus thuringiensis (Bt)is a gram-positive soil bacterium, which contains crystalline toxic protein that is activated upon ingestion by an insect host, binds to the receptor sites in the midgut, and eventually causes insect death. Since the mode of action involves a toxin rather than bacterial infection, several insects developed resistance to Bt pesticides or transgenic crops that contain Bt toxins. However, Bt pesticides are still very popular and used against a variety of lepidopteran (Bt subsp. aizawai and Bt subsp. kurstaki), dipteran (Bt subsp. israelensis and Bt subsp. sphaericus), and coleopteran (Bt subsp. tenebrionis) pests.

Spinosad is a mixture of macrocyclic lactones, spinosyns A and spinosyns D, derived from Saccharopolyspora spinosa, an actinomycete gram-positive bacterium, and is used against dipteran, hymenopteran, lepidopteran, thysanopteran, and other pests. Spinosad products, while naturally derived are registered as chemical pesticides, not as biopesticides.

Insect resistance to spinosad later led to the development of spinetoram, which is a mixture of chemically modified spinosyns J and L. Both spinosad and spinetoram are contact and stomach poisons and act on insect nervous system by continuous activation of nicotinic acetylcholine receptors. However, insect resistance to both spinosad and spinetoram has been reported due to extensive use of these pesticides. Cross-resistance between spinosad and some chemical insecticides has also occurred in some insects.

RESISTANCE TO VIRAL BIOPESTICIDES

Baculovirus infections in lepidoptera have been known for centuries, especially in silkworms. Currently, there are several commercial formulations of nucleopolyhedroviruses (NPV) and granuloviruses (GV). When virus particles are ingested by the insect host, usually lepidoptera, they invade the nucleii of midgut, fatbody, or other tissue cells and kill the host. Baculoviruses are generally very specific to their host insect species and can be very effective in bringing down the pest populations.

However, variations in the susceptibility of certain insect populations and development of resistant to viruses has occurred in several host species. Resistance to different isolates of Cydia pomonella granulovirus (CpGV-M, CpGV-S) in codling moth (Cydia pomonella) populations is well known in Germany and other parts of Europe.

RESISTANCE TO FUNGAL BIOPESTICIDES

There are several fungi that infect insects and mites. Fungal infection starts when fungal spores come in contact with an arthropod host. First, they germinate and gain entry into the body by breaching through the cuticle. Fungus later multiplies, invades the host tissues, kills the host, and emerges from the cadaver to produce more spores. Entomophthoralean fungi such as Entomophthora spp., Pandora spp., and Neozygites spp. can be very effective in pest management through natural epizootics, but cannot be cultured in vitro for commercial scale production. Hypocrealean fungi such as Beauveria bassiana, Isarea fumosorosea, Metarhizium brunneum, and Verticillium lecanii, on the other hand, can be mass-produced in vitro and are commercially available.

These fungi are comparable to broad-spectrum insecticides and are pathogenic to a variety of soil, foliar, and fruit pests of several major orders. Since botanical, bacterial, and viral biopesticides have insecticidal metabolites, proteins, or viral particles that have specific target sites and mode of action, insects have a higher chance of developing resistance through one or more mechanisms. Although fungi also have insecticidal proteins such as beauvericin in B. bassiana and I. fumosorosea and dextruxin in M. anisopliae and M. brunneum, their mode of action is more through fungal infection and multiplication and arthropods are less prone to developing resistance to entomopathogenic fungi.

However, insects can develop resistance to entomopathogenic fungi through increased melanism, phenoloxidase activity, protease inhibitor production, and antimicrobial and antifungal peptide production. It appears that production of detoxification enzymes in insects against fungal infections can also impart resistance to chemical pesticides. Infection of M. anisopliae in the larvae of greater wax moth, Galleria mellonella, increased dexotification enzyme activity and thus resistance to malathion.

These examples show that insects can develop resistance to biopesticides in a manner somewhat similar to chemical pesticides, but due to the typically more complex and multiple modes of action, at a significantly lesser rate depending on the kind of botanical compound or microorganism involved. Resistance to entomopathogenic fungi is less common than with other entomopathogens. Since biopesticide use is not as widespread as chemical pesticides, the risk of resistance development is less for the former.

However, excessive use of any single tool has the potential for resistance or other issues and IPM, which uses a variety of management options, is always a good strategy.

BIOCHEMISTRY BEHIND BIOPESTICIDE EFFICACY

Biopesticides are attracting global attention as new tools to kill or suppress pest populations such as weeds, plant pathogens and insects while posing less risk to people and the environment than synthetic pesticides. Globally, the use of biopesticides has steadily increased by about 10% per year. Biopesticides are typically derived from living organisms, of plant or microbial origin; this paper focusses only on the latter.

More than 225 microbial biopesticides are manufactured in 30 countries in the Organization of Economic Development and Cooperation. The NAFTA countries (USA, Canada, and Mexico) use about 45% of all biopesticides sold, the European Union uses 20%, Oceanic countries use 20%, the South and Latin Americas use 10%, and India and Asia use 5%.

Bacillus thuringiensis has dominated the bioinsecticide marketplace since the 1950s. This organism is used on 75% of the crops and forests in North America to control a wide variety of insect pests. In the 1980s, Colletotrichum gloeosporioides (Penz.) Sacc. f. sp. aeschynomenes, Phytophthora palmivora Butler, and C. gloeosporioides f. sp. malvae were developed as bioherbicides for northern jointvetch, strangler vine, and round-leaved mallow, respectively.

In the 1990s, Streptomyces griseoviridis was among the first biofungicides available for controlling root-infecting fungi in greenhouse crops. Despite the long term success of a few products, there is global pressure for new biopesticides. The rate of commercial success is less than 10%. The challenges facing researchers have usually been biological, environmental, technological and commercial constraints. The question of how we can improve the chances of success remains.

The mode of action of a biopesticide is a critical component in commercial success. It determines efficacy of pest control, efficiency of use, consistency of response, host target and nontarget susceptibility. Mode of action also impacts the choice of manufacturing method and the final cost of production. Typically, there are three broad modes of action for microbial biopesticides:

i. biological or ecological modes of action,
ii. physical means, or
iii. chemical or biochemical processes.

Biological modes of action can include predation or competition. Physical control may involve creating barriers or occupying space. Chemical or biochemical approaches disrupt biochemical, genetic or structural function in the targeted pest. Researchers reported that more successful biopesticides utilize compounds that are produced by the microorganism, rather than relying only on the infection process. Researchers gives examples such as B. thuringiensis and B. subtilis, which produce toxic proteins and agrastatins, respectively. The production of these compounds has been optimized through fermentation. Biopesticides that utilize secondary metabolites may be more akin to synthetic pesticides. However biopesticides utilize multiple modes of action, greatly reducing the risk of resistance developing. For example, a single biopesticide may produce more than one secondary compound that works synergistically with the physiological interaction of the microbe with the host.

In contrast, pest resistance is a major concern associated with synthetic pesticides which often rely on a single mode of action. This review will focus on biopesticides that employ chemical or biochemical modes of action. Specifically, we will survey three examples of biopesticides – a fungal bioherbicide, a bacterial biofungicide and viral bioinsecticide – whose modes of action are at least partially known. Through these three case studies, we will explore the importance of understanding and using chemistry and biochemistry in effective biopesticides.

CASE STUDIES OF BIOPESTICIDES THAT USE CHEMISTRY

The bioherbicide Phoma macrostoma

Phoma macrostoma Montagne is a bioherbicidal fungus, meaning that it can be used to control undesirable weedy plants. P. macrostoma produces photobleaching symptoms in a wide range of broadleaf weeds, such as Canada thistle (Cirsium arvense (L.) Scop.), dandelion (Taraxacum officinale Weber ex F.H. Wigg.), chickweed (Stellaria media (L.) Vill.), and scentless chamomile (Matricaria perforata Merat). Monocots such as turfgrass, wheat and barley are unaffected by P. macrostoma. A critical component of P. macrostoma phytotoxicity is the production of two structurally related secondary metabolites known as macrocidins A and Z.

Macrocidins inhibit carotenoid biosynthesis in susceptible plants, inducing chlorosis, bleaching and eventual plant necrosis. The individual roles of each of the four macrocidins are currently unknown. In addition to producing macrocidins, P. macrostoma enters weed tissues via locations adjacent to root hairs, colonizing intercellularly beside the vascular trachea, thereby interfering with tissue functionality. P. macrostoma also colonizes resistant plants, albeit at a greatly reduced hyphal density and only in the outer layers of the root. The physical colonization of the host plays a quantifiable role in bioherbicide effectiveness. The weed control achieved by a formulation containing both living P. macrostoma and macrocidins is decreased by roughly 15 to 20% by autoclaving which kills the fungus, but does not degrade macrocidins. Identification of macrocidins provided inspiration and a starting point for future research.

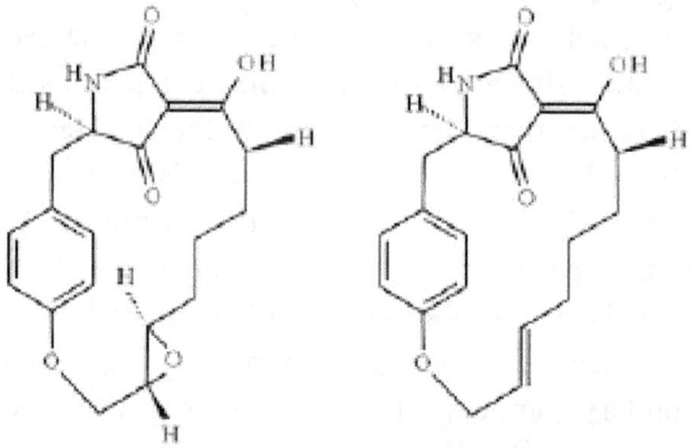

Fig. 1. Two major groups of macrocidins: Macrocidin A (left and Macrocidin Z (right).

For example, the development of sensitive reverse phase liquid chromatography macrocidin detection systems facilitated the work of Bailey et al.. The latter work explored the water-solubility of macrocidins and the impact of potential rainfall events on run-off of macrocidins. Researchers discovered that the susceptibility of macrocidins to leaching and movement to non-target locations is dependent on saturated conditions and soil type. This information facilitates informed decision making about when and where to apply P. macrostoma-based formulations, leading to a more useful bioherbicide and better efficacy.

The realization that higher levels of macrocidins in P. macrostoma formulations correlated with improved weed control point toward new strategies for developing a bioherbicide that would be more effective at lower doses, and hence more cost-efficient. Such strategies include elucidating fermentation and growing conditions under which macrocidin production can be increased, strain selection for naturally-occurring genotypes or genetic/epigenetic manipulation of P. macrostoma to generate a "super-producer".

Modifications of P. macrostoma that avoid the need to release an organism carrying heterologous DNA could be prioritized in order to avoid potential concerns surrounding genetically modified organisms (GMOs). For example, de Crecy et al. have used so-called "directed evolution" to enhance fungal tolerance for high temperatures.

The knowledge that macrocidins work by inhibiting carotenoid biosynthesis could also prove useful. For example, carotenoid-protective ingredients could be avoided in formulations.

Additionally, factors that enhance localization of macrocidins to leaves (where carotenoid biosynthesis occurs) might be considered. A more complete understanding of how macrocidins interfere with carotenoid biogenesis could further expedite formulation development, or at least suggest avenues for future research.

Inhibition of carotenoid biosynthesis in general is a mode of action shared by macrocidins and various synthetic herbicides such as norflurozon, fluridone and diflufenican. However, the way(s) in which macrocidins impact this process appear to be unique from, and more complex than, the modes of action of known synthetic herbicides.

Hence, P. macrostoma could provide a valuable alternative to control weeds with resistance to carotenoid biosynthesis targeting synthetic herbicides.

Phytotoxins produced by some plant pathogenic fungi are harmful to animals and/or humans. Berestetskiy gives examples such as Aspergillus fumigatus, Alternaria tenuissima, Penicillium spp. and Fusarium spp., all of which produce harmful mycotoxins. However, macrocidins and P. macrostoma are safe for humans, animals and the environment, effective against a wide range of weeds and have a long shelf-life. Hence P. macrostoma and its secondary metabolites are a good choice for urban lawn and golf course weed control. In 2011 and 2012, P. macrostoma was registered for broadleaf weed control in turfgrass in Canada and the USA, respectively.

The biofungicide Bacillus subtilis

Bacillus subtilis is a vital component of the biopesticide strategy for crop disease management. A unique characteristic of B. subtilis is its ability to undergo differentiation from a vegetative cell to a highly resistant spore following exhaustion of an essential nutrient i.e. carbon, nitrogen or phosphorus or development of another unfavorable condition.

B. subtilis is a Gram positive rod-shaped bacterium, ubiquitous in nature and produces endospores. These spores can withstand a high degree of exposure to desiccation, radiation (UV and gamma), moist and dry heat and oxidizing agents. The robust nature of the B. subtilis spore and long shelf-life is not lost on the biopesticide industry wishing to commercialize unique isolates with strong antagonistic activity to agricultural phytopathogens.

Bacillus subtilis are isolated from numerous environments including the rhizosphere (a zone enriched with nutrients from plant root exudation), where they effectively compete with diverse microbial population to establish micro-colonies and biofilms on plant roots. Secreted toxic metabolites play a vital role in B. subtilis establishing a niche in the rhizosphere. B. subtilis produce antimicrobial metabolites that target many bacterial and fungal soil inhabitants including plant pathogens, thus providing B. subtilis with a competitive advantage in nutrient limited niches.

B. subtilis are also beneficial to plants in that they promote plant growth through phytohormones production, sequestration of nutrients, stimulation of systemic induced resistance in plants and suppression of plant pathogen activities. Antimicrobial metabolites from B. subtilis have been employed in biocontrol and plant protection by exploiting their antibiosis activity on phytopathogens. These compounds also stimulate the plant's defense systems to pathogens by inducing systemic resistance in bean, Phaseolus vulgarus, and tomato, Solanum lycopersicum.

Hence, B. subtilis can increase plant ecological fitness through sustained colonization of the rhizosphere. The scope of this section will be limited to specific antimicrobial peptides produced by of B. subtilis, proposed function and implications.

Production of antimicrobial peptides appears to be universal among Bacillus spp. B. subtilis isolates are well represented in the literature in terms of discovery, chemical structure and host activity of these products of secondary metabolism. Kunst et al. determined that 4% of the B. subtilis genome was associated with secondary metabolism. Stein estimates that 4-5% of B. subtilis genome was devoted to antibiotic production.

More than two dozen antimicrobial peptides have been discovered from collections of several hundred naturally occurring B. subtilis isolates. These antimicrobial peptides are synthesized by two distinct systems:

1. non-ribosomal peptide synthesis, ex. cyclic lipopeptides, iturin and fengycin families,
2. ribosomal peptide synthesis, which are further subdivided into
   i. post-translationally modified, ex. subtilin,
   ii. non-modified, ex. thurincins, and
   iii. large heat-labile proteins. Antimicrobial peptide synthesis is closely regulated and reliant on both environmental conditions and B. subtilis growth phase.

A large collection of peptide synthetases constitute the non-ribosomal biosynthetic machinery that produces a diverse array of lipopeptides (LPs), such as iturin, fengycin and surfactin families. The LPs are amphiphilic and have cell wall and membrane surface active properties, facilitating the formation of pores in membranes of phytopathogenic bacteria and fungi. Loss of cell membrane integrity causes a disruption in active transport (i.e. nutrient and ion transport) leading to cell death. In vitro studies have demonstrated that surfactins have anti-bacterial activity, while iturins and fengycins have anti-fungal properties.

These in vitro findings support the hypothesis that B. subtilis kills and suppresses deleterious microorganisms in the rhizosphere and provide protection to plants. In a field study, surfactin and iturin were quantified by HPLC-MS from rhizosphere extractions of B. subtilis-colonized cucumber. Levels of surfactin and iturin peaked on roots of 29 day-old cucumber at 60 and 1000 μg/gram fresh weight of root respectively.

In a study of B. subtilis mutants deficient in surfactin production, Blais et al. reported that colonization and biofilm development on Arabidopsis roots was impaired and that infection of Arabidopsis roots by Pseudomonas syringae was not suppressed as compared to roots colonized by wild type B. subtilis capable of biofilm development. Optimum biocontrol activity and colonization of the rhizosphere by B. subtilis requires coordinated production and synergy of complementary LPs.

However, recent studies suggest that surfactin may also be a signal molecule in the quorum-sensing (QS) system in B. subtilis. Surfactin may be involved in initiating biofilm development by sub-populations of the colony and detection of microbial community diversity in the rhizosphere. In general, QS molecules elicit regulation of genes required for microbial survival and act as microbial interspecies communication signals. Therefore B. subtilis surfactins may have a role in communication and survival strategies in addition to their biocontrol activities.

B. subtilis also produce antimicrobial metabolites - bacteriocins - by ribosomal peptide synthesis. Ribosomal syntheses uses the polyketide synthase gene cluster to produce a large assortment of bacteriocins. Researchers propose a classification scheme of the Bacillus spp. bacteriocins that includes post-translationally modified peptides, linear and globular-type lantibiotics, non-modified peptides and large proteins.

Several of the lantibiotics display varying degrees of heat and pH stability, potentially attributable to the disulfide bonds in the individual molecules. B. subtilis produce the well-characterized lantibiotic subtilin, a post-translational modified 32-residue peptide. Subtilin is produced in response to environmental stressors such as decreasing nutrients levels. Subtilin exhibits broad spectrum activity towards Gram-positive bacteria by disrupting membrane function. This differs from activities of bacteriocins produced by Gram-negative bacteria in that the latter have a narrow killing spectrum.

Gram negative bacteria that become targets of B. subtilis lantibiotics undergo subtle changes in the amino acid composition. Conventional pesticide residues on food and in the environment and resistance to them by plant pathogens have promoted interest in the development of alternative means of controlling crop diseases.

B. subtilis isolates are desired for biofungicidal products because of the extensive portfolio of LPs and bacteriocins, status as a safe handling microorganism (generally regarded as safe, GRAS) and long shelf-life. In addition, our understanding of the benefits of B. subtilis in the rhizosphere continues to expand. For example, Researchers illustrated the sophistication of the B. subtilis-Arabidopsis thaliana interaction, showing that B. subtilis colonized and formed biofilms on the roots following secretion of L-malic acid from roots after application of the foliar pathogen, Pseudomonas syringae.

Furthermore, within the context of medical research, additional triggers or stress factors are being sought to increase production of antimicrobial peptides from B. subtilis. Researchers reported that sub-inhibitory levels of extracts of garlic, Allium sativum, induced the production of antimicrobial peptides from B. subtilis. The ability to secrete an array of antagonistic and plant-benefiting compounds into the rhizosphere and form highly resistant spores establishes B. subtilis as a highly desirable biopesticidal product for crop production systems.

THE BIOINSECTICIDAL NUCLEOPOLYHEDROVIRUSES

Insects constitute major pests of food and fiber production in the agricultural and forestry sectors as well as acting as vectors for many disease organisms, infecting plants of agricultural importance, as well as animals and humans. Insect pest control has relied primarily on the use of broad-spectrum chemical pesticides. In addition to negative human health and environmental effects related to chemical insecticides, their heavy usage has led to the emergence of secondary pest problems due to the loss of arthropod natural enemies and the development of resistance in insect pest populations to chemical insecticides.

As a result, more environmentally sustainable pest control strategies have been proposed, notably integrated pest management (IPM). IPM incorporates a range of control tactics including cultural methods, judicious use of chemical insecticides, and biological control agents to suppress insect pest populations below levels that cause economic damage. Biological control agents include microbial pathogens of insects.

Microbial insect pathogens are attractive for IPM systems due to their narrow host range. Currently microbial pest control products constitute only about 1.5% of the world wide pesticide market and microbial insecticides are dominated by Bacillus thuringiensis (Bt) based products. Viruses are associated with insects in a wide range of ecological relationships as pathogens and as specialized symbionts that allow hymenopteran parasitoids to fully develop in their hosts (e.g. polydnaviruses).

In addition, many plant and animal viruses rely on insect vectors for transmission. A number of virus families have been investigated as insect control agents. Viruses in the family Baculoviridae have been shown to have by far the greatest potential as bioinsecticides.

Baculoviruses have long been recognized as potential candidates for biopesticides because of their readily observable symptomology, their ability to produce impressive epizootics in insect populations and because they are detectable by light microscopy as a consequence of being occluded in relatively large protein crystals, referred to as occlusion bodies (OBs), within infected host cells. Currently, a number of baculoviruses are registered as insect control products worldwide.

Most of the registered baculovirus products are for control of lepidopteran and sawfly forest pests but a few have been developed as highly successful biopesticides to control lepidopteran pests in agriculture. Baculoviruses are characterized by rod-shaped enveloped virions and a circular, dsDNA genome of 80–180 kbp. The family Baculoviridae is also defined by 37 core genes that are conserved in all baculovirus genomes.

This set of core genes provides a picture of the common elements required for functional baculoviruses including virion structure, novel polymerases required for virus-specific-DNA replication and gene transcription, and per oral infectivity factors that are essential for infection of insect midgut cells. Their host range is restricted to arthropods and most particularly to insects. Thus baculoviruses are considered to be very safe options for insect pest control.

These viruses are infectious orally and the occlusion-derived virions (ODV) infect midgut epithelial cells after ingestion and dissolution of the OB in the alkaline condition of the midgut. Typically, two virion phenotypes occur in baculovirus infections, ODV and budded virions (BV) that spread the infection to tissues throughout the host. Thus the ODV are responsible for horizontal transmission between insect hosts and BV for systemic spread of infection within a host.

Baculoviruses kill their hosts by virtue of a complex replicative cycle that usurps the metabolic machinery of the host cell and diverts it to the production of baculovirus nucleic acids and structural proteins for virus production. No toxins per se are expressed during replication thus the speed of kill of hosts by baculoviruses is slower than other biopesticides such as typical B. thuringiensis strains which produce □-endotoxins.

However, baculoviruses encode a number of proteins and enzymes that enhance their ability to specifically infect and replicate to high levels in insects and to increase the efficiency of spread and persistence of viruses with in the host insects' environment. A few of these key genes will be reviewed.

All baculoviruses encode an occlusion body protein (polyhedrin or granulin) which forms the bulk of the proteinaceous crystal or OB that occludes the virions in the later stages of virus replication. The OBs are responsible for the highly stable nature of these viruses in the environment. While the occluded form of the virus is susceptible to UV inactivation, in the soil the occluded virions can remain viable for several years. The OB protein, polyhedrin, can make up to 30% of infected cell protein and indeed the powerful promoter element of this gene was the initial basis for the development of baculoviruses as a eukaryotic gene expression system.

Upon ingestion of baculovirus OB by a host insect larva, the OB structure and polyhedrin protein is degraded in the alkaline environment of the insect midgut which also includes a diverse mixture of digestive proteases. The process releases the infectious occluded virions into the midgut lumen. However, to infect the midgut epithelial cells the virions must transit the peritrophic matrix (PM), a chitin and glycoprotein layer which lines the midgut lumen of most insect groups. The PM is a physical barrier that protects the underlying midgut epithelial cells from abrasive food particles, digestive enzymes and pathogens that are infectious per os.

A number of baculoviruses encode a family of metalloproteases referred to as "viral enhancing factor" or "enhancins". These metalloproteases are associated with the OB or are incorporated into the envelope of the occluded virion. Enhancins have been demonstrated to increase the oral infectivity of baculoviruses from 2 to 15 fold when present at normal levels in OB or ODV. The potential mode of action of "viral enhancing factor" (VEF) has been investigated using purified VEF proteins and recombinant viruses expressing VEF.

The major impact observed is the degradation of the structural integrity of the PM. More detailed studies have shown that VEF specifically degrades key structural glycoproteins of the PM, namely the insect intestinal mucins, leading to changes in the PM structure. These degradative changes in the PM lead to more efficient transit of infectious ODV across the PM to the susceptible midgut epithelial cells and thus enhanced infectivity.

While baculoviruses encode a number of proteins and enzymes that interact with host cell machinery to maximize virus replication, one of the most interesting virus encoded enzymes interacts directly with the metabolism of a key insect hormone, ecdysone, which triggers molting in insects. Most baculoviruses infecting Lepidoptera encode an enzyme referred to as ecdysteroid UDP-glucosyltransferase (EGT) (reviewed in Rohrmann). EGT catalyzes the transfer of glucose from UDP-glucose to ecdysone thus inactivating the hormone and blocking the molting process from one larval instar to the next.

Molting in insects is an energy intensive process and as a result pathogen-infected insects often die in the process of molting. By blocking the molt, the action of viral EGT extends the life span of a baculovirus infected larva within an instar and leads to the production of more virus OBs. Indeed, the production of OB in wild-type baculoviruses was shown to be approximately 30% greater than in a recombinant baculovirus in which the EGT gene had been deleted.

Interestingly, the EGT deletion virus killed infected larvae significantly faster than the wild-type virus because molting was not blocked and the insects died in the process of molting. It has been proposed that development of recombinant baculoviruses with the

EGT gene under the control of an inducible promoter which would allow for production of OB in factory produced insects but would be EGT minus under "field" conditions could lead to baculovirus-based bioinsecticides with faster speed of kill.

Two of the observable symptoms associated with baculovirus infection in insect larvae are possibly linked to the evolution of pathobiology that ensures maximal dispersal of the virus within the host's environment, thus increasing the chances of epizootic development in host insect populations. It has long been recognized that baculovirus-infected-larva show an increased predilection to wandering and climbing behaviour. Indeed, some of the earliest reports of baculovirus infections refer to "tree top disease" describing the tendency for infected larvae to climb to the tops of the plant canopies and die.

The infected cadavers hang clinging to plant tissue, become flaccid and begin to "melt". Several viral encoded enzymes play a role in this process which leads to increased dispersion of baculovirus OB in the host's environment. Both baculovirus EGT and a gene encoding a protein tyrosine phosphatase have been implicated in the hyperactivity and wandering behaviour seen in infected lepidopteran larvae.

The evidence for the involvement of these enzymes in the altered behaviours comes from studies comparing wild-type and gene deletion virus infections. The "melting" phenomenon has been linked to two virus encoded enzymes that facilitate the degradation of host tissues late in virus infection. Chitinase genes are found in almost all baculoviruses infecting lepidopteran insects. This gene is expressed late in the baculovirus infection cycle and the enzyme is retained in the endoplasmic reticulum until infected cells begin to lyse.

Chitinase is responsible for the digestion of chitin, one of the major components of insect exoskeletons. In addition to chitinase, a virus encoded protease, cathepsin, is expressed late in baculovirus infection. This protease aids in the breakdown of host cellular structure and eventually the integrity of the infected cadaver, thus maximizing baculovirus OB dispersion.

Evidence for the importance of these two virus encoded enzymes in cadaver liquefaction comes from studies with recombinant viruses in which one or both of these genes have been knocked out. Baculovirus constructs in which either or both genes have been deleted lead to virus infections in which cadavers remain intact for several days after death rather than displaying the typical "melting" symptomology.

THE NEW INTEGRATED PEST MANAGEMENT PARADIGM

The concept of integrated pest management (IPM), a sustainable strategy for managing pests, has been in practice for a long time. Although multiple sources define IPM in different ways, previous models primarily focused on the ecological, and to some extent on the evolutionary, aspects of pest management. A recent IPM pyramid presented by identified a lack of a holistic IPM approach that uses both traditional and modern tools.

However, his conceptual framework mainly dealt with the ecological aspects of pest management with an emphasis on interdisciplinary research approach. Several reports indicated that IPM implementation depends on numerous factors including the level of education, economic and social conditions, environmental awareness, rational thinking, moral values, regulatory aspects, government policies, availability of IPM tools, extension education, consumer preference, and retail marketing. However, there is no IPM model that encompasses all these factors and provides a comprehensive description.

The interpretation of IPM also varies among those who develop, promote, or practice IPM strategies. For example, according to the United States Department of Agriculture-Agricultural Research Service, IPM is a sustainable, science-based, decision-making process that combines biological, cultural, physical, and chemical tools to identify, manage, and reduce risk from pests and pest management tools and strategies in a way that minimizes overall economic, health, and environmental risks.

Several other definitions also focus on minimizing or eliminating the reliance on chemical control options, adopting a number of other options with the emphasis on environmental and human health. However, some practitioners interpret IPM as rotating chemicals from different mode of action groups to maintain pest control efficacy and reduce pesticide resistance with an emphasis on reducing pest damage. These definitions and interpretations represent a variety of objectives and strategies for managing pests including vertebrate and invertebrate pests, diseases, and weeds.

IPM is not a principle that strictly and uniformly applies to every situation, but a philosophy that can guide the practitioner to use it as appropriate for their situation. For example, host plant resistance is effectively used in some crops with pest and disease resistant or tolerant varieties, but not in other crops. Pheromones are widely used for mating disruption, mass trapping, or monitoring of certain lepidopteran and coleopteran pests, but not for several hemipteran pests. Biological control is commonly used for greenhouse pests, but not to the same extent in the field.

Mechanical tools such as bug vacuums are used in high-value crops such as strawberry, but they are not an economical option in non-specialty crops and are not carbon efficient because of fossil fuel consumption. While chemical pesticides should be used as the last resort, in principle, sometimes they are the first line of defense to prevent the area-wide spread of certain endemic or invasive pests and diseases or to protect the seed and transplants from common and persistent pest problems.

Seed treatment with chemical pesticides, e.g., has become a popular prophylactic measure in many crops in recent years. Crop production is an art, a science, and an enterprise, and by adding environmental and social factors, IPM—an approach used in crop production—is also influenced by a number of factors.

Each grower has their own strategy for producing crops, minimizing losses, and making a profit in a manner that is acceptable to the retailer, safe for the consumers, and less disruptive to the environment. In other words, IPM is an approach to manage pests in an economically viable, socially acceptable, and environmentally safe manner.

Keeping this short, but complex, definition in mind and considering recent advances in crop production and protection, communication technology, and globalization of agriculture and commerce, a new paradigm of IPM is presented with its management, business, and sustainability aspects having the following direct or indirect objectives:

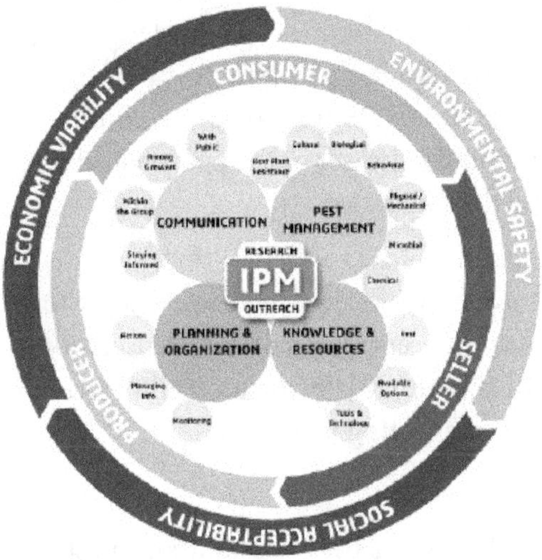

Fig. 2. New IPM paradigm with its various components and influencing factors for economically viable, socially acceptable, and environmentally safe pest management.

- Update the IPM concept as appropriate for modern times and encourage re-evaluation of what is perceived as sustainable
- Build consumer confidence and education in an IPM-based production system that is ideal for all crops and situations, ensures global food security, and eliminates food-based social inequality
- Ensure profitability for the producers while allowing informed consumers, rather than special interest groups or retailers, to make their food choice
- Minimize potential negative impact of the non-IPM-based conventional practices or those perceived to be sustainable alternatives on the environment and challenges associated with managing certain pests

Management Aspect

There are four major components in the new IPM model that address various pest management options, the knowledge, and resources the grower has to address the pest issue, planning and organization of information to take appropriate management actions, and maintaining good communication to acquire and disseminate knowledge about pests and their management.

Pest Management

The concept of pest control has changed to pest management over the years knowing that a balanced approach to managing pest populations to levels that do not cause economic losses is better than eliminating or eradicating (except for newly introduced invasive pests), for environmental and economic reasons.

Although the term control is frequently used in literature and conversations, it generally refers to management. A thorough knowledge of general IPM principles and various management options for all possible pest problems is important as some are preventive and others are curative. Some of the recommended practices may not be practical in all situations and the grower or the pest control professional has to choose the option(s) appropriate for their situation.

It is also essential to understand inherent and potential interactions among these management options to achieve desired control. The following are common control options that can be used at different stages of crop production to prevent, reduce, or treat pest infestations. Each of them may provide a certain level of control, but their additive effect can be significant in preventing yield losses.

Host Plant Resistance

A strategy that involves the use of pest-resistant and pest-tolerant cultivars developed through traditional breeding or genetic engineering (, ,). These cultivars possess physical, morphological, or biochemical characters that reduce the plant's attractiveness or suitability for the pest to feed, develop, or reproduce successfully. These cultivars resist or tolerate pest damage and thus reduce the yield losses. This option is the first line of defense in IPM.

Cultural Control

Adopting good agronomic practices that avoid or reduce pest infestations and damage refers to cultural control. Choosing clean seed or plant material is critical to avoid the chances of introducing pests right from the beginning of the crop production. Adjusting planting dates can help escape pest occurrence or avoid most vulnerable stages. Early planting of cowpea reduced aphid, thrips, and pod bug infestations in Uganda and the legume pod-borer (Maruca vitrata), the legume flower thrips (Megalurothrips sjostedti), and the pod sucking bug (Clarigralla tomentosicollis) in Nigeria.

Plant density or row spacing will also have an impact on pest infestations. High plant density reduced root maggot (Delia spp.) infestations in canola in Canada and aphid infestations in cowpea in Uganda. Modifying irrigation practices, fertilizer program, and other agronomic practices can create conditions that are less suitable for the pest.

Micro-sprinklers are installed on the strawberry beds as a spider mite control strategy especially in organic strawberries in California (personal observation). Strawberry plots with micro-sprinklers also appeared to have less severe powdery mildew (caused by Podosphaera aphanis) and botrytis fruit rot (caused by Botrytis cinerea) infections compared with the plots with overhead aluminum sprinklers. Low potassium content in plants induces jasmonic acid synthesis in plants and helps with plant's ability to withstand insect pests and certain diseases. Increased plant nitrogen can exacerbate arthropod infestations.

High or low nitrogen content in the plant can also contribute to some disease problems. Destroying crop residue and thorough cultivation will eliminate breeding sites and control soil-inhabiting stages of the pest. Sanitation practices to remove infected/infested plant material, regular cleaning field equipment, avoiding accidental contamination of healthy fields through human activity are also important to prevent the pest spread. For example, winter plowing of orchard floors reduced the pistachio psyllids (Agonoscena pistaciae) overwintering in the leaf litter and weeds in Iran. Plowing is also an important control option to destroy the crop residue and expose the soil-inhabiting stages of several vegetable pests.

Sanitation practices such as bagging unmarketable berries or even changing the harvest schedule from every 3 d to 1–2 d reduced spotted-wing drosophila (Drosophila suzukii) infestations. Crop rotation with non-host or tolerant crops will break the pest cycles and reduce their buildup year after year. Crop rotation tactic has been used for insect, disease, and weed management in many cropping systems. Intercropping of non-host plants or those that deter pests or using trap crops to divert pests away from the main crop are some of the other cultural control strategies in IPM.

Biological Control

Natural enemies such as predatory arthropods and parasitic wasps can be very effective in causing significant reductions in pest populations in certain circumstances. Periodical releases of commercially available natural enemies or conserving natural enemy populations by providing refuges or avoiding practices that harm them are some of the common practices to control endemic pests. Biological control has been successfully used in greenhouses and specialty crops such as strawberries grown in the field. To address invasive pest issues, classical biological control approach is typically used where natural enemies from the native region of the invasive pest are imported, multiplied, and released in the new habitat of the pest. The release of irradiated, sterile insects is another biological control technique that has been effectively used against a number of pests.

Behavioral Control

The behavior of the pest can be exploited for its monitoring and control through baits, traps, and mating disruption techniques. Baits containing poisonous material will attract and kill the pests when distributed in the field or placed in traps. Pests are attracted to certain colors, lights, odors of attractants or pheromones. Devices that use one or more of these can be used to attract, trap or kill pests. Pheromone lures confuse adult insects and disrupt their mating potential, and thus reduce their offspring.

Physical or Mechanical Control

This approach refers to the use of a variety of physical or mechanical techniques for pest exclusion, trapping (in some cases similar to the behavioral control), removal, or destruction. Pest exclusion with netting or row covers, handpicking or vacuuming to remove pests, mechanical tools for weed control, traps for rodent pests, modifying environmental conditions such as heat or humidity in greenhouses, steam sterilization or solarization, visual or physical bird deterrents such as reflective material or sonic devices are some examples of physical or mechanical control.

Microbial Control

Using entomopathogenic bacteria, fungi, microsporidia, nematodes, or viruses, and fermentation byproducts of some microbes against arthropod pests, plant parasitic nematodes, and plant pathogens generally come under microbial control.

Chemical Control

Chemical control typically refers to the use of synthetic chemical pesticides. However, to be technically accurate, chemical control should include synthetic chemicals as well as chemicals of microbial or botanical origin. Although botanical extracts such as azadirachtin and pyrethrins, and microbe-derived toxic metabolites such as avermectin and spinosad are regarded as biologicals, they are still chemical molecules, similar to synthetic chemicals, and possess many of the human and environmental safety risks as chemical pesticides.

Chemical pesticides are categorized into different groups based on their mode of action and rotating chemicals from different mode of action groups is recommended to reduce the risk of resistance development. Government regulations restrict the time and amount of certain chemical pesticides and help mitigate the associated risks.

The new ribonucleic acid interference (RNAi) technology where double-stranded RNA is applied to silence specific genes in the target insect is considered as biopesticide application. Certain biostimulants based on minerals, microbes, plant extracts, seaweed or algae impart induced systemic resistance to pests, diseases, and abiotic stressors, but are applied as amendments without any claims for pest or disease control.

These new products or technologies can fall into one or more abovementioned categories of pest management. All the pest management options need careful consideration and application to avoid potential risks. For example, several pests developed resistance to transgenic crops with Bacillus thuringiensis toxic proteins and planting non-transgenic plants along with resistant plants has been recommend, among other strategies, to reduce the resistance development. The western corn rootworm, Diabrotica virgifera virgifera, adapted to the cultural practice of rotating soybean with corn and gut bacteria, among other factors, appears to facilitate this adaptation.

Pests can also develop resistance to botanical and microbial pesticides if they are overused. While mating disruption is successfully used for controlling several pests, factors such as migration of mated females and alternative mate recruitment strategies of some species could affect the efficacy of this technique. Pesticide resistance in arthropod pests is a longtime problem in pest management and one of the key factors for developing IPM strategies.

As required by the crop and pest situation, one or more of the control options can be used throughout the production period for effective pest management. When used effectively, nonchemical control options delay, reduce, or eliminate the use of chemical pesticides.

Although pest management decisions are supposed to be based on economic injury levels and thresholds, in many situations they are either not available, difficult to determine, not applicable to all geographic regions or seasons, or existing ones need revalidation. Some of the established thresholds are also questionable.

Because crop production is highly precise due to modern technologies on one hand, as well as highly variable depending on a myriad of biotic and abiotic factors and the proprietary practices of different farming operations, information management and decision-making parts play a critical role in IPM. One cannot offer a one-size-fits-all solution and the pest control efficacy depends on several factors in addition to the option used.

Knowledge and Resources

The knowledge of various control options, pest biology and damage potential, and suitability of available resources enables the grower to make a decision appropriate for their situation. Conversations with growers in different parts of United States and other countries revealed that IPM implementation is limited by the lack of sufficient knowledge, resources, or immediate economic benefit. It is also evident from many conversations with growers and pest control professionals in California that resistance management seen as rotating pesticides among different mode of action groups is commonly perceived as IPM, although resistance management is only a part of IPM.

IPM implementation is especially a challenge in developing countries or with low-income growers in developed countries. Socio-psychological factors including rational and moral considerations were found to the drivers in IPM implementation in Iran. A survey of vegetable growers in Sri Lanka showed that nearly 50% of them practiced calendar-based chemical pesticide applications before pest or disease occurrence and only 20% had some understanding of IPM. A survey conducted by also showed that a lack of qualified IPM experts and extension educators was an impediment for IPM implementation in developed countries.

PEST

Identification of the pest, understanding its biology and seasonal population trends, damaging life stages and their habitats, nature of the damage and its economic significance, the vulnerability of each life stage for one or more control options, host preference and alternate hosts, predictability of pest occurrence based on the environment, cropping trends, farming practices, and other influencing factors, and all the related information is critical for identifying an effective control strategy.

Available Control Options

Because not all control options can be used against every pest, the grower has to choose the ones that are ideal for the situation. For example, systemic insecticides are more effective against pests that mine or bore into the plant tissue. Pests that follow a particular seasonal pattern can be controlled by adjusting planting dates. Commercially available natural enemies can be released against some, while mating disruption works well against others. Entomopathogenic nematodes can be used against certain soil pests, bacteria, and viruses against pests with chewing mouthparts such as Lepidoptera and Coleoptera, while fungi are effective against a variety of pests.

Although planting alfalfa strips in strawberry fields to attract and vacuum the western tarnished plant bug, Lygus hesperus, has been recommended as an IPM tactic the idea did not take off because growers were not ready for such an intercropping arrangement. Growers are reluctant to lose some of the land for alfalfa and concerned that alfalfa might attract more L. hesperus into strawberry fields. In some areas, where agricultural land is in shortage or there is a continuous demand for specialty crops, certain IPM practices such as allocating space for natural enemy habitats or crop rotation are not always possible. For example, on some farms, strawberries are grown year after year without a rotation with vegetables or cover crops on the California Central Coast (personal observations).

TOOLS AND TECHNOLOGY

A particular pest can be controlled by certain options, but they may not all be available in a particular place, for a particular crop, or within the available financial means. For example, the release of natural enemies may be possible in high-value specialty crops, but not in large acreage field crops. A particular pesticide might have been registered for a pest on some crops, but not on all. Use of netting, row covers, or tractor-mounted vacuums can be effective, but very expensive limiting their availability to those who can afford.

This is an important component where diagnostic and preventive or curative decisions are made based on available and affordable control options. With more regulations on pesticide use and a reduction in the number of active ingredients in some crops, there is a higher emphasis on better understanding of available control options. Regulatory guidelines that limit the use of certain pesticides or promote the use of others can have a major influence on IPM implementation. Many countries have phased out broad-spectrum pesticides and fumigants. Recent concern for pollinator health has also led to restrictions on the use of certain pesticides.

While these factors in the knowledge and resources component cover the implementation part of IPM, having sufficient resources to develop IPM strategies is a critical part in the whole equation that is often ignored. Shortage of IPM specialists, limited financial resources for research and extension, and inherent challenges in conducting time-consuming applied agricultural research are some of the hurdles in developing and disseminating IPM strategies. Too often, many of the IPM tools and technologies do not reach the implementation stage due to practical limitations such as a high cost of commercialization or lack of interest in its adoption.

PLANNING AND ORGANIZATION

This component deals with the management aspect of the new IPM model for data collection, organization, and actual actions against pest infestations.

Pest Monitoring

Regularly monitoring the fields for pest occurrence and spread is a basic step in crop protection. Early detection in many cases can help address the pest situation by low-cost spot treatment or removal of pests or infested or infected plant material. When pest infestations continue to grow, regular monitoring is necessary to assess the damage and determine the time to initiate farm-wide control. Monitoring is also important to avoid calendar-based pesticide applications especially at lower pest populations that do not warrant treatments.

For example, sampling-based fumigation along with improved sanitation and other practices can be an economical alternative to calendar-based fumigation of wheat in elevators. During many conversations with growers and colleagues in extension, it appeared that thorough scouting or decision-making solely based on scouting are not possible due to the lack of resources.

However, drone-assisted aerial imagery for pest detection and identification or to locate areas that are exposed to biotic or abiotic stressors can improve the monitoring efficiency and precision. Some of these tools are already available for commercial use.

Managing Information

A good recordkeeping about pests, their damage, effective treatments, seasonal fluctuations, interactions with environmental factors, irrigation practices, plant nutrition, and other related information from year to year will build the institutional knowledge on a farm and prepares the grower to take preventive or curative actions.

Corrective Actions

Taking a timely action is probably the most important aspect of IPM. Even with all the knowledge about the pest and availability of resources for its effective management, losses can be prevented only when corrective actions are taken at the right time. Good farm management will allow the grower to act in a timely manner. These actions are not only necessary to prevent damage on a particular farm, but also to prevent the spread to neighboring farms. When pest management is neglected on a farm, it can spread to neighboring farms and become an area-wide problem with larger regulatory, social, and economic implications.

Communication

Good communication to transfer the individual or collective knowledge for the benefit of everyone is the last component of the new IPM model. Modern and traditional communication tools can be used for outreach as researchers develop information about endemic and invasive pests, emerging threats, and new control strategies.

Staying Informed

Growers and pest control professionals should stay informed about existing and emerging pests and their management options. Science-based information can be obtained by attending extension meetings, webinars, or workshops, reading the newsletter, trade, extension, or scientific journal articles, and keeping in touch with researchers and other professionals through various communication channels. There are several online resources from universities and other reputed institutions and smartphone applications that provide regular updates.

Well-informed growers can be well prepared to address pest issues. Keeping abreast with pest issues and their management trends is also very important for researchers and extension professionals as they develop and disseminate new strategies.

Communication Within the Group

Educating farm crew through periodical training or communication will help with all aspects of pest management, proper pesticide handling, ensuring worker safety, and preventing environmental contamination. Knowledgeable field crews will be able to identify and monitor pest problems and effectively execute the management strategies.

Communication Among Growers

Although certain crop production and protection strategies are considered proprietary information, sharing knowledge and resources with each other will improve pest control efficacy and benefit the entire grower community. Pests do not have boundaries and can spread to multiple fields when they are not effectively managed throughout the region. Growers in developing countries feel that collective action is required for IPM implementation as there is a lack of sufficient knowledge.

Communication With the Public

Public demand is influenced by retail marketing strategies, concerns for food safety based on pesticide regulations in their region, and a lack of knowledge on food production among other factors. During multiple outreach meetings and field tours organized exclusively for the public in Southern California during the past few years, it was evident that a majority was not aware of farming systems, had misunderstanding about sustainable agriculture, and believed that organic food is pesticide-free (Dara personal observations and unpublished survey data). They indicated a change in their knowledge and a potential change in their behavior in making food choices following the discussions and field tours. Educating public will help their understanding of and preference for organic, conventional, or sustainably produced food as well as influence policy and regulatory decisions in their regions to ensure food safety and security.

Research and Outreach

Research and outreach are an integral part of the IPM model to identify and anticipate pest problems, develop preventive and curative strategies, and effectively disseminate the information through traditional and modern communication tools and strategies.

United States Environmental Protection Agency recognizes education and outreach as the key factors in IPM implementation. A study conducted by involving IPM professionals and practitioners from 96 countries revealed that inadequate training and technical support as major obstacle for IPM implementation.

Identified that science-based solutions and extension services are critical for IPM implementation in vegetables and fruits in New Zealand. Developing science-based information through applied research, effective outreach based on the socioeconomic and demographic structure of the clientele, networking and communication skills of extension educators, and reputation of researchers and extension educators play an important role in educating IPM practitioners.

In addition to the research and outreach foundation and the four components of management, factors that influence profitable, safe, and affordable food production at a larger scale and their implications for global food security should also be included in an IPM model. There are two layers surrounding these four components addressing the business and sustainable aspects of food production.

Business Aspect

Consumers want nutritious, healthy, and tasty produce that is free of pest damage at affordable prices. Growers try to meet this demand by producing food that meets all the consumer needs, while maintaining environmental and human safety, and still being able to make a profit. Sellers evaluate the market demand and strategize their sales to satisfy consumers while making their own profit to stay in the business. In an ideal system, consumer, producer, and seller would maintain a harmonious balance of food production and sale. In such a system, food is safe and affordable to everyone, there will be food security all over the world, and both growers and sellers make a good profit with minimal risk to the environment in the process of food production. However, this balance is frequently disrupted due to

1. consumers' misunderstanding of various food production systems, their demand for perfectly shaped fruits and vegetables at affordable prices, or their willingness to pay a premium price for food items that are perceived to be safe,
2. growers trying to find economical ways of producing high-quality food while facing with continuous pest problems and other challenges, and
3. sellers trying to market organic food at a higher price as a safer alternative to conventionally produced food.

If growers implement good IPM strategies to produce safe food and consumers are aware of this practice and have confidence in food produced in an IPM-based system, then sellers would be able to market what informed-consumers demand. Extension traditionally focused on educating the growers and those involved in food production, but public education on the importance of IPM can have a significant influence on the way food is produced.

SUSTAINABILITY ASPECT

IPM is an approach to ensure economic viability at both consumer and producer level (seller is always expected to make a profit), environmental safety through a balanced use of all available pest control options, and social acceptability since IPM-based food is safe and affordable.

Although organic food production is generally perceived as safe and sustainable, the following examples can explain why it is not necessarily true. Organic food production is not pesticide-free and some of the pesticides used in an organic system are as harmful to humans and nontarget organisms as some chemical pesticides.

For example, pyrethrins are highly toxic to honey bees, fish, and aquatic invertebrates. Certain organically accepted pesticides have toxins or natural chemical molecules that are very similar to those in synthetic pesticides. In fact, some synthetic pesticides (e.g., synthetic pyrethroids vs pyrethrins and neonicotinoids vs nicotine) are manufactured imitating the pesticidal molecules of natural origin. Pests develop resistance to biopesticides just as they develop resistance to chemical pesticides.

Arthropod resistance to abamectin, B. thuringiensis formulations or its cry toxins in genetically modified crops spinosad, and other biopesticides are well documented. Kaolin particle films used in organic farming might control some pests, but can negatively impact natural enemies and cause other pests to proliferate. Organic farming practices might encourage natural enemy populations, but a higher number of natural enemies does not always result in pest suppression. Mechanical pest control practices such as vacuuming or tilling utilize fossil fuels and indirectly have a negative impact on the environment.

For example, diesel-powered tractors are operated for vacuuming western tarnished bug in strawberry 2–3 times or more each week while a pesticide application typically requires the use of tractor once every 7–14 d. To control certain pests, multiple applications of organic pesticides might be necessary with associated costs and risks, while similar pest populations could be controlled by fewer chemical pesticide applications. It is very difficult to manage certain plant diseases and arthropod pests through nonchemical means in some crops.

Inadequate control not only leads to crop losses, but can result in their spread to larger areas making their control even more difficult. Manure commonly used in organic farms can have a bigger carbon footprint, and nitrate shortage in the root zone and nitrate leaching are common in organic farms. A meta-analysis of European research by identified both positive and negative impacts of organic farming in terms of nutrient management and emphasized the need to improve both organic and conventional systems for reducing negative environmental impacts and increasing yields.

Many growers prefer a good IPM-based production to an organic production for the ease of operation and profitability. However, they continue to produce organic food to stay in business. Multiple conversations with the strawberry and vegetable growers in the Central Coast region of California indicated that they produce organically to meet the market demand, but they prefer a system where they can use nonorganic options when and as needed. Growers in Bangladesh, Haiti, Moldova, and Myanmar, during training programs organized by the U.S. Agency for International Development, shared that organic farming is more challenging than conventional farming especially for pest management, but they produce organically because of high returns.

A small study conducted in the United Kingdom reported that organic growers had ecocentric reasons for protecting the environment while conventional growers had anthropocentric reasons to ensure food security. However, a more recent review article thoroughly compared organic and conventional farming approaches at the global level. When the productivity, biodiversity, water quality, off-site environmental impacts, carbon footprint, climate change, and other aspects were compared, carefully run conventional farming appears to be more sustainable than organic farming. Compared to organic farming or a conventional farming with non-ecocentric approach, a conventional system based on IPM principles and focused on sustainability is safe, profitable, and practical.

While middle and upper-class consumers may be willing to pay higher prices for organically produced food, many of the low-income groups in developed and underdeveloped countries cannot afford such food. Organic food production can lead to social inequality and a false sense of well-being for those can afford.

Nutritional quality of organic foods is more for certain nutrients, while it is more for conventional foods for others. Although subjective well-being is reported from organic food consumption, the evidence of health benefits from the nutritional quality of organic foods is yet to be validated. Food security for the growing world population is necessary through optimizing input costs, minimizing wastage, grower adoption of safe and sustainable practices, and consumer confidence in food produced through such practices.

14

NOVEL BIOPESTICIDE FORMULATIONS FOR ENVIRONMENTAL SUSTAINABILITY

Biopesticides have been developed as environment friendly alternate to chemical pesticides. Pesticidal properties of neem leaves have been known in literature. A novel biopesticide using whole neem fruit in combination with other ayurvedic herbs in micro fine dust form has been developed. The formulation particles create a sustained release effect since the ingredients are gradually released and thus a long term protection from insects is achieved. The product is economical and efficient in the field.

Biopesticides are biological pest control agents that are viewed as safer alternatives to the synthetic chemicals that dominate the global insecticide market. A major constraint on the wider adoption of biopesticides is their susceptibility to the ultraviolet (UV: 290–400 nm) radiation in sunlight, which limits their persistence and efficacy. Here, we describe a novel formulation technology for biopesticides in which the active ingredient (baculovirus) is micro-encapsulated in an ENTOSTAT wax combined with a UV absorbant (titanium dioxide, TiO_2). Importantly, this capsule protects the sensitive viral DNA from degrading in sunlight, but dissolves in the alkaline insect gut to release the virus, which then infects and kills the pest.

We show, using simulated sunlight, in both laboratory bioassays and trials on cabbage and tomato plants, that this can extend the efficacy of the biopesticide well beyond the few hours of existing virus formulations, potentially increasing the spray interval and/or reducing the need for high application rates. The new formulation has a shelf-life at 30 °C of at least 6 months, which is comparable to standard commercial biopesticides and has no phytotoxic effect on the host plants.

Taken together, these findings suggest that the new formulation technology could reduce the costs and increase the efficacy of baculovirus biopesticides, with the potential to make them commercially competitive alternatives to synthetic chemicals.

Biopesticides, used to suppress pest population, consists of living organisms or products derived from these organisms. Biopesticides decompose more quickly in the environment and are generally less toxic towards non- specific target species. Thus they can often help suppress resistant pathogens and can be applied in alternation with other pesticides. Biopesticides generally contains low volatile organic chemicals (VOC) content and thus reduces air pollution caused by high-VOC chemicals compared to chemical pesticides. In other words, these beneficial microorganisms are generally obtained from aerial or underground parts of plants that are naturally less or not at all affected by a pathogen that devastates a neighboring group of the same plant species.

Using microorganisms which are a part of our biome as pesticide eliminates the need to incorporate any chemical substance in the ecosystem and thus doesn't disturb the ecological balance. The use of non- degradable chemicals in farming is harmful as it accumulates in tropic levels, and disturbs the balance of the ecosystems. Phyto-stimulation is a direct promotion of plant growth through the modulation of the plant's hormonal balance. Several microorganisms are capable to produce and excrete a variety of plant hormone-like compounds including auxin, gibberellins, cytokines etc. and also enzymes that degrade a precursor of ethylene thus promoting plant growth especially under stress conditions.

Biopesticides are effective in very small quantities and can decompose quickly, largely avoiding the pollution problems that are caused by conventional chemical pesticides, when used as a component of Integrated Pest Management (IPM) programs, can greatly decrease the use of conventional pesticides, without decreasing the crop yields. Moreover, a microorganism that possesses a combination of these growth-promoting activities and bio control potential offers the advantage to supply the crop in one application with a bio pesticide and a bio fertilizer.

MICROORGANISMS WITH BENEFICIAL BIOCONTROL PROPERTIES SELECTED

- Bacillus thuringiensis a ubiquitous grampositive, spore-forming bacterium that forms parasporal crystal during the stationary phase of its growth cycle responsible for the control of certain insect species among the orders Lepidoptera, Diptera, and Coleoptera. There are more recent reports of B. Thuringiensis isolates active against other insect orders (Hymenoptera, Homoptera, Orthoptera, and Mallophaga) and against nematodes, mites, and protozoa. They are also found to produce fungicidal enzyme chitinase

- Bacillus subtilis, possess several characteristics that enhance its survival in the rhizosphere and is also found to be effective as a biopesticide. They can display various mechanisms of biocontrol and biostimulation (produces auxin) and known to secrete chitinase which inhibits pathogenic fungal growth in rhizosphere.

- Beauveria bassiana produce some secondary metabolites in which beauvericin have insecticidal activity, cytotoxic and ionophoric property. It parasitizes a wide range of arthropods Such as termite, thrips, aphids, whiteflies, rice pests, corn borer, sucking pests and different types of beetles.
- Trichoderma viride colonizes in root and produces growth promoter gibberellins. They also secrete cellulase, glucanase and other enzyme which can dissolve the hyphae of invading mycelial fungi, thus reduce their virulence and the plants are protected from harmful fungus.
- Paecilomyces lilacinus, a common soil hypomycetes with a cosmopolitan distribution, parasitizes eggs of root knot causing agents such as nematode Meliodogyne incognita, Chitwood, Macrophomina phaseolina etc.

ISOLATION AND PURIFICATION OF CULTURE

Suitable colony from the above incubated plates were taken and streaked on PDA (for fungi) and NA (for bacteria) slants then incubated for 24 hours.

Characterization of the organism:

Bacteria from the pure culture were subjected to Gram staining and fungi from the pure culture was stained with lacto-phenol cotton blue stain and observed under light microscope.

Interaction Study:

All possible combinations between bacteriabacteria, fungi-bacteria as well as fungi- fungi were studied to check to determine their interaction behavior. In case of bacteria and fungi, the growth was checked on both Nutrient agar as well as Potato dextrose agar (PDA).

Fungal- Bacterial interaction

With the help of a sterilized cork borer, disc was cut out from PDA plates containing fungal pure culture and was placed in the center of the plate with bacterium streaked along four sides of the plate, which were then allowed to incubate at 37°C and 25°C in duplicate sets for 48 hrs.

Fungal - Fungal interaction

With the help of separate sterilized cork borers, discs were cut out from plated containing pure culture and then the two fungal discs were placed equidistant from the central line of the PDA plate and incubate at 25°C in duplicates for 72 hrs.

Bacterial-Bacterial interaction

Two individual bacteria from pure culture were Tstreaked on NA plates and allowed to incubate at 37°C for 24 hrs.

FORMULATION OF LIQUID PESTICIDES

For bacterial pesticide (SXCPB1)

1% peptone and 0.5% dextrose was added in sterile distilled water followed by addition of 1.5% rice bran oil to prepare an emulsion. Then Bacillus thuringiensis and Bacillus subtillis spore suspension was inoculated in 1:1 ratio. The formulation was incubated at 37°C for 24 hours.

For fungal pesticide 1 (SXCPF1)

2% rice bran oil was added to potato decoction to prepare an emulsion followed by the addition of 10% Beauveria bassiana spore suspension and 72hrs incubation at 25°C.

For fungal pesticide 2 (SXCPF2)

2% rice bran oil was added to potato decoction and mixed well to prepare an emulsion, 5% then 1:1 Trichoderma viride and Paecilomyces lilacinus spore suspension was added and incubated at 25°C for 48 hrs.

A mixture of charcoal powder and oil cake was made in 3:7 proportion followed by addition of 2% w/w, 1:1 Trichoderma viride and Paecilomyces lilacinus crushed mat and thorough mixing. The formulation was then packed in zip- lock bags and incubated at 25°C for 48 hrs.

APPLICATION OF FORMULATIONS ON PLANT SET UPS

After 2 weeks of germination, products were applied in desired amount to the plants accordingly. Six pots of control plants with base soil were only watered without any supplementary addition and the other 6 control pots were supplemented with renowned chemical pesticide. 0.8g of SXCFP3 was added in rhizosphere of the experimental set at a regular interval of 2 weeks, to prevent soil borne pathogenic attack of the plants. And of the other three types of formulation (SXCBP1, SXCFP1, SXCFP2) were sprayed on emergence of aphids or flies. Capsicum annum was selected as model plant to determine effects of biopesticide application, due to its fast growth rate and suitability to climate of Eastern India.

Chlorophyll assay

Acetone extract of 10g fresh leaves from each set up was centrifuged at 5000 rpm for 5minutes in room temperature and absorbance of chlorophyll content was measured at 645nm and 663 nm using UV-Vis spectrophotometer.

Shelf life check

The shelf life of formulated products were checked after 3 months of formulation by spread plate technique.

GROWTH MEASUREMENT OF HOST

After application of formulated product developmental of the plants- root length and shoot length were measured. The plant setups were observed to get prevented from pest and no negative impact on their growth or death was observed. All the sets were growing green and healthy. Moreover the growth promoting factors produced by B. subtilis such as Auxin. and by other members promoted increased shoot length and root length in SXCBP1 applied plants compared to other set ups where increased shoot length, root length was not observed in a drastic level but a gradual increase similar to the control was noted (Fig 1).

Fig. 1. Experimental set up after application of formulated pesticide

Chlorophyll assay

Chlorophyll assay of the experimental hosts were done to check developmental status. Chlorophyll assay was performed samples were spectroscopically measured. It was observed that the chl a: chl b ratio for all the treated plants lay nearby the expected range. Especially in case of SXCBP1 applied plants, quite a good level of chlorophyll content in the host plants was observed.

Shelf life check

On checking after 3 months, SXCBP1 was 2X10P 8 PCFUs/litre formulation, SXCFP1 was 1×10w spore/ml and SXCPF2 was 6X10y spore/ml.

UNIQUE REQUIREMENTS AND CHALLENGES FOR BIOPESTICIDE FORMULATION

The biological activity or mode of action of a pesticide is determined by its biologically active microorganism or metabolite. The active ingredient in its raw form requires formulation to facilitate its handling, storage and application.

A key difference between the formulation of conventional (chemical) pesticides and the formulation of biopesticides is the living nature of the biopesticide, and their biological viability is sensitive to storage conditions and environment.

A challenge for the formulation of microbial biopesticides is the presence of additional fermentation materials besides the microbes themselves. These materials may adversely react with formulation ingredients.

Another challenge facing microbial biopesticide formulators is the tendency for gram-positive bacteria such as Bacillus spp. to produce a slime layer in connection with endospore formation. Suspension of these bioformulations for spraying can lead to excessive sliminess and the blockage of sprayer filters and nozzles. While the slime layer is removable e.g. by centrifugation, it may contain valuable components such as lipopeptides, which may contribute to the biopesticidal effect.

Shelf life (storage stability)

Other challenges to biopesticide formulation include the shorter shelf life of biopesticides compared to conventional pesticides. Significant formulation efforts are being targeted at the long-term stabilization of microbial and other biopesticides.

Shelf life measured in years is normal for conventional plant protection products, while for biopesticides a shelf life of 1 to 2 years is desirable. For many biopesticides, however, shelf life is measured in weeks, and the correct storage of these products is vital. This is especially important for non-spore-forming, gram-negative bacteria.

Shelf life and viability may be extended by reducing storage temperature through freezing or refrigeration. While refrigeration is not an option in most cases, exceptions do exist for high-value crops, and there appears to be a growing tendency for growers to treat biopesticides as perishable high-value products.

Microorganisms may be frozen or refrigerated in culture broth or an appropriate buffer. However, depending on the microorganism, storage in culture broth may impact long-term cell viability, and extensive testing and development in this area is currently a priority.

Besides storage, longer shelf life may be obtained by increasing the number of microorganisms in the product, to ensure viability despite a subsequent decline in their population.

Fermentation parameters such as the rate of dehydration and the physiological state of the bacteria on harvest from the medium will also affect the shelf life of the final formulation.

Semiochemicals

Semiochemicals – natural compounds such as pheromones – are difficult to formulate due to their volatility. Their encapsulation in traditional polymer matrixes or natural biological microcapsules such as yeast cells may reduce their volatility and degradation by facilitating their controlled release to the environment.

However, as microcapsules are about the same size as pollen grains, microencapsulated biopesticides may pose a hazard to foraging bees. These may carry the microcapsules back to their hives where the biopesticide is released, with the risk of harming the bees.

Delivery

Due to the particulate nature of most biopesticides, agitation of the spray suspension is vital to prevent precipitation in the spray tank. Biopesticides are formulated for use with standard application equipment, and additional levels of filtering and straining as well as the use of large-orifice nozzles are recommended to prevent clogging.

Biopesticides are exposed to detrimental environmental conditions when applied in the field, including sunlight (UV), fluctuating moisture conditions, plant physiological and biochemical responses as well as competition by microorganisms already present on the leaf surface. These conditions may lead to shorter persistence once the biopesticide is applied to crops (field persistence).

UV-stability may be addressed by the addition of protectants (sunscreens such as oxybenzone and light blockers, such as lignin), the inclusion of antioxidant-rich natural products, and through the encapsulation of the active metabolite or microorganism.

Future developments are expected to include expression of genes coding for enzymes involved in DNA repair. In addition, an understanding of crop physiology may be used to reduce the exposure of biopesticides to sunlight, e.g. by utilizing the physiology of maize leaf whorls to funnel biopesticide granules to the feeding sites of larvae.

Since many biopesticides are slow to control the pest or pathogen, growers may feel that the product has insufficient effect and reapply unnecessarily. Diligent reading of labels and an understanding how the product works, as well as appropriate scouting is necessary for the optimal application of these products, allowing growers to reap the full benefits of biopesticides and attain consistent results.

Formulation types

Due to differences in environmental conditions, biopesticides may be formulated in different forms, each designed for a specific market. Many of the formulation types described in the previous articles on conventional pesticide formulation are relevant for biopesticides.

Most biopesticides are formulated as dry (solid) formulations. Dry formulations of dormant bacterial spores tend to have a longer shelf life compared to most liquid formulations. Dry formulations may, however, have several disadvantages, including lower field performance and the risk of nozzle blockage.

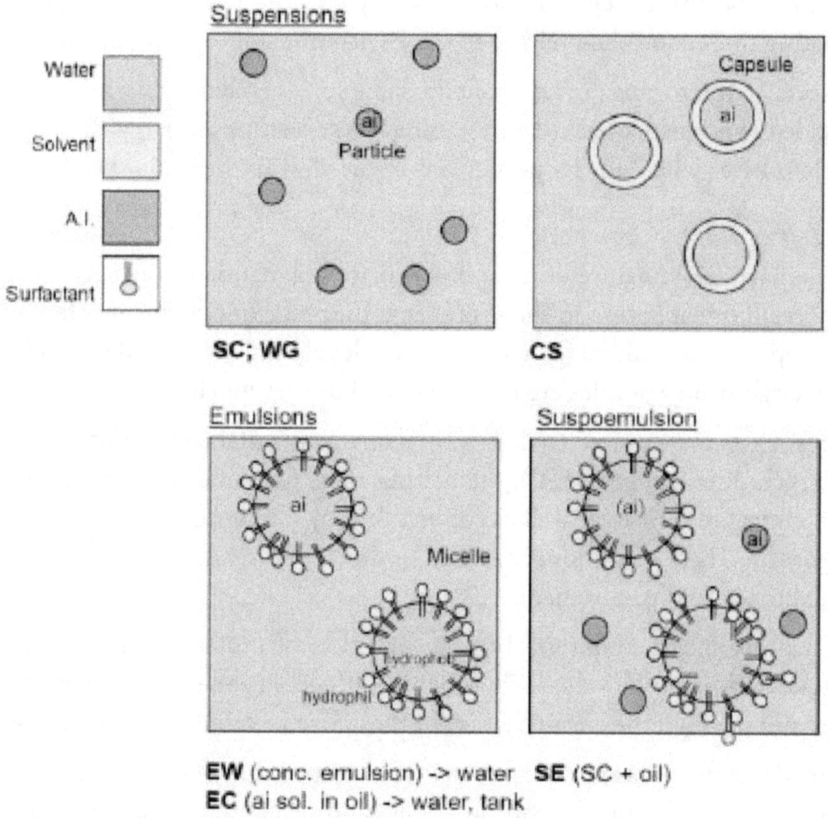

Fig. 2: Summary of selected formulation types in aqueous spray solutions.

Dry products include: dusts (to be applied dry), wettable powders (WP; intended to be applied as an aqueous suspension) and granules (G), which may be applied directly or mixed as an aqueous suspension (water dispersible granules; WG or WDG).

Liquid biopesticide formulations typically comprise biopesticidal metabolites or microorganisms suspended in a medium containing adjuvants, protectants and nutrients.

They may be formulated as conventional formulations such as suspension concentrates (SC) or as cell suspensions in water, oils or combinations of both (emulsions; EC or EW) and microcapsules (CS).

Liquid biopesticide formulations address issues regarding shelf life and field efficacy, and may generally be applied using existing field equipment without issue. The latter is of importance for the acceptance of biopesticides by growers, who usually are not willing to invest in new equipment specifically for the application of biopesticides.

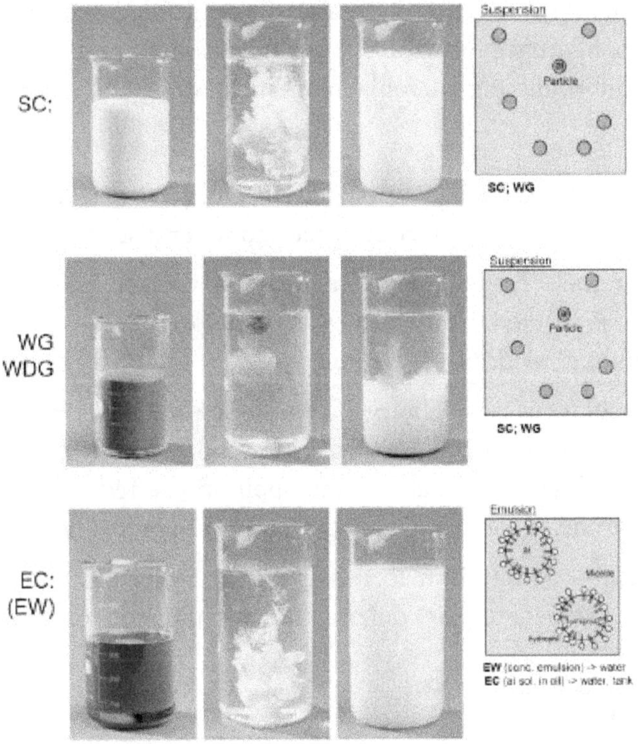

Fig. 3: Suspensibility (SC; WG) and solubility (EC; EW) of pesticide formulations.

Inerts

Formulation components other than the active ingredient or microorganism are called inerts. Inerts may contribute to improving product performance by optimising contact to leaf, soil or insect surfaces, by reducing desiccation, by increasing shelf life or by increasing solubility or suspensibility in water.

Inerts can include dry carriers (mineral: e.g. clay, or organic: e.g. flour) or liquid carriers (e.g. vegetable oils) as well as binders, desiccants, thickeners, surfactants, stickers and UV protectants. Formulation carriers can affect survival and viability, by affecting both drying-survival and storage stability.

To ensure microbial survival and efficacy after delivery, formulations may contain carbon sources (nutrients) such as molasses or peptone and moisture-retaining polymers. Hydrogels consist of a network of hydrophilic polymer chains which can hold large amounts of water.

Hydrogels may be used in slow-release biopesticide formulations to improve nutrient and moisture retention as well as reduce degradation, leading to improved biopesticide viability and efficacy.

For biopesticides, it is important that formulation inerts have low toxicity, both from an environmental and health perspective, but also with respect to the viability of the microorganism being formulated. Chitosan is often included in biopesticide formulations (for example, in hydrogel formation), and its possible antibacterial effects must be considered when formulating microbial biopesticides.

Trends

The trend in conventional and biopesticide formulation is to move away from dusts, wettable powders and suspension concentrates to water dispersible granules.

Controlled release formulations are being developed to optimise effect, while nanotechnology is expected to provide new types of formulations such as nanoemulsions, nanosuspensions and nanocapsule formulations.

Another area of focus is the choice of appropriate adjuvants for optimal biopesticidal activity. Here, cooperation between biologists and chemists is facilitating the development and introduction of new formulations and applications, reducing the inconsistency of field performance and encouraging growers to implement this new technology.

Attention is also being given to delivery and application practices, specifically to understanding the impact of plant defence induction kinetics on application timing and placement.

ADJUVANTS

Adjuvants are ingredients added to a formulation (formulation adjuvants) or tank mix (spray adjuvants) to improve the efficacy or safety of pesticide and biopesticide formulations.

For this discussion of pesticide and biopesticide adjuvants, our attention will primarily focus on foliar application:

Adjuvants fall into two groups: Activator adjuvants (improving the activity of the pesticide) and Modifier or Utility adjuvants (those improving the handling and application of the pesticide).

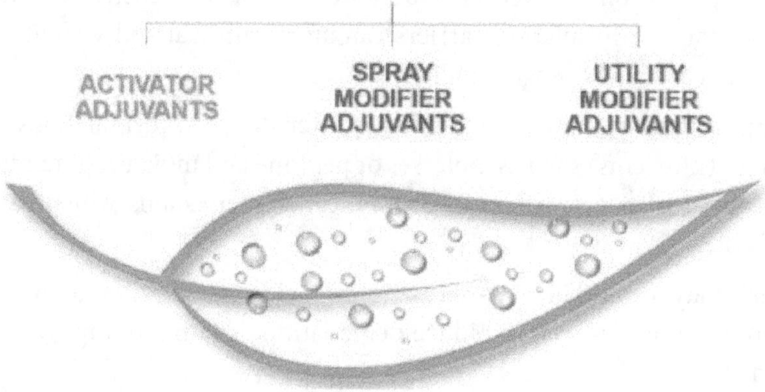

Fig 4. Adjuvant groups.

Modifier adjuvants may in turn be grouped into those affecting the physical properties of the spray solution (spray modifier adjuvants) and those affecting their handling and application (utility modifier adjuvants).

Together, adjuvants modify the spray solution to improve handling, application and efficacy in the case of conventional pesticides. For biopesticides, adjuvants may be selected to optimise contact to leaf, soil or insect surfaces, to reduce desiccation and to protect against UV degradation, to increase shelf life or to increase solubility or suspensibility in water.

Activator Adjuvants

Activator adjuvants command the greatest market share, and their function is to modify spray mixture properties to optimise pesticide and biopesticide retention and uptake by improving their spreading and penetrating properties, thereby improving the efficacy of the product.

For herbicides, care should be taken to evaluate possible crop phytotoxicity (loss of weed:crop selectivity) due to an increase in herbicide efficacy by activator adjuvants.

Wetters / spreaders

Surfactants or surface acting agents – are most commonly used to increase spray coverage on leaf surfaces, by reducing the surface tension of spray droplets and allowing them to spread beyond their initial contact area. Surfactants may be charged (non-ionic) or charged (anionic or cationic).

Non-ionic surfactants are most commonly used in pesticide and biopesticide applications, are compatible with most pesticides and are typically used with systemic products.

Anionic surfactants are typically used to improve the efficiency of contact insecticides to control small, relatively immobile pests such as mites, thrips and aphids, while cationic surfactants are rarely used due to potential phytotoxicity risks.

Organosilicones are silicon-based surfactants which have an extreme spreading ability ("super-spreaders"). Super spreaders can facilitate the movement of pesticides deep into the crop canopy, or into the other difficult-to-reach areas such as leaf whorls.

While super spreaders offer enhanced coverage their use may lead to increased rates of spray droplet evaporation (with potentially lower pesticide uptake rates), or coalescence and potential run-off from the leaf surface.

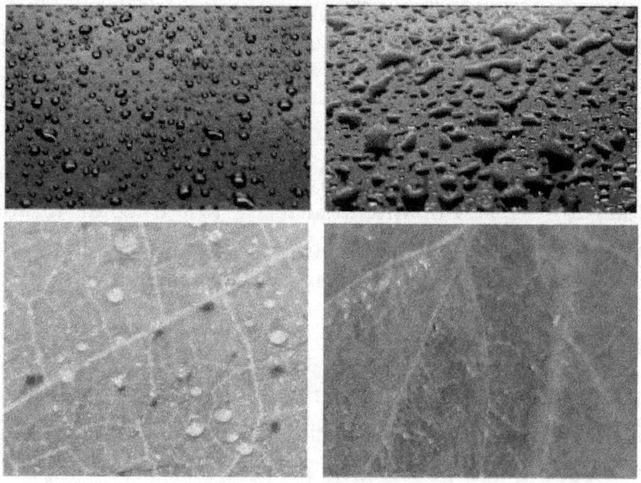

Fig. 5: Spray droplets without adjuvants (top left) and with added wetter adjuvant (top right). In the absence of adjuvants, spray droplets "pearl" and can be seen to sit on leaf hairs (bottom left) without contacting the leaf; with the addition of a superspreader, the spray drops spread and cover the entire surface of the leaf.

Penetrants / translocators

Penetrants are generally used to optimise active ingredient uptake in the grasses as well as broadleaf plants with a thick cuticula. Plants growing under hot, dry conditions may develop a thicker cuticula to reduce desiccation, and the inclusion of penetrants in spray solutions may be particularly beneficial under these conditions.

In a previous article in this series, the partition coefficient (logP) and dissociation constant (pKa) of a pesticide active ingredient were shown to be determinants of pesticide mobility in plants.

By affecting uptake across the cuticula as well as intracellular or extracellular transport within the plant, adjuvants are an additional factor deciding whether chemical and biochemical active ingredients have apoplastic (extracellular), symplastic (intracellular) or translaminar (from adaxial to abaxial leaf surface) activity.

Crop oil concentrates (COC; petroleum or vegetable-based oil with surfactants to emulsify the oil) and methylated seed oils (MSO; modified vegetable-based oil) function as penetrants (facilitating active ingredient uptake by softening or dissolving leaf cuticular waxes) and translocators (facilitating movement within cell walls and membranes), and may act as humectants.

Nitrogen based fertilizers are generally used in combination with a surfactant and a crop oil concentrate to improve the uptake and efficiency of systemic pesticides.

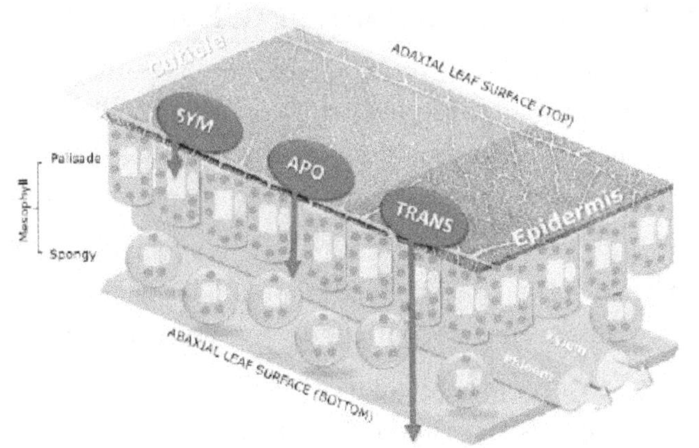

Fig. 6: apoplastic (extracellular), symplastic (intracellular) or translaminar (from adaxial to abaxial leaf surface) active ingredient pathways.

Modifier (Utility) Adjuvants

Modifier adjuvants address specific conditions that may affect the spray solution or its application, typically by modifying the physical characteristics of the spray solution, facilitating handling, safety and application.

Spray Modifiers

Spray modifiers affect the physical properties of the spray solution and include stickers, drift retardants (thickeners, typically polymers which minimise small droplet formation) and humectants.

Humectants extend the drying time of spray droplets, allowing more time for cuticular absorption and epidermal uptake to occur.

Stickers are ingredients that increase droplet or particle adherence to the leaf surface and are often combined with wetting or spreading agents to improve coverage. Stickers prevent run-off on application as well as wash-off during heavy rainfall.

Utility Modifiers

Utility modifiers address handling and application issues. Buffers can be used to maintain a slightly acid pH in spray solutions, preventing alkaline hydrolysis of susceptible pesticides, and may improve the uptake of certain herbicides.

AMS (ammonium salts) may be added to reduce hard water problems in spray solutions, while antifoam agents suppressed foam that can cause pump and spray problems.

Suspension agents contribute to maintaining particles in suspension, and are particularly relevant for microbial biopesticides as well as conventional pesticides prone to precipitation.

CONSIDERATIONS FOR BIOPESTICIDE ADJUVANTS

While many conventional adjuvants are suitable for biopesticide applications, special consideration must be given for the living nature of biopesticide products. The objectives of adjuvants in biopesticide formulations are to stabilise the active metabolite or microorganism, to optimise delivery to the target, to protect the biopesticide during storage and in the field, and to enhance the efficacy of the biopesticide.

Due to the particulate nature of most biopesticides, suspension agents as well as agitation of the spray suspension is required to prevent precipitation in the spray tank.

Biopesticides are sensitive to sunlight. UV-stability may be addressed by the addition of protectant adjuvants (sunscreens and light blockers).

To prevent desiccation and ensure microbial survival and efficacy after delivery, formulations may contain moisture-retaining polymers (e.g. hydrogels) and carbon sources (nutrients) such as molasses or peptone.

When choosing adjuvants to optimise microbial growth, it may be necessary to consider whether they also stimulate and support pathogen germination and infection.

Finally, it is important that adjuvants for biopesticides have low toxicity, both from an environmental and health perspective, but also with respect to the viability of the microorganism being formulated.

CONCLUSION

Until recently, slow progress in formulation and delivery research has been a major hurdle to biopesticide development and acceptance. Current research efforts in bioformulation technology are focused on promoting the bioefficacy post-application, especially with regard to application timing and placement of the formulated product.

Formulation is one of the key factors in determining the commercial success of a biopesticide candidate. There are significant costs involved in solving the biological issues of biopesticide formulations (including storage and field stability as well as efficacy consistency), but the benefits in terms of market access and the acceptance of biopesticides as viable commercial alternatives or supplements to conventional plant protection products warrant this effort.

GLOSSARY

Accuracy (of GPS receivers): The measure of closeness of an object's actual (true) position to the position obtained with a GPS receiver. Accuracy levels are used to rate the quality of GPS receivers.

ACE: Agriculture in Concert with the Environment.

ACS: Alternative conservation system.

Active ingredient: In any pesticide product, the component that kills, or otherwise controls, target pests. Pesticides are regulated by the EPA primarily on the basis of active ingredients. See also Inert ingredient.

ADI: Acceptable daily intake (the amount of a substance per 1 kg of body weight that can be ingested daily without health risk). This value is determined on the basis of the no observed adverse effect level obtained from long-term toxicity studies and other studies.

Aerial Imagery: Photos taken from airplanes, satellites, or UAVs to assist growers to determine variations within an area of interest such as a field.

Aerial remote sensing images: Photographs taken from the air, usually done by drones, to help farmers determine the status of the field.

Agribusiness: Agriculturally related businesses that supply farm inputs (such as fertilizer or equipment) or are involved in the marketing of farm products (such as warehouses, processors, wholesalers, transporters, and retailers). Farms are not usually included when the term agribusiness is used.

Agricultural chemicals: Chemicals used in agriculture, including fertilizers, herbicides and pesticides.

Agricultural data: Data from farm activities and conditions. Data can include soil analysis, nutrition information, seed selection, plant population, and yield data.

Agricultural pollution: Wastes, emissions, and discharges arising from farming activities.

Agricultural product: The term agricultural products is often used as a collective expression and like agricultural commodity, is defined in different ways to meet different circumstances. Agricultural protection zoning: Local zoning codes that include provisions such as large lot size requirements and use limitations to separate farming and related activities from other land uses

Agricultural production: A series of agricultural activities and the resulting agricultural products will be sold in the market.

Agriculture 4.0: An analogy to Industry 4.0. It is the integrated internal and external networking of farming operations as a result of the emergence of smart technology in agriculture.

Agriculture as a Service: Known as FaaS, this refers to the way farmers use subscriptions or pay-per-use for technology and data. For small to medium farms, this is ideal for data-driven decision making to increase productivity and efficiency, including farm management solutions, production assistance and market access.

Agro-chemicals: Chemicals used in agriculture, which include fertilizers, herbicides, and pesticides.

Agronomic data: Data derived from activities and conditions on farm fields. The data can include soil analysis, nutrient information, hybrid selection, plant populations, and yield data.

Agronomy: The science of field-crop production and soil management.

Air pollution: Contamination of the atmosphere by substances that, directly or indirectly, adversely affect human health or welfare

Algorithm: A mathematical formula that can be used to control variable rate applications.

Algorithm: A mathematical formula that may be used to control variable rate applications.

Animal biologics: Vaccines, bacterins, antigens, diagnostic kits and other products of biologic origin.

Aquifer: An underground geological formation, or group of formations, containing usable amounts of groundwater that can supply wells or springs for domestic, industrial, and irrigation uses.

Arid: A relatively dry climate in which annual precipitation is less than 10 inches, which generally is insufficient for crops to be grown without irrigation. Such areas usually are the focus of debate over federal water policies.

Artificial intelligence: Computers simulate intelligent behaviour, such as decision making. AI can refer to anything from computer programs that play chess to speech recognition systems like Amazon Alexa, which are used to interpret and respond to speech.

Auto-Steer – a GPS guidance system that steers agricultural equipment with centimeter accuracy. This level of accuracy requires real time kinematic (RTK) correction of GPS signals. Auto-steer is an add-on component for equipment. It includes both the GPS system to receive and process the signals, software and hardware to allow the input of control maps and the mechanical equipment to actually steer the tractor. Some new tractors are available "auto-steer ready."

Base Map: a simple map that shows the boundaries of a field or section and information about any unique feature (sinkholes, or streams).

Base Station: The RTK-GPS receiver and radio that are placed in a stationary position, functioning as the corrections source for roving tractor units in an area. These stations can be either portable or permanently installed systems and their coverage can range from 5 to 10 miles depending on topographic conditions, antenna height, and radio-transmit power.

Big Data Analytics: Examine a large number of different data sets (called big data) to reveal hidden patterns, unknown correlations, market trends, and customer preferences that can be used for decision making.

Big Data: Information collected in a large number of different data sets through IoT sensors and devices.

Biochemicals: Biochemical pesticides are naturally occurring substances that control pests by non-toxic methods. Biochemicals include insect sex pheromones that disrupt mating and plant extracts that lure pests into traps. Our copper fungicides are considered biochemicals. Copper hydroxide (the Kocide products) and copper octanoate (Cueva) control fire blight and bacterial disorders. Potassium bicarbonate (Carb-O-Nator) is a biochemical used for control of powdery mildew and Botrytis and peroxyacetic acid (Rendition) is a bactericide/fungicide often used in greenhouses to sanitize surfaces.

Biocontrols: Previously a biocontrol was known as the release of a beneficial species that is a predator of a specific pest, such as lady bugs, lacewings or six spotted thrips to control mites. Today, biocontrols also include the use of live organisms that can be tank mixed and applied via sprays or chemigation. The live organism Isaria fumosorosea (PFR97-) controls aphids, whiteflies, thrips, mites and psyllids. Paecilomyces lilacinus (MeloCon) is a fungus that when it is applied to and watered into soil infects and kills nematodes.

Biofuels: Fuels made from biomass, which in the United States largely include corn-based ethanol (blended into gasoline and called gasohol) and soybean-based biodiesel. Biofuels are a subset of renewable fuels, which are a subset of alternative fuels.

Biological control: The practice of using beneficial natural organisms to attack and control harmful plant and animal pests and weeds is called biological control, or biocontrol.

Biological monitoring: Using living organisms to test the quality of either effluent to be discharged into receiving waters, or waters downstream from a discharge.

Biological oxygen demand (BOD): See Biochemical oxygen demand.

Biologics: Immunization vaccines, bacterins, antigens, and antitoxins and other preparations made from living organisms and their products, intended for use in diagnosing, immunizing, or treating humans or animals, or in related research. The Animal and Plant Health Inspection Service has responsibility for approving some animal biologics. See Veterinary biologics.

Biopesticide: A pesticide that is biological in origin (i.e., viruses, bacteria, pheromones, natural plant compounds) in contrast to synthetic chemicals. Transgenic Bt cotton and corn are biopesticides because Bacillus thuringiensis (Bt) is a naturally occurring soil bacterium that has been genetically engineered into the plants. Biopesticide is a broad term for pesticides that are derived from a natural source and used as an insecticide, miticide, fungicide and/or an herbicide. The naturally occurring sources, or active ingredients, can be derived from animals, plants, bacteria and certain minerals. According to the EPA, as of April 2016, there are 299 registered biopesticide active ingredients and 1401 active biopesticide product registrations. Some examples of biopesticide active ingredients include azadirachtin (Neemix 4.5), an insect growth regulator that is extracted from the seeds of the neem tree. Azadirachtin controls leafminers, whiteflies and aphids. Spinosad (Seduce) is used in an insecticidal bait to control ants, earwigs and cutworms. Spores of Bacillus amyloliquefaciens strain D747 (Double Nickel 55) control diseases such as powdery mildew, Botrytis, bacterial disorders and foliar diseases.

Biopharmaceutical (biopharm) crops: Crops that are bioengineered to produce pharmaceuticals, industrial chemicals and other medical and industrial products (sometimes called plant-made pharmaceuticals).

Biostimulants: A biostimulant product has an active ingredient that causes a plant response, such as a yield increase or other aspect of growth. A more technical definition comes from the Biological Products Industry Alliance (BPIA): Plant biostimulants contain substance(s) and/or microorganisms whose function when applied to plants or the rhizosphere is to stimulate natural processes to enhance/benefit nutrient uptake, nutrient efficiency, tolerance to abiotic stress and crop quality. There are many examples of biostimulants available to us from literally hundreds of companies that sell biostimulant products today. This wide-ranging category of active ingredients include acids, such as humic acids, fluvic acids, amino and fatty acid products. Biostimulants can be extracts of seaweed/kelp, carboxyls, allelochemicals and other organic matter extracts, including sugar beets extracts. There are microbial biostimulants that include Trichoderma, mycorrhizae and other beneficial fungi, such as Gliocladium virens (SoilGard) which controls damping off and other soil-borne diseases. The biostimulant active ingredient in Certis USA's AgSil products potassium silicate stimulate plants to withstand stress, drought, salinity and metal toxicity. Bacillus mycoides isolate J (LifeGard), a biological plant activator that manages fungal and bacterial diseases, is also a biostimulant. Its spores trigger an internal plant response to "switch on" resistance genes that cause a cascade of plant metabolic responses that limit infection and disease development.

Biotechnology: Broadly defined as transferring genes from one living entity (plant, animal, or microbe) into another (or into a synthetic compound) using advanced recombinant DNA technology (also called genetic engineering (GE)).

Boom/section Controller – An electronic device that is capable of turning on/off sections of a chemical application boom manually or automatically in combination with GPS positioning and area mapping.

CAN-Bus (in tractors and implements): CAN-Bus is a high-speed, wired data network connection between electronic devices. The hardware/wiring of CAN-Bus networks are generally the same, while the protocols for communication can be different and vary depending on the industry where they are used. These networks are used to link multiple sensors to an electronic controller, which can be linked to relays or other devices on a single set of wires. This reduces the amount of wires needed for a system and allows for a cleaner way to connect additional devices as system demands change.

Contour Line: a line used to represent the same value of an attribute (elevation or yield).

Contour Map: yield map that combines dots of the same intensity/yield level by interpolating (or kriging).

Contouring: interpolation method used to distinguish between different levels of an attribute (elevation, fertility, yield).

Control Segment: the network of tracking stations that monitor and control GPS satellites.

CORS (network): Continuously Operating Reference Station. A network managed by the U.S. office of National Ocean Service (NOAA) to provide GNSS data consisting of carrier phase measurements throughout the United States.

Data Layer (in GIS): A layer of information on a GIS map. A map can have many layers to present different types of information. For example, the first layer of a map may be a satellite image of an area. The next layer may have only lines that represent roads or highways. The next layer may contain topographic information and so forth.

Database: a collection of different pieces of georeferenced information (yield, soil type, fertility) that can be manipulated (layered) in a GIS model.

Differential Correction: correction of a GPS signal that is used to improve its accuracy (to less than 100 m/~330 ft) by using a stationary GPS receiver whose location is known. A second receiver computes the error in signal by comparing the true distance from the satellites to the GPS measured distance.

Differential Global Positioning System (DGPS): a method of using GPS which attains the position accuracy needed for precision farming through differential correction.

Flow Sensor: a sensor that measures the amount of flow through an enclosure (tube, pipe or housing) per unit of time.

Geographic Data: data that contains information about the spatial location (position) and the attribute being monitored (yield, seed population, etc.). Also referred to as spatial data.

Geographic Information System (GIS): a computer-based system used to input, store, retrieve, and analyze geographic data sets. The GIS is usually composed of map-like spatial representations called layers which contain information on a number of attributes such as elevation, land ownership and use, crop yield and soil nutrient levels.

Georeferencing: the process of adding geographic data to yield data or other field attributes either in real-time (on-the-go) or by post-processing or the process of associating data points with specific locations on the earth's surface.

Geo-Stationary Satellite: an orbital path of a satellite that is synchronized with the earth's orbit or space vehicles in an orbit which keeps them over the same location on the earth at all times.

Global Positioning System (GPS): A system using satellite signals (radio-waves) to locate and track the position of a receiver/antenna on the Earth. GPS is a technology that originated in the U.S. It is currently maintained by the U.S. government and available to users worldwide free of charge. There are 30 satellites in the GPS constellation.

Global Positioning System: A system that uses satellite signals to locate and track the position of the receiver/antenna on the Earth.

Global warming: See Climate change and/or Greenhouse gases.

GLONASS (GLObal naya NAvigatsionnaya Sputnikovaya Sistema): Russian version of the American GPS satellite system. It is a radio-based satellite navigation system operated for the Russian government by the Russian Space Forces with a constellation of 24 operational satellites in 2010.

GNSS (Global Navigation Satellite System): Is the standard generic term for satellite navigation systems that provide geo-spatial positioning with global coverage using time signals transmitted from satellites. The United States GPS and the Russian GLONASS are the only two fully operational GNSS. Top of the line GNSS receivers can communicate with both GPS and GLONASS satellites effectively doubling the available reference satellites at any given time.

GPS Antenna: The device that receives satellite signals from space. On most hand-held GPS devices, the antenna is integrated into the receiver device. For machine GPS systems, the antenna is typically an external device that can be mounted on top of the vehicle, away from the receiver.

Grid map: Soil or plant samples can be obtained to analyze predetermined location information in the field. This information can be used to assess fertility requirements and to determine the approximate location of different fertilizer applications.

Grid Mapping: Predetermined locations in a field where soil or plant samples may be obtained for analysis. The test information can be used for assessing fertility needs and determining approximate locations for varying fertilizer and lime applications.

Grid Sampling: the collection of samples from small, uniform-sized cells based on a systematic grid laid out across a field. Grid location in the field is used to develop a field map for the attribute measured.

Ground Control Points: stationary objects/areas on the earth's surface that provide georeferenced points in a remote sensing image/aerial photograph.

In an inter-species relationship in nature in which one preys on the other specifically, the predator is referred to as "a natural enemy" of the prey.

Interpolation: mathematical procedure for estimating unknown values from neighboring **known data.**

ISOBUS: ISOBUS standard 11783 is a communication protocol for the agricultural industry that is used to specify a serial data network for control and communications on forestry or agricultural tractors and implements. ISOBUS-compliant tractors and implements come with round 9-pin connectors.

Kriging: an interpolation method for obtaining stastically unbiased estimates for field attributes (yield, nutrients, elevation) from a set of neighboring points.

LANDSAT (LAND SATellite): US satellites used to study the earth's surface using remote sensing techniques.

Latitude: A global standard coordinate used to identify a position on earth given in degrees, minutes and seconds, indicates the north/south position above/below the equator, positive is in the northern hemisphere and negative is in the southern hemisphere.

LC50: median lethal concentration. A concentration of the poisonous substance in air or in liquid, which kills 50% of the subject animals within the set amount of time. The higher the value, the less toxic.

LD50: median lethal dose. A dose level that kills 50% of the subject animals within the set amount of time. This value varies depending on inter-species variation and administration route. The higher the value, the less toxic.

Lightbar (in machine guidance): A device connected to a GPS receiver typically consisting of a row of led lights to provide the tractor operator with a visual guide, day or night. The lightbar does not automatically steer the tractor or machine, rather it aids the operator in driving on the imaginary reference line.

Longitude: A global standard coordinate used to identify a position on earth given in degrees, minutes and seconds, indicates the east/west position around the globe from a reference point which overlays Greenwich, England. Negative values are east of Greenwich and positive values are west.

Machine data: Information about how the device works. This can include fuel consumption, machine health indicators, diagnostic codes, and engine performance.

Machine learning: Machine learning is a sub-area of artificial intelligence. The core principle is that the machine acquires data and learns by itself.

Microbial Pesticides: These products contain a microorganism (bacterium, fungus, virus or protozoan) as its active ingredient. Specificity is a hallmark of the microbial pesticides. Typically, the active ingredient is specific for its target pest. For example, Cydia pomonella GV (CYD-X HP) is a virus that infects and kills codling moth larvae and no other pest. The most widely used microbial pesticides are the Bacillus thuringiensis (Bt) products. Certis USA's Bt products include Deliver and Javelin that control Lepidoptera caterpillar pests, but are of low risk to beneficial insects, workers and the environment.

NDVI Image: The Normalized Difference Vegetation Index (NDVI) is a simple graphical indicator that can be used to analyze remote sensing measurements and assess whether the target being observed contains live green vegetation or not.

No observed adverse effect level: NOAEL. NOAEL refers to a level that can be ingested by animals daily over a lifetime without adverse effect on the health of the animals, which is obtained form toxicity studies. Unit: mg/kg/day

Phenome: The set of physical and biochemical traits belonging to an organism.

Phenomics: The measurement of phenomes.

Ppm: Parts per million. By analogy, one gram in a 1-ton small truck.

Precision Farming: managing crop production inputs (seed, fertilizer, lime, pesticides, etc.) on a site-specific basis to increase profits, reduce waste and maintain environmental quality.

Prescribed Application: The dispensing of a material or chemical into the field on a prescribed or predetermined basis. A prescription map is generated by an expert (grower and/or agronomist) based on information about the field in use before an application. The prescription determines how much of something will be applied.

Prescription File: A computer generated GIS file that assigns a value to a given geographical area. Example: Nitrogen application rate.

Raster Format: format for storing GIS spatial data in which the data is stored in cells which are addressed by rows and columns of the cells.

Rate Controller: An electronic device that varies the amount of chemical/plant nutrient applied to a given area.

Real-Time Correction: correction of a GPS signal by simultaneously transmitting the differential correction information to a mobile receiver.

Real-Time Kinetic (RTK) – procedure whereby carrier-phase corrections are transmitted in real time from a reference receiver to the user's receiver. Depending on local availability, RTK corrections can be delivered by radio modem from an on-site base station or a state's CORS network, or even over the internet using Wi-Fi.

Receiver (in GPS hardware) – A computer-radio device that receives satellite information by way of radio waves to determine the position of its antenna relative to the earth's surface. The antenna can be integrated into the receiver or connected externally with a cable.

Remote sensing: Scanning an object by satellite or high-flying aircraft in order to obtain information without direct contact between the sensor and the object.

Satellite: A communications vehicle orbiting the Earth. Satellites typically provide a variety of information from weather data to television programming. Satellites send time-stamped signals to GPS receivers to determine the position on the Earth.

Semi-variance: a measure of the difference between two neighboring data points (defined as 1/2 the square of the difference between two values).

Site-Specific Crop Management (SSCM) – the use of yield maps, grid sampling and other precision tools to manage the variability of soil and crop parameters and aid decisions on production inputs.

Soil Map: a map that indicates differences in soil properties (texture, fertility, organic matter, pH, etc.) within a field.

Spatial Data: Data that contains information about the spatial location (position) and the attribute being monitored such as yield, soil properties, plant variables, seed population, etc. Synonymous with geographic data.

Spatial Resolution: the size of the smallest unit that can be identified by a remote sensing **observation.**

Spatial Variability: differences in observed attribute(s) that are noted between locations in a field.

Spectral Resolution: the ability of sensing system to differentiate between electromagnetic radiation of different wavelengths.

Speed Sensors: sensors that measure the rotational speed of a shaft or the reflection of radio or sound waves off the ground to determine forward speed.

Sustainable intensification: The goal of agriculture is to increase the production of existing farmland while minimizing damage to the environment, conserving biodiversity, and protecting the land for future crops.

TGAI: Technical grade of active ingredient. This is a chemical compound produced in the chemical industry, which an agricultural chemical contains as an active ingredient. TGAI includes byproducts of synthesis, non-reactive raw materials, and decomposed matter, as well as the principle component of the agricultural chemical.

The Bottom Line: No matter what terminology is used, biological products offer sustainable alternatives, resistant pest management and harvest solutions, and low pesticide residues for market flexibility and export accessibility. They are valuable additions to our industry's pesticide offerings, no matter how you spell them.

UAV: Stands for Unmanned Aerial Vehicle and is an aircraft piloted by remote control or onboard computers. Commonly known as a drone.

Unmanned Aerial Vehicles (UAVs): An unmanned aerial vehicle (UAV), commonly known as a drone and also referred by several other names, is an aircraft without a human pilot aboard. The flight of UAVs may be controlled either autonomously by onboard computers or by the remote control of a pilot on the ground or in another vehicle. In agriculture, UAVs are typically used to survey crops. The available two types of UAVs: fixed-wing and rotary-wing: are both equipped with cameras and are guided by GPS. They can travel along a fixed flight path or be controlled remotely.

Variable irrigation: An emerging technology used in conjunction with a central hub irrigation system.

Variable-Rate Application (VRA): adjustment of the amount of crop input such as seed, fertilizer, lime or pesticides to match conditions (yield potential) in a field.

Variable-Rate Technology (VRT): system of sensors, controllers and agricultural machinery used to perform variable-rate applications of crop production inputs.

Vector Format: a format for storing and displaying GIS spatial data that is stored as points, lines or areas to create a map object. By using a nearly continuous coordinate system, vector data can be more accurately georeferenced than raster data.

Wireless Communication: Data transfer and voice communications using radio frequencies or infrared light.

Yield Map: a map that indicates differences in crop yield within a field. Data is usually collected by a yield monitor on a combine over one to three second intervals.

Yield Monitor: electronic device that continuously measures and records crop yields and moisture on-the-go.

Zone Management: The information-based division of large areas into smaller areas for site specific management applications.

BIBLIOGRAPHY

Allan S, Shi B, Adkins SW. Impact of parthenium weed on human and animal health. Parthenium Weed Biol Ecol Manag. 2018;7:105.

Appleby AP, Müller F, Carpy S. Weed control. In Ullmann's Encyclopedia of Industrial Chemistry, (Ed.). https://doi.org/10.1002/14356007.a28_165

Aqeel-ur-Rehman and Z. A. Shaikh, "Smart agriculture," in Application of Modern High-Performance Networks, pp. 120–129, Bentham Science Publishers Ltd, 2009.

Arad, B., Balendonck, J., Barth, R., Ben-Shahar, O., Edan, Y., Hellström, T., et al. (2020). Development of a sweet pepper harvesting robot. J. Field Robot. 13–1. doi: 10.1002/rob.21937

Arad, B., Kurtser, P., Barnea, E., Harel, B., Edan, Y., and Ben-Shahar, O. (2019). Controlled Lighting and Illumination-Independent Target Detection for Real-Time Cost-Efficient Applications. The Case Study of Sweet Pepper Robotic Harvesting. Sensors 15–1 ,19. doi: 10.3390/s19061390

Arefi, A., Motlagh, A. M., Mollazade, K., and Teimourlou, R. F. (2011). Recognition and localization of ripen tomato based on machine vision. Aust. J. Crop Sci. –1144 5 1149.

B. Basnet and J. Bang, "The state-of-the-art of knowledge-intensive agriculture: a review on applied sensing systems and data analytics," Journal of Sensors, vol. 2018, Article ID 3528296, 13 pages, 2018.

Bac, C. W., Hemming, J., and van Henten, E. J. (2014a). Stem localization of sweet-pepper plants using the support wire as a visual cue. Comput. Electr. Agricult. –111 ,105 120.

Bac, C. W., Henten, E. J., Hemming, J., and Edan, Y. (2014b). Harvesting robots for high-value crops: state-of-the-art review and challenges ahead. J. Field Robot. 31:888e911.

Bargoti, S., and Underwood, J. P. (2017). Image Segmentation for Fruit Detection and Yield Estimation in Apple Orchards. J. Field Robot. 1060–1039 ,34.

Barriuso, A., G. Villarrubia González, J. de Paz, Á. Lozano, and J. Bajo, "Combination of multi-agent systems and wireless sensor networks for the monitoring of cattle," Sensors, vol. 18, no. 2, p. 108, 2018.

Barth, R., Hemming, J., and Van Henten, E. J. (2019). Angle estimation between plant parts for grasp optimisation in harvest robots. Biosyst. Eng. 46–26 ,183.

Bechar, A. (2010). Robotics in horticultural field production. Stewart Postharvest Rev. ,6 11–1. doi: 10.1111/plb.12914

Birrell, S., Hughes, J., Cai, J. Y., and Iida, F. (2019). A field-tested robotic harvesting system for iceberg lettuce. J. Field Robot. 21–1 ,37. doi: 10.1002/rob.21888

Blas, M. R., and Blanke, M. (2011). Stereo vision with texture learning for fault-tolerant automatic baling. Comput. Electr. Agricul. 168–159 ,75.

Blok, P. M., van Boheemen, K., van Evert, F. K., IJsselmuiden, J., and Kim, G. (2019). Robot navigation in orchards with localization based on Particle filter and Kalman filter. Comput. Electr. Agricult. 269–261 ,157.

Borghi, E., Avanzi, J.C., Bortolon, L., Junior, A.L. and Bortolon, E.S. 2016. Adoption and use of precision agriculture in Brazil: Perception of growers and service dealership. Journal of Agricultural Science 8 (11): 89.

Boselin, P.S.R., Pradeep, M. and Gajendran, E., 2017. An analysis of smart irrigation system using wireless sensor network. Star International Journal 5 3(3): 12-17

Bramley, R., and Trengove, S.A.M., 2013. Precision agriculture in Australia: Present status and recent developments. Engenharia Agrícola 33 (3): 575-588.

Bréant, D., Jézéquel, K. and Lebeau, T., 2002. Optimization of the cell release from immobilized cells of Bacillus simplex cultivated in culture media enriched with Cd2+: influence of Cd2+, inoculum size, culture medium and alginate beads characteristics. Biotechnology Letters 24: 1237–1241.

Brown, G. K., and Schertz, C. E. (1967). Evaluating Shake Harvesting of Oranges for the Fresh Fruit Market. Trans. ASAE 578–577 ,10.

Buemi, F., Massa, M., Sandini, G., and Costi, G. (1996). The AGROBOT project. Adv. Space Res. 189–185 ,18.

Bulanon, D. M., Burks, T. F., and Alchanatis, V. (2010). A multispectral imaging analysis for enhancing citrus fruit detection. Environ. Control Biol. 91–81 ,48.

Bulanon, D. M., Kataoka, T., Okamoto, H., and Hata, S. (2004). "Development of a real-time machine vision system for the apple harvesting robot," in Proceedings of the

SICE Annual Conference in Sapporo, Hokkaido Institute of Technolgy, Japan, 595–598.

Bulanon, D. M., Kataoka, T., Ota, Y., and Hiroma, T. (2002). AE—automation and emerging technologies: a segmentation algorithm for the automatic recognition of Fuji apples at harvest. Biosyst. Eng. 412–405 ,83.

Butzen, S., 2013. Common nitrogen fertilizers and stabilizers for corn production. In: Crop Insights. vol. 23.

Ceres, R., Pons, J. L., Jiménez, A. R., Martín, J. M., and Calderón, L. (1998). Design and implementation of an aided fruit-harvesting robot (Agribot). Indus. Robot ,25 346–337.

Chen, S. W., Shivakumar, S. S., Dcunha, S., Das, J., Okon, E., Qu, C., et al. (2017). Counting apples and oranges with deep learning: a data-driven approach. IEEE Robot. Automat. Lett. 788–781 ,2.

Chen, Y., Lee, W. S., Gan, H., Peres, N., Fraisse, C., Zhang, Y., et al. (2019). Strawberry yield prediction based on a deep neural network using high-resolution aerial orthoimages. Remote Sens. 11:1584.

Comba, L., Biglia, A., Aimonino, D. R., and Gay, P. (2018). Unsupervised detection of vineyards by 3D point-cloud UAV photogrammetry for precision agriculture. Comput. Electr. Agricult. 95–84 ,155.

Cubero, S., Aleixos, N., Albert, F., Torregrosa, A., Ortiz, C., García-Navarrete, O., et al. (2014). Optimised computer vision system for automatic pre-grading of citrus fruit in the field using a mobile platform. Precision Agricult. 94–80 ,15.

d'Grand, E., Rabatel, A. G., Pellenc, R., Journeau, A., and Aldon, M. J. (1987). Magali: A self-propelled robot to pick apples. Am. Soc. Agricult. Eng. Paper 358–353 ,46.

Daudelin, J., Jing, G., Tosun, T., Yim, M., Kress-Gazit, H., and Campbell, M. (2018). An integrated system for perception-driven autonomy with modular robots. Sci. Robot. 6–1 ,3.

De la Concepcion, A. R., R. Stefanelli, and D. Trinchero, "A wireless sensor network platform optimized for assisted sustainable agriculture," in IEEE Global Humanitarian Technology Conference (GHTC 2014), pp. 159–165, San Jose, CA, USA, October 2014.

Dey, D., Mummert, L., and Sukthankar, R. (2012). "Classification of plant structures from uncalibrated image sequences," in Proceedings of the 2012 IEEE Workshop on the Applications of Computer Vision (WACV), Breckenridge, CO, 329–336.

Edan, Y., Han, S., and Kondo, N. (2009). Automation in Agriculture, Springer Handbook of Automation. Berlin: Springer Berlin Heidelberg, 1128–1095.

Edan, Y., Rogozin, D., Flash, T., and Miles, G. E. (2000). Robotic melon harvesting. IEEE Trans. Robot. Automat. 835–831 ,16.

Ehud, B., Rotem, M., and Ohad, B. (2016). Colour-agnostic shape-based 3D fruit detection for crop harvesting robots. Biosyst. Eng. 70–57 ,146.

Fischer, J. W., Greiner, K., Lutman, M. W., Webber, B. L., and Vercauteren, K. C. (2019). Use of unmanned aircraft systems (UAS) and multispectral imagery for quantifying agricultural areas damaged by wild pigs. Crop Protect. 125:104865.

Fu, L., Feng, Y., Majeed, Y., Zhang, X., Zhang, J., Karkee, M., et al. (2018). Kiwifruit detection in field images using Faster R-CNN with ZFNet. IFAC Papers On Line ,51 50–45.

Gago, J., C. Douthe, R. E. Coopman et al., "UAVs challenge to assess water stress for sustainable agriculture," Agricultural Water Management, vol. 153, pp. 9–19, 2015.

Ivanova, E., Teunou, E. and Poncelet, D., 2005. In: Gruev, B., Nikolova, M., Donev, A. (Eds.), Alginate based macrocapsules as inoculants carriers for production of nitrogen biofertilizers. Proceedings of the Balkan Scientific Conference of Biology in Plovdiv (Bulgaria), pp. 90–108.

Jorquera, M., Herna´ndez, M., Rengel, Z., Marschner, P. and Mora, M.L., 2008. Isolation of culturable phosphobacteria with both phytatemineralization and phosphatesolubilization activity from the rhizosphere of plants grown in a volcanic soil. Biology and Fertility of Soils 44: 1025–1034.

Kalubhai, P., 2015. Present Status and Prospects of Micro Irrigation System in Mahisagar District of Gujarat. Doctoral dissertation, Anand Agricultural University, Anand, India.

Kennedy, I.R., Choudhury, A.T.M.A. and Kecske´s, M.L., 2004. Non-symbiotic bacterial diazotrophs in crop-farming systems: can their potential for plant growth promotion be better exploited? Soil Biology and Biochemistry 36: 1229–1244.

Keskin, M., and Sekerli, Y.E., 2016. Awareness and adoption of precision agriculture in the Cukurova region of Turkey. Agronomy Research 14 (4): 1307-1320.

Kim, S.Y., Kim, M., Lee, S., Lee, J. and Ko, G., 2012. Survival of microorganisms on antimicrobial filters and the removal efficiency of bioaerosols in an environmental chamber. Journal of Microbiology and Biotechnology 22 (9): 1288–1295.

Kumar R, Katiyar R, Kumar S, Kumar T, Singh V. Lantana camara: an alien weed, its impact on animal health and strategies to control. J Exp Biol Agric Sci. 2016;4:321–37.

Kurtulmus F, Lee WS, Vardar A (2014) Immature peach detection in colour images acquired in natural illumination conditions using statistical classifiers and neural network. Precis Agric 15(1):57–79.

Linker R, Cohen O, Naor A (2012) Determination of the number of green apples in rgb images recorded in orchards. Comput Electron Agric 81:45–57.

Liu W, Anguelov D, Erhan D, Szegedy C, Reed S, Fu C-Y, Berg AC (2016) Ssd: single shot multibox detector. In: European conference on computer vision. Springer, Berlin, pp 21–37.

Maher, A., E. Kamel, F. Enrico, I. Atif, and M. Abdelkader, "An intelligent system for the climate control and energy savings in agricultural greenhouses," Energy Efficiency, vol. 9, no. 6, pp. 1241–1255, 2016.

Mohammed, G., F. Trolard, M. Gillon, A. L. Cognard-Plancq, A. Chanzy, and G. Bourrie, "Combination of a crop model and a geochemical model as a new approach to evaluate the sustainability of an intensive agriculture system," Science of The Total Environment, vol. 595, pp. 119–131, 2017.

Naz, M.Y., Sulaiman, S.A., 2016. Slow release coating remedy for nitrogen loss from conventional urea: a review. Journal of Controlled Release 225(10): 109–120.

Neethirajan, S., 2017. Recent advances in wearable sensors for animal health management. Sensing and Bio-Sensing Research 12: 15–29. doi:10.1016/j.sbsr.2016.11.004.

Ni, B., Liu, M. and Lu, S., 2009. Multifunctional slow-release urea fertilizer from ethylcellulose and superabsorbent coated formulations. Chemical Engineering Journal 155: 892–898.

Park, D. H. and J. W. Park, "Wireless sensor network-based greenhouse environment monitoring and automatic control system for dew condensation prevention," Sensors, vol. 11, no. 4, pp. 3640–3651, 2011.

Park, M., Kim, C., Yang, J., Lee, H., Shin, W., Kim, S. and Sa, T., 2005. Isolation and characterization of diazotrophic growth promoting bacteria from rhizosphere of agricultural crops of Korea. Microbiological Research 160: 127-133.

Paustian, M., and Theuvsen, L., 2017. Adoption of precision agriculture technologies by German crop farmers. Precision Agriculture 18(5): 701–716.

Perez-Ruiz, M., P. Gonzalez-de-Santos, A. Ribeiro et al., "Highlights and preliminary results for autonomous crop protection," Computers and Electronics in Agriculture, vol. 110, pp. 150–161, 2015.

Puri, M., Solanki, A., Padawer, T., Tipparaju, S.M., Moreno, W.A. and Pathak, Y., 2015. Introduction to Artificial Neural Network (ANN) as a Predictive Tool for Drug Design, Discovery, Delivery, and Disposition: Basic Concepts and Modeling; Elsevier Inc.: Amsterdam, the Netherlands, pp. 3–13. .

Rakun J, Stajnko D, Zazula D (2011) Detecting fruits in natural scenes by using spatial-frequency based texture analysis and multiview geometry. Comput Electron Agric 76(1):80–88.

Stajnko D, Lakota M, Hoèevar M (2004) Estimation of number and diameter of apple fruits in an orchard during the growing season by thermal imaging. Comput Electron Agric 42(1):31–42

Universal Robot Support. https://www.universal-robots.com/download/. Accessed 23 Oct 2019.

Xiaofeng, L., Linlin, Q., Linjian, L. and Gang, W. 2015. Design and implementation of modern greenhouse remote monitoring system based on the Android system. In: 34th Chinese Control Conference (CCC), pp. 5742–5746.

Xu, Z., Chen, J., Wang, Y. and Fan, Z., 2016. A remote monitoring system for greenhouse based on the internet of things. In: MATEC Web of Conferences, EDP Sciences 77:1-4

Yadav T, Chopra NK, Chopra NK, Kumar R, Soni PG. Assessment of critical period of crop-weed competition in forage cowpea (Vigna unguiculata) and its effect on seed yield and quality. Indian J Agron. 2018;63:124–7.

Yan-E, D., 2011. Design of intelligent agriculture management information system based on IoT. In: Proceedings of the 4th International Conference on Intelligent Computation Technology Automation ICICTA 2011, Vol. 1, pp 1045–1049

Zahid, M., Abbasi, M.K., Hameed, S. and Rahim, N., 2015. Isolation and identification of indigenous plant growth promoting rhizobacteria from Himalayan region of Kashmir and their effect on improving growth and nutrient contents of maize (Zea mays L.). Frontiers in Microbiology 6(207): 1–10.

Zhang, X., Zhang, C., Fang, J., Yu, X. and Liang, J., 2009. Smart sensor nodes for wireless soil temperature monitoring systems in precision agriculture. Transactions of the Chinese Society for Agricultural Machinery 40: 237-240.

INDEX

L

M

N

O

P

R

S

T

W

www.ingramcontent.com/pod-product-compliance
Lightning Source LLC
LaVergne TN
LVHW062116200926
848484LV00007B/1682
9781787150560